Integrated Water Resources Management in the 21st Century: Revisiting the Paradigm

Integrated Water Resources Management in the 21st Century: Revisiting the Paradigm

Editors

Pedro Martinez-Santos &
Maite M. Aldaya
Complutense University of Madrid, Spain

M. Ramón Llamas
Complutense University and Water Observatory at the Marcelino Botín Foundation, Madrid, Spain

Editorial assistant

Laurens A.D.R. Thuy
Utrecht University, Utrecht, The Netherlands
Water Observatory, Marcelino Botín Foundation, Madrid, Spain

CRC Press
Taylor & Francis Group
Boca Raton London New York Leiden

CRC Press is an imprint of the
Taylor & Francis Group, an **informa** business
A BALKEMA BOOK

CRC Press/Balkema is an imprint of the Taylor & Francis Group, an informa business

Typeset by V Publishing Solutions Pvt Ltd., Chennai, India
Printed and bound in The Netherlands by PrintSupport4U, Meppel

Published by: CRC Press/Balkema
P.O. Box 11320, 2301 EH Leiden, The Netherlands
e-mail: Pub.NL@taylorandfrancis.com
www.crcpress.com – www.taylorandfrancis.com

Library of Congress Cataloging-in-Publication Data

Integrated water resources management in the 21st century: revisiting the paradigm / edited by Pedro Martinez-Santos, Maite M. Aldaya, Ramón Llamas.
p. cm.
Includes bibliographical references and index.
ISBN 978-1-138-00143-5 (hardback)
1. Integrated water development--Case studies. 2. Water-supply--Management--Case studies. 3. Water--Government policy--Case studies. I. Martinez Santos, Pedro.

TC409.I5585 2014
333.91–dc23

2013045285

ISBN: 978-1-138-00143-5 (Hbk)
ISBN: 978-1-315-79409-9 (eBook PDF)

Contents

Foreword

Two years ago, the Water Observatory of the Botín Foundation launched a book entitled '*Water, agriculture and the environment in Spain: can we square the circle?*'. Based on the Spanish experience, this volume showed how emerging paradigms have changed the water management picture, and how water management is subject to an increasing number of variables that transcend the watershed scale. Take for instance the concepts of soil storage (*green* water) or virtual water flows, which trigger essential changes to the way water balances have been traditionally computed; or the role of food trade on water use patterns, which presents strong implications for the conventional notion of food security. More than ever, these have made water policy a global issue.

The present book takes the idea one step further. While keeping an eye on the Spanish experience, we attempt to draw from the experience of international institutions and specific case studies to single out some of the aspects that will constrain the way we manage our water resources in the future. Readers will find that some of the chapters present a more traditional outlook, while others are more innovative. All of them however show that water management is subject to the same rapid change that permeates most aspects of our daily lives, and that adapting is perhaps the main task at hand.

This book compiles the papers presented by sixteen international experts at the 6th Botín Water Workshop, held in Madrid, Spain, in November 2012. It is divided in three sections. Section one provides an introduction to the prevailing concepts of integrated water resources management (IWRM). Thus, chapters one to three showcase the history of the IWRM concept and its guiding principles. They also acknowledge some of the emerging issues that water managers will need to take into account in coming years, as well as critical views as to what IWRM can achieve and what lies beyond its means. Chapters four and five present views from members of the Organization for Economic Co-operation and Development and the World Trade Organization, two international organizations which are likely to play increasingly important roles in the manner in which water embedded in food and manufactured products flows around the globe.

Section two builds on the Water Observatory's ongoing research on Spain and Latin America. Chapters six to ten draw overarching conclusions from issues such as the role of virtual water trade, the state of freshwater ecosystems, the relevance of groundwater resources and urban water supply or the role of water institutions. These set the scene for section three. Chapters eleven to sixteen present a series of selected case studies dealing with the reality of integrated water resources management in countries so diverse as Peru, Spain, Chile or China. All of them showcase the beauty of the concept when applied to real-life situations, as well as its practical limitations. Overall, most authors highlight the importance of IWRM as a *useful utopia*, whose value lies more with the steps that need to be taken than with reaching an ever-elusive end goal.

We hope that this book will provide a valuable contribution to the current debate on the world's water resources.

The Editors

About the Water Observatory of the Botín Foundation

Over the last two decades the Marcelino Botín Foundation has maintained the Observatory of Trend Analysis, a private research body whose purpose is to investigate the drivers that catalyze change in today's society.

The Water Observatory, one of its branches, is Spain's first interdisciplinary think tank on water. It was created in 1998 and specializes in analyzing water-related issues of national and global importance (http://www.fundacionbotin.org/water-observatory_trend-observatory.htm). The Observatory is known for its scientific rigor, which combines natural science with economic and social science perspectives to offer new ideas that may contribute to underpin decision-making in the field of water resources.

Since its early days, the Observatory has provided a forum for the exchange of ideas among water experts, facilitating the transfer of new knowledge to water decision makers and to society as a whole. In more recent times, the Water Observatory has focused on developing and disseminating water footprint, virtual water trade and water and food security concepts, particularly in the context of Spain and Latin America.

As per the proposal of Board Member Emilio Botín O'Shea, the Botín Foundation launched a multidisciplinary analysis of Spain's groundwater resources in 1998. Over the ensuing four years, Professor M.R. Llamas led this project, which culminated with a presentation at the 2003 Kyoto World Water Forum. Its outcomes remain an influential benchmark for groundwater policy in Spain.

Between 2004 and 2008 work focused on the organization of international workshops on water resources. The main findings of these meetings were compiled in two books, entitled *Water Crisis: Myth or Reality?* and *Water Ethics*, both which can be freely downloaded from the website. Since 2008 the Water Observatory has also published twelve monographs (PAV and SHAN papers), a special issue of the *Water Policy* journal and three books, entitled *Water Footprint and Virtual Water Trade in Spain* and *Water, Agriculture and the Environment is Spain: can we square the circle?* and *Rethinking Water and Food Security*, respectively. Most of these works have been presented at major international meetings, including the Stockholm Water Week or the World Water Forum.

Towards the end of 2005, Prof. Llamas delivered a speech in the opening session of the academic year of the Royal Academy of Sciences of Spain. This conference dealt with new water management paradigms, including the *colors of water* and the role of virtual water flows. Research on these new ideas crystallized in the formal constitution of the Water Observatory as a think-tank to promote innovation in water management practices. Thus, the Water Observatory was established as an ongoing collaboration agreement between the Botín Foundation, the Universidad Complutense de Madrid and the Universidad Politécnica de Madrid.

Research staff is recruited through these universities, which also provide the venue for the Secretariat and offices. Ramón Llamas, Emeritus Professor of the Universidad Complutense, is the serving Director, while Alberto Garrido, professor of Agricultural Economics and Natural Resources at the Universidad Politénica of UPM, and Director of the CEIGRAM, is Deputy Director.

The activities of the Water Observatory have led to outstanding policy impacts. Take for instance the effect of the groundwater project on the groundwater management regulations of Spain's 2001 National Water Plan; or the decision by the Ministry of Environment to include the water footprint in Spain's River Basin Management Plans (BOE, no. 229, September 22nd 2008), in line with the European Union (EU) Water Framework Directive (WFD) (2000/60/EC).

Section 1

Introduction and international perspectives

Chapter 1

Integrated Water Resources Management (IWRM): The international experience

Mohamed Ait Kadi
Chair of GWP Technical Committee, Global Water Partnership, Stockholm, Sweden

ABSTRACT: Water is now seen as a central plank of sustainable natural resources management, it is embedded in all aspects of development – food security, health, and poverty reduction – it is essential for economic growth, and it sustains the natural ecosystems on which everything else depends. Implicit in all this is the need for integration. This is already at the heart of Integrated Water Resources Management (IWRM) and this chapter argues that it is timely to revisit IWRM as an approach that can facilitate and lead the process of 'greening' the world's economies. There are skeptics, but evidence is presented from the 2012 UN survey of 134 countries that 82% have embarked on reforms to improve the enabling environment and integrate approaches to water resources management, 65% have developed IWRM plans, and 34% say they are at an advanced stage of implementation. It is argued that IWRM is no longer just an idea; it is a reality for many. It is truly 'fit for purpose' – a process whose time has come.

1 INTRODUCTION

Integrated water resources management is a concept that few would argue against. Surely, it is a given that if we wish to manage scarce water resources effectively and avoid wastage then we must approach this together in a coordinated manner. Indeed, the idea of taking a 'silo' or fragmented approach to resource planning would seem archaic today. This essentially simple idea has growing international acceptance, yet implementing it in reality has proved to be far from simple and some argue that it is an ideal that is largely unattainable in practice. Biswas (2008) is just one skeptic who wonders why it has not been possible to properly implement a concept that has been around for at least two generations.

But there is another reality. Across the world, economies are expanding, cities are spreading, populations are growing, and many are enjoying better living standards. But what is not growing are the natural resources that underpin all this economic and social development. Water, in particular, is a limiting resource in many countries. We do not yet face a global water crisis, that many are predicting over the next 30–50 years. But there is little doubt that recent global events – increasing food and energy prices, severe droughts and floods, worries about water and food security, and the threat of climate change – have heightened our concerns over water availability.

Water is now seen as a central plank of sustainable natural resources management, it is embedded in all aspects of development – food security, health, and poverty reduction – as it is an essential part of sustaining economic growth in agriculture, industry, and energy generation, and it sustains the natural ecosystems on which everything else depends. The Stockholm Statement (2011) described water as the 'bloodstream of the green economy'. But if the world continues to use water at current rates it is estimated that demand could outstrip supply by as much as 40% by 2030.

Water management and water scarcity have rightly come to preoccupy many different sectors in many different societies as never before. There is now a worldwide movement and new enthusiasm towards sustainable resource use. And implicit in all this is integration. This has always been at the heart of IWRM and so it is now timely to re-visit this concept as an approach to development that can facilitate and lead the process of 'greening' the world's economies. How has thinking about IWRM shifted in recent years? Is there a more universal understanding of what IWRM means? Are the drivers in place to make it more attainable? What are the experiences of those who seek to practice IWRM? Is it now a process that is truly 'fit for purpose' – a process whose time has come?

2 OUR JOURNEY

The origins of IWRM are now part of water resources history. Snellen & Schrevel (2004) provide an excellent account of this. They cite establishing the Tennessee Valley Authority (TVA) in 1933 as an early example of how to bring together the different facets of water use, such as navigation, flood control, and power production, for economic development. But the modern ideas of IWRM – the need for integration in and across the water sector – have their roots in the 1977 international water conference, which led to the Mar del Plata Action Plan. In 1992 IWRM was incorporated in what has now become known as the 'Dublin Principles' which was a precursor to including IWRM in Agenda 21 of the United Nations Conference on Environment and Development (UNCED). This is about improving water resources management by connecting the many different water services and providing good governance, appropriate infrastructure, and sustainable financing. In 1996 the Global Water Partnership (GWP) was founded to foster IWRM and provide a global forum for dialogue among corporations, governmental agencies, water users, and environmental groups to promote stability through sustainable water resources development, management, and use. In 2002 the World Summit on Sustainable Development (UN, 2005) in Johannesburg again called for the development of 'IWRM and all countries agreed to develop IWRM and water efficiency plans'.

The advent of IWRM marked a fundamental shift away from the traditional top-down, supply-led solutions to water problems that were dominated by technology (McDonnell, 2008). When water was plentiful and abstractors few, the rules of water sharing in most societies were few and basic. But as water use increased, shortages occurred and awareness grew of the impact this had on the environment, more complex institutions were needed to negotiate and coordinate water allocations

Box 1: Integrated water resources management: Definition and approaches

The Global Water Partnership (2000) defined IWRM as:

> 'IWRM is a process which promotes the coordinated development and management of water, land, and related resources in order to maximize the resultant economic and social welfare in an equitable manner without compromising the sustainability of vital eco-systems'

This definition is the most authoritative definition on IWRM. These are often referred to as the IWRM 'principles' (GWP, 2000).

IWRM is defined as a process; it does not offer a 'blueprint' approach to water management. Water resources are different from place to place and so too are development priorities and social and economic issues. Country or water basin planning may differ but IWRM provides a common approach and experience shows that there are features common to all. These include a strong enabling environment; sound investments in infrastructure, clear robust and comprehensive institutional roles; and effective use of available management and technical instruments. These are the practical elements of implementing IWRM (Muller & Lenton, 2009).

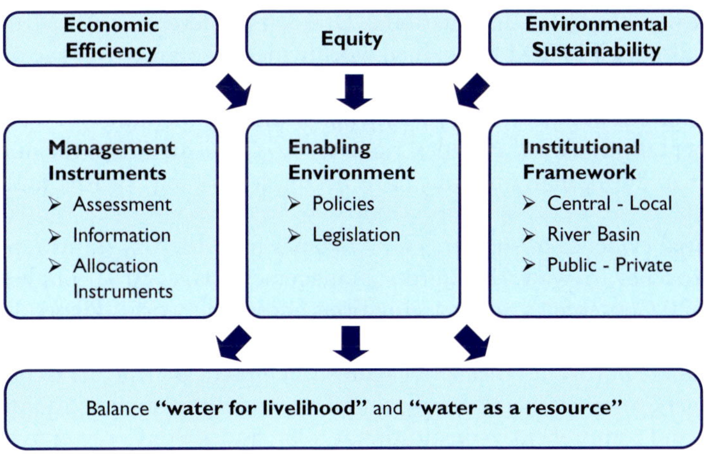

Figure 1 Common IWRM features.

Since its inception in 1996 GWP has driven a worldwide movement towards IWRM. It has helped countries around the world to (1) recognize basic principles that underpin good water management; (2) develop a stronger enabling environment of policies and laws; (3) build more appropriate institutional frameworks; and to (4) share, adopt and adapt management instruments and tools (Ait Kadi & Arriens, 2012).

among different users. Administrations responsible for developing and managing water resource infrastructure had to pay more attention to managing and protecting the resource (Muller & Lenton, 2009).

At first 'integration' meant bringing together water resources with engineering and economic driven solutions. But there was increasing awareness that the way land is managed impacts water resources and vice versa and that water quantity could not be managed in isolation from water quality. As water demands increased it became clear that bridges were needed between human and natural systems and between the water sector and the economy. 'Vertical' bridging was also needed across different levels of decision-making from local, provincial, and national to river basins and transnational scale. So the idea of integration grew to include more decentralized approaches to water management in a more holistic context together with an appreciation of local ideas and demand management (McDonnell, 2008). But integration did not mean that everything needed to be together and managed under 'one roof' or that sectoral decision-making should be abandoned. On the contrary, this was considered to be both undesirable and unworkable. What was clear though was that integration meant increasing complexity and this has undoubtedly contributed to the concerns about fully achieving it.

3 THE STATUS OF IWRM IN THE WORLD

Concerns about the reality of implementing IWRM have been raised many times. Most recently in the 2011 International Dresden conference, IWRM experts concluded that, although IWRM has gained worldwide acceptance over the past 20 years and is now included in national policies, strategies, and laws, actual implementation is lagging behind (Borchardt, 2011). So what is the current state of IWRM planning and implementation? Are the critics right to be sceptical? Since the Johannesburg commitment in 2002, many nations have begun developing 'IWRM and water efficiency plans'.

Substantial evidence to support IWRM comes from the UN status report on Integrated Approaches to Water Resources Management (UN, 2012) published in time for the Rio+20 Conference. Some 134 nations across the world (Figure 2) responded to the survey carried out to determine progress towards sustainable water resources using integrated approaches measured against the practical elements of implementing IWRM, namely, a strong enabling environment; sound investments in infrastructure, clear robust and comprehensive institutional roles; and effective use of available management and technical instruments (Muller & Lenton, 2009). Here we review some of the highlights of this survey.

3.1 Enabling environment: Policy, Laws & Plans

Over the past 20 years, since the 1992 UNCED conference, most countries in the survey have observed increases in water-related risks and competition for water use across all sectors (UN, 2012). In high HDI countries (HDI is the Human Development Index which is a composite index that measures health, knowledge, and income according to which countries are categorised in four bands: low, medium, high, and

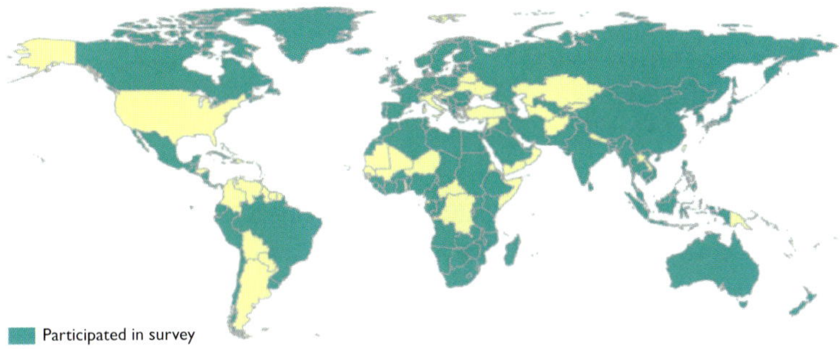

Figure 2 Countries participating in the UN IWRM survey (UN, 2012).

very high) water for the environment is seen as the main priority while in low HDI countries the focus is on domestic water supplies, particularly for growing cities, followed by water for agriculture.

Encouragingly, 82% of countries have embarked on reforms to improve the enabling environment and integrated approaches to water resources management. Some 65% have developed IWRM plans and 34% say they are at an advanced stage of implementation. But 25% report obstacles to implementation such as weak and conflicting legal frameworks and inadequate or non-existent strategic planning. In Albania, the legal system has little in common with the European Union yet they would like to move towards the EU's Water Framework Directive (WFD). In Peru and Samoa national laws conflict with traditional practices and customs. Azerbaijan is concerned with challenging transboundary water issues while countries such as Portugal, Denmark, and Germany, have difficulties in balancing internal agricultural, industrial, and environmental water interests.

Integration is working well in high HDI countries and is driven by water scarcity and environmental concerns (Australia, Spain) and by international agreements such as the EU Water Framework Directive. Even countries with complex federal and state structures have managed to agree and implement sweeping reforms (Brazil, Australia). Many high HDI countries have the added advantage of a legacy of 'easy hydrology' – low rainfall variability, rain distributed throughout the year and perennial rivers sustained by groundwater base flows (Grey & Sadoff, 2007).

Progress is much slower in low and medium HDI countries, which includes most of sub-Saharan Africa and south Asia. Most have large rural populations that rely on subsistence agriculture and are exposed not just to the vagaries of uncertain and unpredictable seasons but also to the 'difficult hydrology' of absolute physical water scarcity and severe flood risk, usually at different times but often in the same place (Grey & Sadoff, 2007). More difficult hydrology usually means more costly infrastructure to control and manage water.

One indicator used in the survey of how well water is integrated, is the inclusion of water in planning documents, such as Poverty Reduction Strategies, National Strategy for Sustainable Development, and National Environmental Action Plan. Almost 80% of countries included water in these plans.

Box 2: Examples of IWRM in selected countries (UN, 2012)

Some are moving ahead:

> 'Better governance, public participation and dissemination of the planning process and the implementation of the Policy instruments are recognized as major advances in the legal and institutional structure' (Brazil).

Some are stalled:

> 'new policy formulation.... contested by the major stakeholders in the water sector, civil society, organizations and the political parties..... not only due to content of the policy document rather due to the approach adopted for consultation' (Sri Lanka)

A surprising 54% of countries are engaged in implementing transboundary water agreements. They recognise that huge differences exist in the purpose and details of such agreements and the additional challenges this brings to IWRM. GWP (2012) is currently addressing this issue as it examines the role of international law as a facilitator of transboundary water cooperation.

Overall, climate change was seen as an important driver of change but less so in low HDI countries compared to other water development threats. The report rightly points out that there is no quick fix for sustainable water management and so national and international leaders must demonstrate their commitment for the long haul if it is to succeed.

3.2 Developing infrastructure

The relationship between water sector reforms and support for infrastructure is a weak but positive one. However, there is little evidence of a fully integrated approach to infrastructure development, although awareness of the need is increasing. Brazil offers one of the few encouraging examples. Over 65% of countries say their infrastructure implementation is advanced but less so for flood management, irrigation, rainwater harvesting, and natural ecosystems.

3.3 Finance

More than 50% of the countries say they have seen a notable increase in funding for water resources development over the past 20 years. Though a few (10–15%) have seen a decline. Many countries finance water resources development and management from a range of government and aid sources. More than 60% of low HDI countries indicate that grants and loans for water projects have increased. Others face obstacles to raising finance but most recognise that more investment is needed. While many countries have adopted policies and laws that water users and polluters should pay

Box 3: Infrastructure and IWRM

Multi-purpose infrastructure:

> 'Nowadays, an integrated approach to the development of multipurpose projects and the incorporation of climate change impacts in the design of infrastructure can increasingly be observed' (Mexico).

Cross sector funding benefits:

> 'Further mainstreaming of water considerations into other sectors' plans (like environment, agriculture) will assist with increasing the available financing through cross-sector activities' (Jordan)

Coordinating infrastructure:

> 'In 2007, the Federal Government launched the Program for Accelerating Development (PAC) based on strong coordination of public expenditure priorities, including actions to enlarge the water supply, sanitation, irrigation and energy infrastructures, as well as other water resources-related actions, among others' (Brazil).

for water benefits and the costs involved, there is little progress in getting users and polluters to pay for water resources and ecosystem services. Therefore, most recognise there is still much to do. Indeed, low revenue generation is cited as one reason for the lack of financial resources for investment and for Operations and Management (O&M) and the need to rely on governmental budgets.

3.4 Institutional improvements

Survey results show that over 70% of countries now manage their water resources at river basin level and processes to achieve this are well advanced in over 50% of countries. Managing water at the lowest appropriate level (subsidiarity) is a fundamental tenet of IRWM which means taking a basin approach and decentralising decision-making. Countries that have made progress with legal and policy reforms are also most likely to have made progress with improving their water management institutions.

Stakeholder participation is also a key component of IWRM and although all countries consider this to be important, progress is slow in all but the very high HDI countries. The latter have usually implemented stakeholder access to information. They see this as an essential condition for active engagement and have a generally positive view of the process though it requires careful management to avoid excessive costs. Given the importance of gender in water resources management it is surprising that 22% of countries thought that gender mainstreaming for water management was not relevant to them.

Box 4: Management instruments in IWRM

'Several management instruments (e.g. issuing of permits, licensing, and monitoring) have been developed and introduced, however; human capacity remains a major challenge. This results in limited compliance to permit conditions and inadequate pollution control. Only 50–60% of permit holders report their abstraction figures' (Namibia)

Box 5: Morocco's experiences in IWRM

In Morocco water security has always been a key component in economic and social development. The growing demand for good quality water stems from increased industrialization and a rapidly growing population, accentuated by a progressive shift from rural to urban living.

Over the last four decades the emphasis was to maximize the sustainable development of the country's surface water resources and their optimal use for irrigated agriculture, potable water supplies, industrialization, and energy generation. This meant significant capital investment in infrastructure to control rivers and to capture and use about two-third of the country's surface water. More major infrastructure projects are at advanced stages of planning and construction to capture most of the remaining surface water potential by 2020.

As Morocco nears the end of the infrastructure development phase, emphasis is shifting to the more sophisticated and difficult task of ensuring socially, technically efficient, and sustainable allocation of existing water resources among competing consumer groups. Morocco's water economy is now characterized by sharply rising costs of supplying additional water and more direct and intense competition among different types of water users and uses. Better management of existing supplies is the most rational response to this growing water scarcity. As irrigation is the predominant water user, serious questions were directed towards reducing wastage in this sector by restructuring production and consumption patterns away from wasteful and low-value crops to more effective water use for high-value crops.

To meet these challenges, Morocco has adopted an IWRM approach through mutually reinforcing policy and institutional reforms as well as the development of a long-term investment program. The major policy reforms adopted include:

- Adopting a long-term strategy for integrated water resources management. The National Water Plan will be the vehicle for implementing the strategy and will serve as the framework for investment programs until 2020;
- Developing a new legal and institutional framework to promote decentralized management and increase stakeholder participation;
- Introducing economic incentives in water allocation decisions through rational tariff and cost recovery;

– Taking capacity enhancing measures to meet institutional challenges for the management of water resources;
– Establishing effective monitoring and control of water quality to reduce environmental degradation.

A new water law was promulgated in 1995 which provides a comprehensive framework for integrated water management.

Investment efforts in the water sector continue to capture most of the remaining potential and develop the accompanying hydropower infrastructure to reduce energy imports; meet the government's objective for the potable water supply sub-sector which is to supply virtually the entire rural and urban population by 2020; and continue the on-going expansion of modern irrigation, which is expected to bring the total irrigated area to 1.35 million hectares by 2020. This is the country's full potential given the available water resources.

Morocco's water reform experience offers a range of useful features covering mainly the new institutional arrangements governing the water sector with the reinforced role of the High Water and Climate Council as an apex body on national water policies and programs and the creation of river basin agencies. At the sub-sectorial level, the Moroccan irrigation agencies are unique as they integrate the provision of production services to farmers with water supply – an approach that is so crucial for enhancing water productivity and farm output.

Although Morocco has developed a comprehensive water sector reform agenda, the pace of implementing the reforms remains slow. It is indeed complicated by the place water occupies within the set of interdependencies among the country's different physical, economic, and social systems. Thus, reforms cannot be seen in isolation from socio-economic development policy, urban policy and urban design, land use, and agricultural policies etc. They should also be considered from the wider political and cultural changes occurring within the country mainly with regard to the progress of democracy and distributive governance.

The Government is continuing its efforts towards completing IWRM reforms mainly through improving sector governance, strengthening policy coordination bodies, and improving sector organization. It has completed the issuance of all the decrees needed to enact the 1995 Water Law. River Basin Agencies are being progressively empowered to enact decentralized and participatory resource planning, co-fund resource conservation and protection projects, enforce 'user pays/polluter pays' policies and aquifer management strategies.

We can draw a number of key lessons from this experience:

> Completing and enforcing IWRM reform is key to restoring sustainability in Morocco's water management and promoting efficient use and allocation of water; achieving a successful, sustainable, cooperative river basin is clearly a challenge; benefit can be taken from other countries' experiences but avoid 'one-size-fits-all' solutions to water resources planning and management; and think political economy (nature and pace of change).

3.5 Applying management instruments

The survey encouragingly shows progress in implementing water management instruments in all HDI groups and this is seen as a good indicator of improving water management performance and integration. Management instruments (tools and methods) help decision-makers to make rational and informed choices between alternative actions and support institutional sustainability. Although there is no accepted measure of water resources management performance, implementing management instruments, which requires both financial resources and human capacity, can be a strong indicator of improving water management performance and integration. The survey also shows that there is a strong correlation between improving a country's enabling environment – policy, laws, and plans – and improving the adoption and use of these instruments.

Over 60% of countries have implemented water resources monitoring and assessment systems but these tend to be high and very high HDI countries. Over 50% of low and medium HDI countries do not have systems in place. A lack of knowledge sharing mechanisms in all but the very high HDI countries constrain integration. These included scarcity and transparency of information, a lack of research, poor data handling and management, and a lack of implementing capacity.

Less than 10% of low and high HDI countries have reached an advanced stage of implementing financial instruments as a means of cost recovery.

4 COMING OF AGE

Over the past five years, people around the world have faced a daunting array of new and increasingly inter-connected crises (food, energy, financial, climate and water) that impact heavily across scales and boundaries, threatening the security of households, communities, natural resources, and national economies. These all hit the poor hardest.

The concurrence of these crises during the last decade has challenged humankind's optimistic vision of continuing progress in development, as well as the validity of the current world economic models. Many more people have come to realise that the earth's resources cannot, in the long run, meet the demand of the world's growing population following the existing models of industrialised societies.

We have also become conscious of the fact that the problems are increasingly and tightly interconnected. The problems we face with climate change for example are embedded in the problems of water security, food security, and energy security (Figures 3 & 4). These interconnections are often ignored when policy-makers devise partial responses to individual problems (Ait Kadi, 2010). They call for broader public policy planning tools with the capacity to encourage legitimate public/collective clarification of the trade-offs and the assessment of the potential of multiple uses of water to facilitate development and growth.

The success or failure of a particular policy initiative or strategic plan is largely dependent on whether the decision-maker truly understands the involved interactions and complexities. It is not surprising that the 'intuitive' or 'common sense' approach to policy design often falls short, or is counter-productive to desired outcomes.

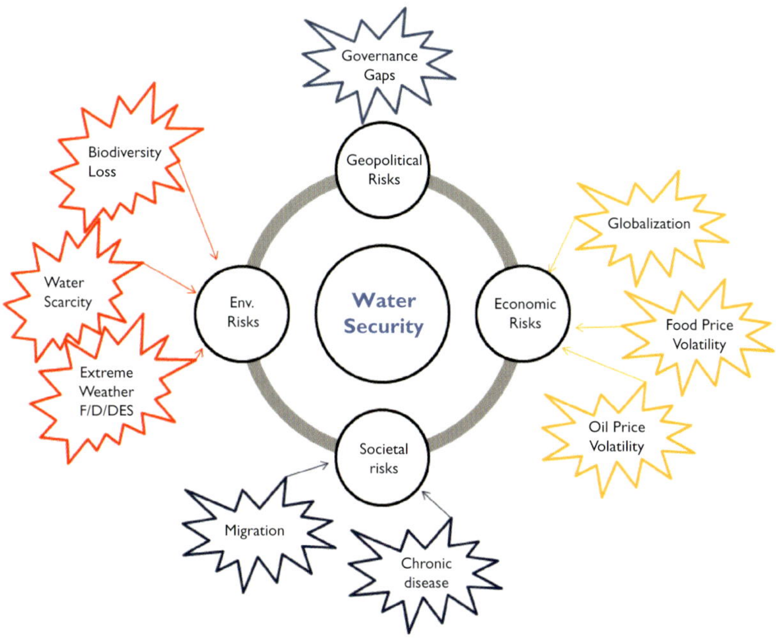

Figure 3 Water Security and Interconnected Global Risks (Ait Kadi, 2010).

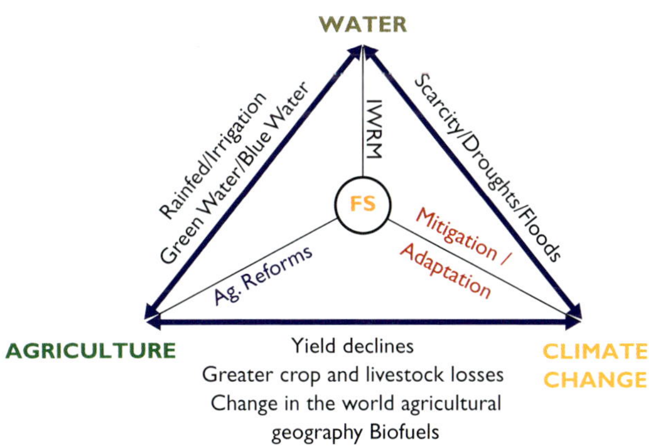

Figure 4 Water – Food Security – Climate Change Nexus (Ait Kadi, 2010).

Understanding the challenges and their interconnections is a critical step in effective policy design, policy implementation, and consensus building. IWRM provides a lens through which the many interlinked drivers and potential consequences of economic, social, and environmental changes can be identified and coordinates actions formulated to holistically achieve economic efficiency, social equity, and environmental sustainability based on publicly available and transparent information .... Its implementation in the new context will need categorizing to optimize spinoffs and plans of action for the short, medium, long term. It requires:

- Policy instruments that promote complementarities (economic, social, environmental) and leverage change;
- Fiscal instruments that give a price to environmental goods;
- Strengthened institutional arrangements that function within increasing complexity, cutting across sectoral silos and sovereign boundaries;
- A new generation of financial instruments that share risk between governments and investors and make new technology affordable;
- Skills development that supports the emerging green sectors in the economy;
- Transparent information and monitoring: set targets, define trajectories and gather the right information to monitor progress (e.g., on water/energy efficiencies);
- Education and awareness raising.

5 CONCLUSION

If we look at the past 20 years of IWRM experience, we see that it can be regarded as a development process towards adaptive solutions to water-related problems. In any location, the water management actions of today will build on the achievements and experiences of the past. But where IWRM was promoted as a rigid formula or prescription it has often failed to deliver the desired benefits.

GWP with others has for many years supported countries in developing an integrated approach to manage and develop water resources. This approach has always suffered from its technical and scientific origins within the water sector and has not been easy to explain or 'sell' to politicians and other non-water decision-makers. Nevertheless, despite the usual naysayers, the principles of IWRM are now becoming well accepted and incorporated in policies throughout the world as is indicated by the UN survey. Indeed, the idea of taking a non-integrated or fragmented approach would seem archaic today. IWRM has however been difficult to apply in practice as it requires considerable coordination and information sharing between multiple sectors and different layers of authority. Administrations are structured principally by economic sectors whereas water, as a natural resource, impacts on and is impacted by these sectors but it has no institutional home. Water resources are thus easily exploited and polluted by users due to its weak management and/or regulation.

In essence IWRM is a means for solving problems and not a formula as it has been some times promoted. It deserves constant reinforcement through technical and political requirements.

Managing water resources now means meeting both human and ecosystem needs. It is clear that IWRM requires fundamental changes in values, beliefs, perceptions and political positions, not just in water management institutions but also in the stakeholders themselves. Progress may be slow and questions complex but there really is no alternative to IWRM (Snellen & Schrevel, 2004).

REFERENCES

Ait Kadi, M. (2010) Exploring the role of water security in regional economic development. In: *GWP Consulting Partners Meeting: Water security and regional economic development, August 2010, Stockholm, Sweden.*

Ait Kadi, M. & Arriens, W.L. (2012) *Increasing Water Security: A Development Imperative.* GWP Perspectives Paper: February.

Biswas, A.K. (2008) Integrated Water Resources Management: Is It Working? *Water Resources Development*, 24 (1), 5–22.

Borchardt, D. (2011) *Message from the Dresden International Conference on Integrated Water Resources Management. Management of Water in a Changing World: Lessons Learnt and Innovative Perspectives.* [Online] Dresden International Conference on IWRM. Available from: www.bmbf.iwrm2011.de/_media/Dresden_to_Bonn_Message_IWRM_2011_11_01.pdf [Accessed May 2012].

Grey, D. & Sadoff, S.W. (2007) *Sink or Swim? Water Security for Growth and Development.* Water Policy, 9, 545–571.

GWP, Global Water Partnership (2000) *Integrated Water Resources Management.* [Online] TAC Background Papers. Number: 4. Available from: http://www.gwp.org/Global/GWP-CACENA_Files/en/pdf/tec04.pdf [Accessed September 2012].

GWP (2012) *Water in the Green Economy.* [Online] GWP Perspectives Paper: June. Available from: http://www.gwp.org/Global/The%20Challenge/Resource%20material/Perspectives%20Paper_Green%20Economy_FINAL.pdf [Accessed May 2012].

McDonnell, R. (2008) *Challenges for Integrated Water Resources Management: How Do We Provide the Knowledge to Support Truly Integrated Thinking?* International Journal of Water Resources Development, 24 (1), 131–143.

Muller, M. & Lenton, R. (2009) *Integrated Water Resources Management in Practice.* Earthscan, London.

Snellen, W.B. & Schrevel, A. (2004) *IWRM for sustainable use of water: 50 years of international experience with the concept of integrated water management.* [Background document]. FAO/Netherlands Conference on Water for Food and Ecoystems. Ministry of Agriculture, Nature and Food Quality, Netherlands.

UN (2005) *Plan of Implementation of the World Summit on Sustainable Development.* [Online] New York, UN. Available from: www.un.org/esa/sustdev/documents/WSSD_POI_PD/English/WSSD_PlanImpl.pdf [Accessed September 2012].

UN (2012) *Status Report on the Application of Integrated Approaches to Water Resources Management.* [Online] UN Water Report. Available from: http://www.unwater.org/rio2012/report/index.html [Accessed October 2012].

Chapter 2

Integrated Water Resources Management: State of the art and the way forward

Pedro Martínez-Santos
Departamento de Geodinámica, Universidad Complutense de Madrid, Madrid, Spain

Maite M. Aldaya
Departamento de Geodinámica, Universidad Complutense de Madrid, Madrid, Spain

M. Ramón Llamas
Water Observatory, Marcelino Botín Foundation, and Royal Academy of Sciences of Spain, Madrid, Spain

ABSTRACT: The concept of Integrated Water Resources Management (IWRM) has gradually gained recognition from water officers, users and academics over the last twenty years. This is largely because IWRM advocates a coordinated approach for managing water resources in a way that balances social and economic needs with care for nature. Whilst these general principles are difficult to argue with, IWRM is both controversial and elusive in practice. Critics argue that IWRM is too vague a concept to be meaningful, and that it lacks a sufficiently clear series of steps for its practical implementation. Not the least of its shortcomings is the fact that many decisions affecting water are made outside the world of water planning. Based on the Spanish experience, this chapter examines the role of water accounting, food trade, environmental externalities and intangible values as key aspects of water management which still need to be resolved in practice.

Keywords: IWRM, intangibles, food trade, virtual water, environmental flows, water pollution

1 INTRODUCTION

Over the last two decades, Integrated Water Resources Management (IWRM) has become a prevailing paradigm among academics, policy-makers and practitioners within the water sector. As a management approach, rather than an end in itself, IWRM encompasses a broad range of aspects, promoting a responsible use of water resources by taking into account cross-sectoral dependencies. Precisely because IWRM is a broad concept, it can be defined in a number of ways. Perhaps the most widely accepted definition is provided by the Global Water Partnership, which states that

IWRM is *'a process which promotes the coordinated development and management of water, land and related resources in order to maximize economic and social welfare in an equitable manner, without compromising the sustainability of vital ecosystems and the environment'* (GWP, 2000). In other words, IWRM is based on the understanding that water resources are an integral component of the ecosystem, a natural resource, and a social and economic good. IWRM recognizes that freshwater is essential for human livelihoods and economic activities, as well as for the environment. Finite and vulnerable, water is dealt with as a valuable resource which needs protection (Foster & Ait-Kadi, 2012). Suitably, IWRM advocates participatory approaches that involve users, planners and policy makers at all levels.

The United Nations report several case studies with on-site implementations of IWRM (UN, 2009). These include examples from places so diverse as Malaysia, Chile, the U.S. or China. Rogers & Leal (2010) also present a series of case studies which could loosely fall under the umbrella of integrated water resources management. Take for instance integrated water supply and wastewater disposal in the Orange County, California. Admittedly, some of the latter are too local or too restricted in scale to be considered full-flight examples. However, they do showcase real-life situations where win-win solutions were adopted to accommodate all water users while also enhancing environmental protection, which is, in fact, the promise of IWRM.

A relatively recent survey by the Global Water Partnership reveals that only twenty out of ninety-five countries surveyed had implemented IWRM at the policy level (GWP, 2006). A more recent undertaking showed that this percentage had grown to the point that 65% of countries presented some degree of IWRM implementation by 2012. However, the same report mentions that it had slowed down or regressed in countries with a low or medium human development index since 2008 (UNEP, 2012a).

Despite some success stories, the practical implementation of IWRM has proven elusive. In this sense, it could be argued that the breadth of the IWRM concept – its main advantage – is also its own undoing. At least, this is the underlying rationale behind most critical voices. Conca (2006), for instance, only recognizes IWRM to be useful as a policy-framing discourse, but highlights that it is also an ambiguous and potentially contradictory notion. As Molle (2008) puts it, IWRM is too vague of a concept to be practical. As such, it can be assimilated to a platonic ideal, much like a 'nirvana' state, one of whose main features is that the sense of progress attached to any shift in its direction is enough to make it an attractive focal point. Moreover, IWRM is essentially hydro-centric whilst, in fact, most decisions that affect water policy are made outside the water sector (UN, 2009).

Based on experiences from different countries, Ioris (2008) contends that water management is essentially the art of choosing between equally important demands. Since every basin is different, this author concludes that there is no blueprint to suit them all. Within this context, claims for wide-ranging integration are unable to offer much help when dealing with specific management questions.

All in all, IWRM is controversial. This chapter deals with some of the aspects which often escape the debate on the practical implementation of integrated water resources management. These include, but are not restricted to, water accounting, food trade, green water and water pollution. The chapter also deals with the intangible value involved in water allocation. These are often underpinned by ethical values – that may be different from country to country – or by the interest of different social groups, whose influence may be of the greatest practical relevance in many cases. Most insights have been drawn from specific dealing with the Spanish experience (De Stefano & Llamas, 2012).

2 THE WORLD'S WATER RESOURCES

The Earth is often known as the 'Blue Planet' for a reason. The hydrosphere contains approximately 1400 million km³ of water. However, only a small amount can be used directly by human beings. Indeed, about 97.5% corresponds to saline waters and is stored in seas and oceans. The main share of this fresh water (69%) is in the form of ice and permanent snow cover in the mountain regions and the poles. In turn, about 30% exists within aquifer systems. Only 0.3% of the total amount of fresh waters on the Earth are concentrated in lakes, reservoirs and river systems (Shiklomanov, 1998). In absolute values, the largest volume of water resources are those of Asia and South America, while the lowest correspond to Europe and Oceania.

Rivers, lakes and aquifers, however, are the principal sources of fresh water to support life necessities and economic activities. Shiklomanov (1998) calculates the difference between precipitation and evaporation from the land surface (119,000–74,200 = 44,800 km³/year) as the total runoff of the Earth's rivers (42,700 km³/year) and direct groundwater runoff to the ocean (2100 km³/year). These figures tend to downplay the role of green water – that which is trapped in the soils and is available for vegetation –, and differ slightly from those shown in Table 1, but they have been considered sufficiently representative for practical purposes.

Although the available figures are subject to uncertainty (Gleick *et al.*, 2011), some authors estimate the world's water withdrawals to be between 3500 and 4000 km³ each year (Shen *et al.*, 2008). Agriculture is by far the most important user, with 70% of the total withdrawal, while industry accounts for 20%. Domestic use only comprises 10%. In this sense, it is important to distinguish consumptive uses from withdrawals. Not all water withdrawals result in consumptive water use. This is because

Table 1 Approximate amount of annual precipitation, evaporation and runoff per continent in relation to the water footprint (Australian Bureau of Meteorology, n.d., and UNEP, 2012a).

Land mass	*Surface (*10³ km²)*	*Population (*10⁶)*	*Avg. rainfall (mm)*	*Total rainfall (mm)*	*Avg. Evap. (mm)*	*Total Evap. (mm)*	*Runoff (km³)*	*WFP[1] (m³/person)*	*WFP (km³)*	*WFP as % rain*
Asia	43,820	4216	650	28,500	410	18,000	10,500	1150	4850	17.0
Africa	30,370	1072	740	22,500	630	19,000	3500	1300	1400	6.2
North America	24,490	346	800	19,500	470	11,500	8000	2800	970	5.0
South America	17,840	596	1600	28,500	900	16,000	12,500	1900	1130	4.0
Europe	10,180	740	820	8400	590	6000	2400	1700	1250	15.0
Oceania	9010	37	440	4000	400	3500	500	1150	45	1.1

Note: Water-related figures have been rounded. Avg. stands for average, Evap. for evaporation and WFP for Water Footprint.
1 Numbers obtained or estimated from figure 4.4 in UNEP (2012a).

a large share of withdrawn waters goes back to the cycle in the form of wastewater or irrigation return flows. On the other hand, not all consumptive uses stem directly from withdrawals. Rainfed agriculture, for instance, represents a significant fraction of the total water use without being responsible for any direct extraction from the water cycle.

This is one of the reasons why global consumption is difficult to estimate. While Shiklomanov (1998) reports that consumptive uses amount to about half the volume of water withdrawn from the cycle – i.e. slightly less than 2000 km^3 per yearmore recent calculations yield significantly different results. Take for instance water footprint estimates, which render a global water consumption per person of about 1400 m^3/yr (Mekonnen & Hoekstra, 2011). Water footprint figures by continent range widely, from North America's 2800 to the Asia and Pacific region's 1150 m^3 per person and year (UNEP, 2012a). Based on these, the world's total water footprint amounts to 9100 km^3 per year, which greatly exceeds the 2000 km^3 estimate provided above.

Part of this variability can be attributed to the 'colors' of water, i.e. blue, green and grey. The blue water footprint is computed as the volume of freshwater in rivers, lakes and aquifers that is consumed to produce the goods and services. On the other hand, the green water footprint is the volume of rainwater stored in the form of soil moisture, which plays a key role in natural ecosystems and rainfed agriculture. Finally, the grey water footprint is the volume of polluted water that associates with the production of goods and services (Hoekstra *et al.*, 2011). The grey water footprint is controversial as the scientific community is yet to reach an agreement as to how it should be calculated. Currently, many authors consider it to be the volume of water that is required to assimilate pollutants to such an extent that the quality of the water remains at or above agreed water quality standards. However, such standards change from country to country and from pollutant to pollutant, and are dependent of the hydrological conditions. For this reason, a good number of authors do not calculate the grey water or simply consider these numbers as a rough indicator.

The water footprint estimates consider green, blue and grey water, whilst Shiklomanov's estimates the blue component and omit green water as a consumptive use. Both examples showcase how global water figures can change depending on how they are computed.

Agriculture accounts for 92% of the total global water footprint. Livestock and related products alone account for 27 percent, whereas industrial and domestic uses only comprise 4.4% and 3.6%, respectively (Mekonnen & Hoekstra, 2011). In absolute terms, about 74 percent of the total water footprint represents rainwater stored in soil (green water), 11 percent represents the consumptive use of surface and groundwater (blue water), and 15 percent represents the freshwater required to assimilate pollution from all sources (grey water). In other words, agriculture is by far the main water consumer, while the role of industrial and urban uses is marginal. However, these two uses play a crucial role in all the water policies due to its political and economic relevance and to the leverage of the public or private corporations that control these activities.

If consumption amounts to roughly 9000 km^3/yr, human beings are only consuming 8.5% of the precipitation that falls on emerged lands (Table 1). This average, however, is not representative for all continents. Take for instance Asia. With a total population of 4200 million people in 2010 and an average water footprint of 1150 m^3 per person and year, the largest and most populated continent on Earth yearly uses about 4850 km^3 of water. This is 17% of its total rainfall. In contrast, Oceania and South America use up to 1.1% and 4% of their rainfall, respectively (Figure 1).

Looking at these numbers one would think that water scarcity at the world scale is hardly a matter of concern. In this regard, and leaving aside the philosophical issues raised in the previous section, there are several issues that need to be considered for integrated management purposes. First and foremost, water is irregularly distributed and accounted for. This implies that these figures are not representative of all basins across each continent. Some regions be subject to greater pressures and confronted with water scarcity sooner or later. In this regard, it becomes obvious that global figures are only useful as a first approximation, and that numbers need to be dealt

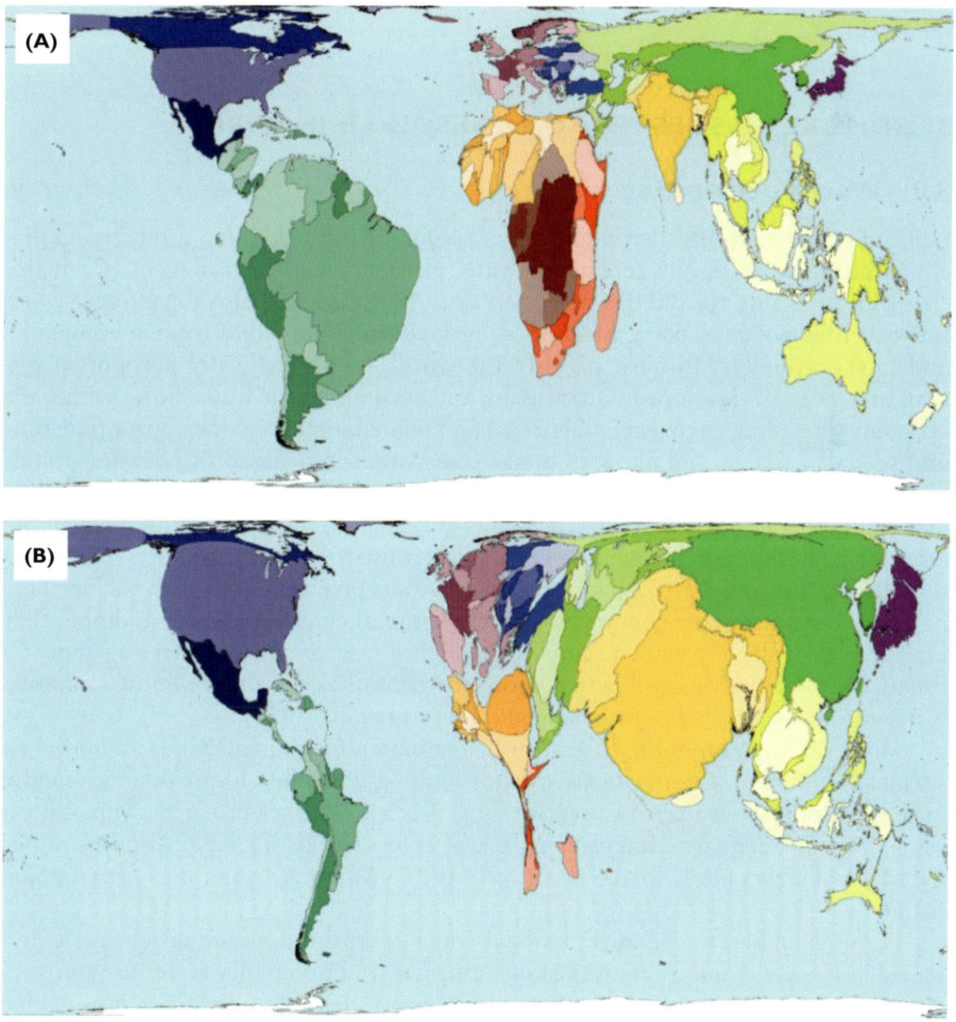

Figure 1 (A) Total rainfall (1961–1990); (B) Water use. Country areas are deformed as a function of total rainfall and water use, i.e. the larger the country is represented, the larger its proportional share over the total of all countries (González-Vallvé, 2012).

with at the basin or sub-basin scale. Besides, not all water resources are destined for human consumption.

An important corollary is that, given the role of agriculture as the world's main water consumer, food trade may provide a valuable means to balance the irregular distribution of water across regions. Secondly, overall water figures tend to overlook the problem of water pollution. While techniques to clean up our water resources are becoming increasingly available, water pollution – particularly diffuse agricultural contamination – is perhaps the greatest challenge to deal with. Contamination is not the only environmental challenge, though. Environmental flows and green water are also essential to underpin ecosystems. Finally, there are a series of intangible values – social, economic, political and cultural – that constrain the implementation of the IWRM principles at the basin scale. The following sections will delve further into these aspects.

3 SOME KEY ISSUES YET TO BE RESOLVED IN IWRM

3.1 Water accounting

In a context of unevenly distributed water resources and increasing drought and precipitation in some regions, enhanced water efficiency and management is a major challenge, not only for direct water users and managers, but also for indirect water users such as policy makers, businesses, agricultural commodity trading companies and final consumers. In most parts of the world, consistent water accounting systems are yet to be developed. Quantifying and accounting for water flows within the economy (including environmental needs) and related impacts in the appropriate time and spatial scales, would allow to attain transparent information to develop robust allocation and management systems that underpin a green economy (UNEP, 2012b).

Traditional national water use accounts only refer to the water withdrawal within a country. They do not distinguish between water use for making products for domestic consumption and water use for producing export products. They also exclude data on water use outside the country to support national consumption. In addition, they include blue water use only, excluding green and grey water. In order to support a broader sort of analysis and better inform decision-making, the traditional national water use accounts need to be extended (Hoekstra *et al.*, 2011).

However, sometimes the lack of sufficient data about climate, soils and growing seasons is in many cases the greatest factor limiting the ability to provide meaningful information on the water consumptive uses of crops or other products. This is most often due to inadequate databases, or lack of access to existing data. And this causes a cascade of errors in the final estimation of the consumptive water uses per crop and surface.

As an example, the Spanish national water footprint accounting scheme is presented in Figures 2 and 3 versus the governmental traditional blue water use statistics from the non-consumptive perspective, in Tables 2 and 3. The large differences in blue water consumptive and non-consumptive uses in Spain, depending on the data source, reflects the need of systematic, standardized and comprehensive water accounting systems (UNSD, 2012).

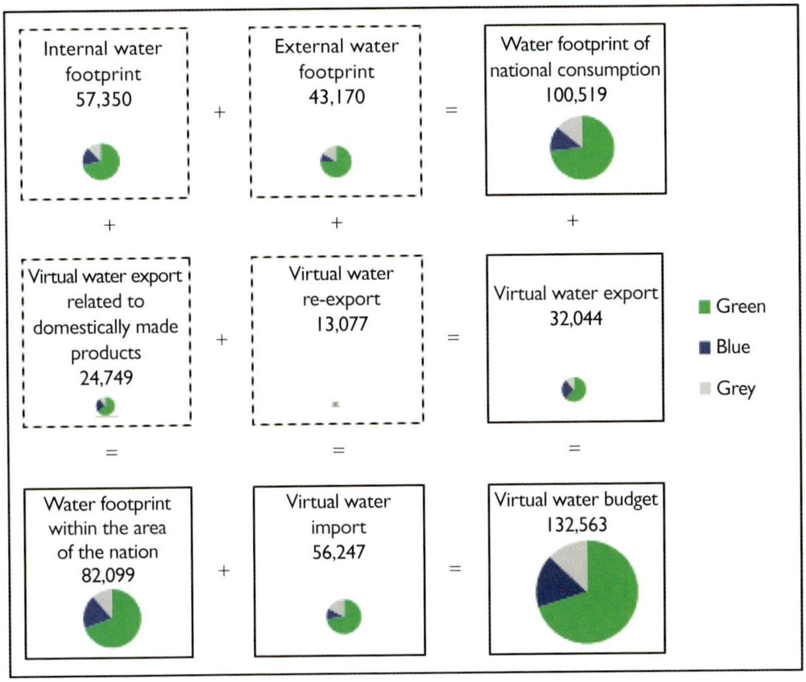

Figure 2 The Spanish water footprint accounting scheme per type of water (green, blue, grey). The accounting scheme shows the various balances that hold for the water footprint related to national consumption (WFcons, nat), the water footprint within the area of the nation (WFarea, nat), the total virtual water export (Ve) and the total virtual water import (Vi) (Mm^3/year) (Mekonnen & Hoekstra, 2011).

3.2 Food trade as a means to balance the irregular temporal and spatial distribution of water

There is enough fresh water to produce food for the global population now and in the future. However, world leaders must take action now by embracing transparency, removing perverse subsidies and making WTO agreements fairer. According to the McKinsey report, in a business-as-usual scenario, under an average economic growth scenario and if no efficiency gains are assumed, global demand to withdraw water would outstrip currently accessible water supplies by 40% by 2030 (2030 Water Resources Group, 2009). There is a need to improve water efficiency and productivity around the world, both in humid and arid regions. This water embedded in commodities can be then redistributed through trade. The virtual water trade, at national, regional and global scales, can be a tool for improving overall efficiency by capitalizing on the comparative advantages of certain water uses in particular regions. Virtual water trade could be therefore a significant part of the solution of the global water challenge.

Traditionally, water resources management has been dealt with from the local, river basin or national perspective. Even if it is increasingly recognized that water

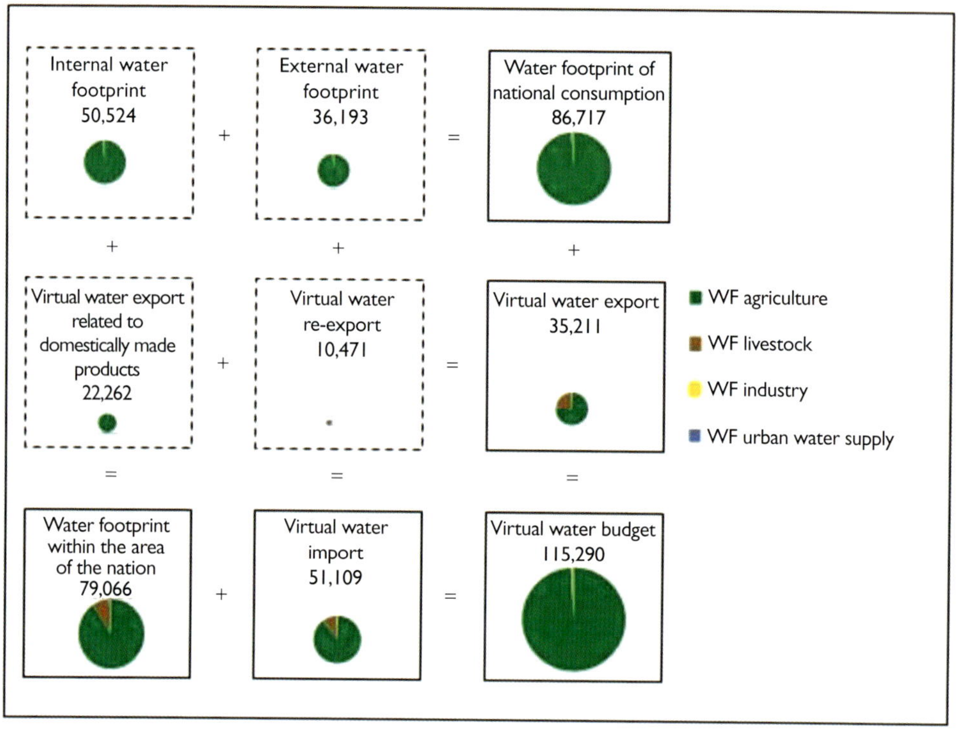

Figure 3 The Spanish water footprint accounting scheme per sector considering green and blue water. The accounting scheme shows the various balances that hold for the water footprint related to national consumption (WFcons, nat), the water footprint within the area of the nation (WFarea, nat), the total virtual water export (Ve) and the total virtual water import (Vi) (Mm^3/year) (Mekonnen & Hoekstra, 2011). Note: When there is no brown legend, livestock is included in the WF agriculture.

Table 2 National blue water accounting in Spain (Mm^3/year).

	Blue water withdrawals			*Blue water consumption*	
Source	MIMAM, (2001)*	OSE (2010) [in Fundación Mapfre, 2011]	MIMAM (2007)	Garrido *et al.*, (2010)	Mekonnen & Hoekstra (2011)
Urban water supply	4667	4941	3619	4042	479
Agriculture	24,094	16,211	11,897 (crop) 259 (livestock)	16,178 (crop) 260 (livestock)	14,136 (crop) 750 (livestock)
Industry	1647	1772	1268	1700	330
Cooling	4915	6795	–	–	–

*Actual consumption has been evaluated in MIMAM (2001) in 20,783 Mm^3/year and the volume of returns in 14,539 Mm^3/year.

Table 3 The Spanish water footprint accounting scheme per type of water (green, blue, grey) (Mekonnen & Hoekstra, 2011).

Item	*Green*	*Blue*	*Grey*	*Total*
Internal water footprint	41,039	9485	6826	57,350
Virtual water export related to domestically made products	16,052	6210	2487	24,749
Water footprint within the area of the nation	57,091	15,695	9313	82,099
External water footprint	32,561	3632	6977	43,170
Virtual water re-export	8349	2122	2606	13,077
Virtual water import	40,910	5754	9583	56,247
Water footprint of national consumption	73,600	13,116	13,803	100,519
Virtual water export	19,524	9049	3471	32,044
Virtual water budget	93,124	22,165	17,274	132,563

Note: The accounting scheme shows the various balances that hold for the water footprint related to national consumption (WFcons, nat), the water footprint within the area of the nation (WFarea, nat), the total virtual water export (Ve) and the total virtual water import (Vi) (Mm^3/year).

governance has a global dimension, the linkages between international trade and freshwater resources are rarely analyzed.

An obvious effect of international trade in water-intensive commodities is that it generates water savings in the countries that import those commodities. This effect has been discussed since the mid-1990s (Allan, 2011; Hoekstra, 2003). By 'importing' virtual water embodied in agricultural commodities, a nation 'saves' the amount of water it would have required to produce those commodities domestically. Import of water-intensive commodities reduces national water demand and potential impacts. The importing country can then use the domestic water for more valuable purposes, such as ecosystem services, urban water supply or tourism. Virtual water trade – the water embedded in traded products ranging from crops to manufactured goods – between or within nations could also mitigate drought cycles and be an alternative to inter-basin water transfers. Virtual water trade can be a means of reducing water scarcity. This is particularly relevant to arid or semi-arid countries with scarce water resources. It also allows access to global water, usually green, to these countries with scarce water resources. Virtual water trade, thus, has the potential to relieve water stress and improve water security.

The other side of international trade in water-intensive commodities is that it takes water in the exporting countries, which can no longer be used for other (domestic) purposes (Chapagain *et al.*, 2006). National water losses may be positive or negative from an economic perspective depending on the context. Water losses are positive in economic sense if the benefit in terms of foreign earnings they provide is outweighed by the opportunity costs of water use and the negative externalities left at the production site. The social and environmental costs that are often associated with water use remain in the exporting countries; they are not included in the price paid for the products by the consumers in the importing countries. Even if there is a net global water loss from an exchange, there might be a saving of blue water at the cost of a greater loss of green water or vice versa.

About one-fourth of the global water footprint is related to production for export (Table 4). The global virtual water trade for agricultural and industrial products totaled 2320 km^3 per year between 1996 and 2005, with crops contributing 76 percent and animal and industrial products each contributing 12 percent. Within agriculture, it is mainly low-economic value and high virtual water trade commodities that are traded: Oil crops (43%), staple food (cereals, 17%) followed by industrial products (12%) and stimulants (8%) (Figure 4) (Mekonnen & Hoekstra, 2011).

International trade can save water globally when a water-intensive commodity is traded from an area where it is produced with high water productivity (low water input per unit of output) to an area with lower water productivity (high water input per unit of output) (Mekonnen & Hoekstra, 2011). However, at the global level, the overall water saving estimate is generally a small percentage, about 5% of the global agricultural water consumption, which is estimated at about 8400 km^3/year. The order of magnitude of this percentage is similar to the range of error in the calculations of the water footprint volumes by commonly used methods, when these are estimated

Table 4 International virtual water flows (1996–2005) (Mekonnen & Hoekstra, 2011).

Item	*Volume (billion m^3/yr)*	*Percentage (%)*
Crop products	1766	76
Farm animal products	272	12
Industrial products	282	12
Total	2320	100

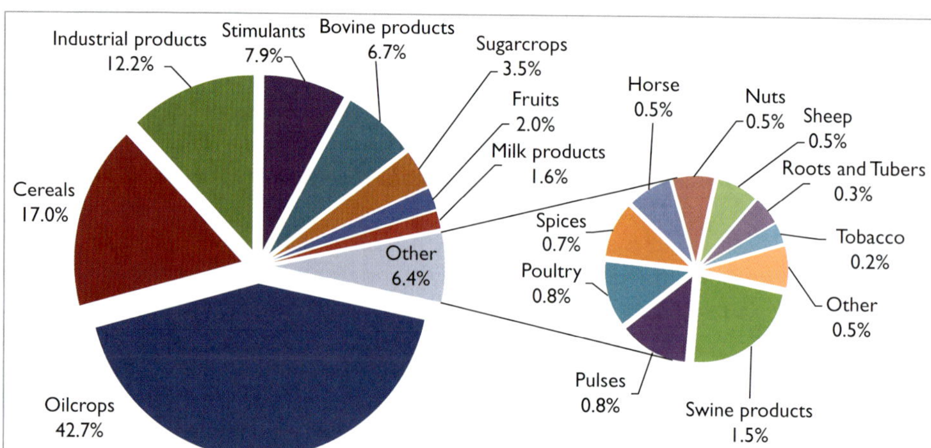

Figure 4 Contribution of different product categories to the global sum of international virtual water flows (Mekonnen & Hoekstra, 2011).

for national or bigger areas. Water footprint calculations, when applied to large scale regions may have significant uncertainties and limitations (Liu *et al.*, 2009).

Recent studies quantitatively corroborate that main international virtual water flows are based on green water (Aldaya *et al.*, 2010b; Mekonnen & Hoekstra, 2011) (Figure 5). Major exporters produce under relatively favorable productive rainfed conditions while most importers would have relied (at least partially) on their blue water resources. Virtual water trade, thus, can reduce irrigation water demand and play a role in ensuring water and water dependent food security in water-short countries. The virtual water trade can efficiently redistribute global water and partially help to address the impacts of production.

At present, however, this option is far from being fully exploited due to the absence of a more water friendly international trade regime with equal access to global markets. Other obstacles are formed by the inadequacy of water pricing structures worldwide and the agricultural subsidies in the EU and USA (Aldaya *et al.*, 2010b).

The current global virtual water trade is primarily among the countries above the low-income level in the World Bank country classification (Yang *et al.*, 2006). Countries with low-income levels are minor participants. As Allan (2006) points out, socio-economic development is a prerequisite to access virtual water in the global system. Besides, other factors are also interrelated with global green water trade such as availability of land, labor, technology, the costs of engaging in trade, the potential for further increases in the productivity of soil water and irrigation water, national food policies and international trade agreements. There are factors that can also contribute to increased food demand and to increased water use for food production, such as population growth, changes in diets and the use of cereals and oilseeds for biofuel production (De Fraiture *et al.*, 2007; Gerbens-Leenes *et al.*, 2009).

In the future, in a context of greater water demand, green virtual water trade will probably become increasingly important from a global perspective. Rainfed agriculture, with some of the highest yields in several regions, hold great underexploited potential for increasing water productivity through better water management practices – gaining more yield and value from water. In this context, the socio-economic development of poor economies in humid regions, such as the case of Sub-Saharan

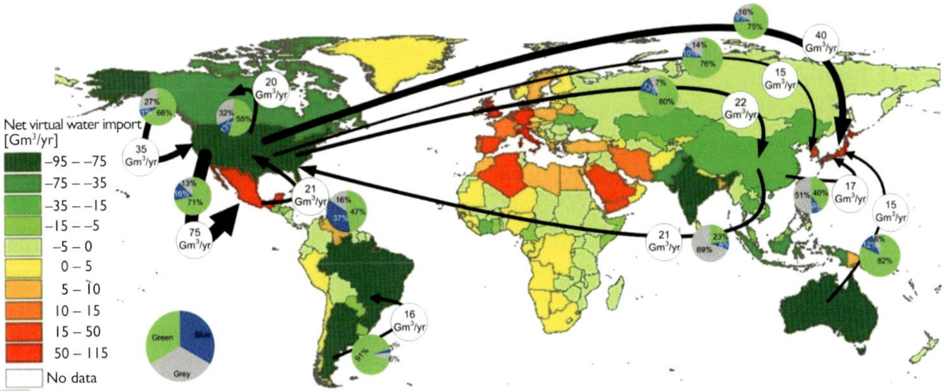

Figure 5 National virtual water balances. Arrows show gross virtual water flows >15 Gm³/yr (1 Gm³ = 1 km³) (Mekonnen & Hoekstra, 2011).

Africa, could let these economies enter the international market and promote the virtual water trade solution, as Argentina did earlier. The importance of international green virtual water trade and its contribution to water and food security in the future will, though, depend on factors such as the productivity of blue and green water, international trade agreements, and the nature of domestic economic objectives and political considerations.

3.3 Environmental needs and water pollution in the context of IWRM

Assessment of water availability, water use and water scarcity has been the subject of increasingly intensive research over the past years (Millennium Ecosystem Assessment, 2005). However, the requirements of ecosystems for green and blue water have rarely been considered explicitly in such assessments (Aldaya *et al.*, 2010a; UNEP, 2012a).

This is a very important issue from a global perspective. Falkenmark & Rockström (2004) estimated the water required to maintain ecosystem goods and services as 75 percent of the total rainfall, while direct human water use represented 25 percent of the total. These figures include both blue (groundwater and surface water) and green water (water stored in the soil).

However, more accurate figures are needed in relation to the water quantity that is accessible and reliable for human use considering the water requirements for the ecosystems, which is in principle a smaller quantity than the absolute raw water available in nature. Estimations of this type are rare and generally without enough factual evidence.

Although many goals in the Johannesburg Plan of Implementation acknowledge the importance of marine and coastal ecosystems (WSSD, 2002), there is less recognition of water needs to support ecosystems, which are themselves legitimate water users (UNEP, 2012b). Ecosystems, which provide life-supporting goods and services, need for water of adequate quantity and quality as well as appropriate timing. Although the importance of formally recognizing the environment as a legitimate water user is increasing, it remains on a relatively small scale in practice, with many aquatic ecosystems still at risk (Garrick *et al.*, 2009). This includes not only environmental flow requirements – or aquatic ecosystem water requirements or ecosystem blue water requirements, including groundwater and surface water, but also ecosystem green water requirements – or terrestrial ecosystem water requirements (De Stefano & Llamas, 2012).

Few are the studies including both the aquatic and terrestrial ecosystem water requirements (Aldaya *et al.*, 2010c; Salmoral *et al.*, 2011; Willaarts, 2012) (Figure 6), even if recent studies imply that this might be crucial. Latest research suggests that forests might be by far the major water consumers in semi-arid climates, above agriculture, and that changes in land use may have a greater impact compared to climate change on runoff reduction downstream (Willaarts, 2012). In this kind of climate, the green water-based grasslands, man-made agrosilvopastoral systems, has enhanced the maintenance of an extraordinarily high biodiversity.

If computing environmental requirements is difficult, water contamination poses another major challenge. Accounting for water quality and related impacts on water

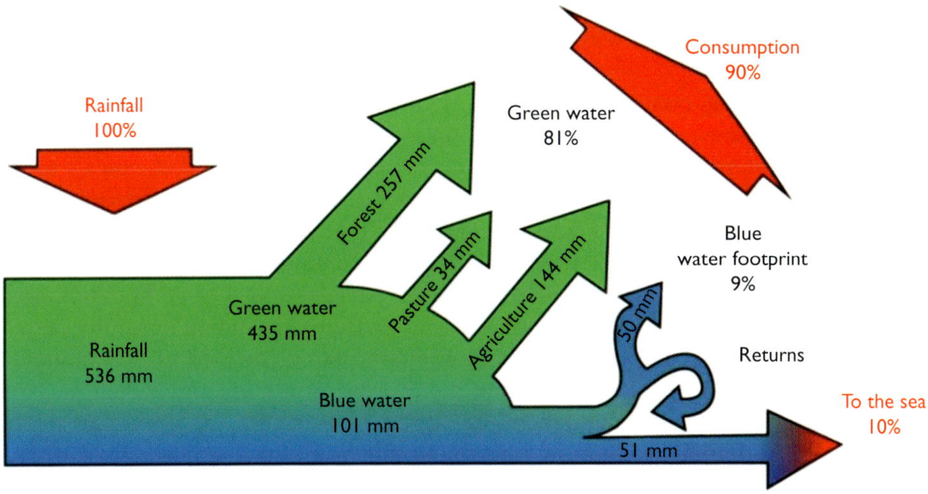

Figure 6 Water balance in the Guadalquivir river basin (Salmoral *et al.*, 2011).

resources is problematic. This is due to many factors, including the various different types of pollutants coming from industrial facilities and agriculture; the interactions among pollutants; the variety of ways water quality can be compromised (i.e. contaminant loads, temperature, odor, turbidity), and the various approaches to accounting for the resulting impacts to ecosystems and communities.

A wide range of human and natural processes affect the biological, chemical, and physical characteristics of water, and thus impact water quality. Contamination by pathogenic organisms, trace metals, and anthropogenic and toxic chemicals, the introduction of non-native species, and changes in the acidity, temperature, and salinity of water can all harm aquatic ecosystems and make water unsuitable for human use and ecosystem support (UNEP, 2010).

Solving water quality problems requires strategies to prevent, treat, and remediate water pollution. As a first order intervention, pollution can be prevented before it enters waterways; second, wastewater can be treated before it is discharged; and third, the biological integrity of polluted watercourses can be physically restored through remediation.

In most industrialized nations, water quality improvement efforts focus on two approaches: The construction of centralized or on-site water-treatment facilities and wastewater plants, and regulations aimed at individual point source-polluters who discharge water pollution, including both direct dischargers who discharge effluents into receiving waters and indirect dischargers who release pollutants into sewer systems that flow into treatment plants. As a result, the technology used for cleaning urban and industrial waters has improved dramatically over the last decades. Today, most effluents can be treated at a relatively low cost (US\$0.5 to 1.5 per m^3). Hence, the issue of water quality in rivers has become a matter of political will and social awareness.

The flipside of the coin is diffuse pollution, which is particularly critical in aquifer systems. This is the result of precipitation runoff from many diffuse sources including fertilizers, nutrients, and pesticides from agriculture, and oil, grease and toxics from urban settlements. These diffuse pollutants from multiple sources are not easily regulated, which largely explains why the efforts to address point sources have been more successful to date.

Finally, emerging contaminants pose an increasingly important threat to the health of water bodies, particularly in developed regions. A wide range of chemicals, designed for use in industry, agriculture, and as consumer goods and chemicals unintentionally formed or produced as by-products of industrial processes or combustion, are potentially of environmental concern. The term 'emerging contaminants' does not necessarily correspond to 'new substances', i.e. newly introduced chemicals and their degradation products/metabolites or by-products, but also refers to compounds with previously unrecognized adverse effects on the ecosystems, including naturally occurring compounds. Therefore, 'emerging contaminants' can be defined as contaminants that are currently not included in routine monitoring programs and which may be candidates for future regulation, depending on research on their (eco)toxicity, potential health effects, public perception, and on monitoring data regarding their occurrence in the various environmental compartments (Petrovic *et al.*, 2008).

3.4 Intangible values and IWRM

Intangible values are defined as those elements that should be considered in the process of implementing IWRM but which are not easily quantifiable. This concept was first included in the Declaration on the Ethics of Freshwater Uses by UNESCO's World Commission on the Ethics of Science and Technology (Delli Priscoli *et al.*, 2004). The idea of intangibles, also known as 'externalities' or 'intrinsic values', is recurrent in water-related debates. This is only to be expected, as moral and ethical values ultimately underpin most human decisions (Llamas, 2012).

In recent times, globalization has triggered a renewed interest in water ethics. Our planet is home to seven billion people distributed in two hundred countries and at least as many different cultures. Developments carried out within a given region can have an impact on others, what often raises moral dilemmas. Since ethics are intrinsically intertwined with the local cultures and religions, regulating these activities poses a significant challenge. Some authors consider that regulations should stem from the universal principles of natural law, whilst others advocate a binding international consensus. Overall, it is difficult to approach the issue of intangible values in a manner that encompasses all relevant aspects. Perhaps the most illustrative way for the purpose of this chapter is to focus on a specific geographic setting and to reflect on the interests of different social collectives. Thus, the remainder of this section will deal mostly with the Spanish experience, presenting an overview of the aspects involved with politics, farming, local corporations, conservationist groups, the academia, water managers and society at large.

Although sometimes overlooked by the average citizen, water management is crucial in national-scale politics. Take for instance those United Nations Millennium Development Goals that pertain drinking water and basic sanitation and the recent decision to declare these as an obligation for every country.

Political parties are usually content if urban water supply is satisfied. This is relatively simple to achieve – provided that there is enough funding –, as the technical solutions and expertise are readily available. It is also understandable in the sense that urban agglomerations are the key to both economic development and election polls. However, urban water only represents 8% of the total consumptive uses in Spain. This means that a large share of the available human and economic resources are devoted to managing a small part of the water. In contrast, agricultural water management receives comparatively little attention despite being the main consumer. This becomes particularly clear in the case of groundwater resources, where illegal use is rampant across the country (Llamas & Martínez-Santos, 2005; Martínez-Santos *et al.*, 2008).

Water issues are frequently used as a political weapon during election time. Take for instance the Ebro water transfer. In 2001, Spain's conservative party passed a decree for a 1000 km pipeline to take water from the Ebro river to the rest of Spain's Mediterranean coast. The socialist party, together with farmers from the Ebro basin and conservation lobbies opposed the transfer strongly. Large demonstrations – at times attended by several hundred thousand people – took place in different cities of Spain, as well as in Brussels. As a result, water allocation became a major issue in the months prior to Spain's 2004 general election. The Ebro transfer was finally overruled following a democratic overturn of the conservative government.

The emotional side of things tends to overrule technical expertise in such debates. Consequently, some water managers view IWRM with skepticism. A leading author of Spain's White Book on Water (MIMAM, 2000), who was also the Chief Planning Officer of one of Spain's River Basin Authorities for several years, told the authors of this chapter that the efforts to obtain reliable data to underpin the implementation of IWRM are largely useless because politicians are likely to ignore whichever information that does not suit their agenda (Cabezas, 2012, personal communication).

While this may be the case in current times, it is expected that things should change in the long run. After all, sound theories tend to permeate the collective mind and eventually evolve into common sense. In this regard, relevant personalities from Spain's two main political parties have advocated the need for an overall agreement – a 'pact' – on water issues. A recent book by the Water Observatory of the Botín Foundation shows that the current climate of economic duress may be beneficial for this purpose (De Stefano & Llamas, 2012).

Farming – and farmers – also bring important intangibles to water debates. As a matter of fact, farmer lobbies have a great leverage in most countries (Mukherji, 2006), from the USA where they only represent 2% of the working force, to India, where they represent 60%. In Spain farmers only account for 4% of the working population, with differences from less than 1% in Balearic Archipelago to 12% in Extremadura. Relevant changes in water policy almost necessarily need to take into account the position of farmer lobbies. Win-win outcomes must be sought over zero-sum solutions, as the latter are unlikely to be translated into practice. In this sense, technology may prove beneficial in integrating the needs of ecosystems in water planning. Spain, whose agricultural sector is highly technified, provides a good example of how the classic 'more crops and jobs per drop' motto is gradually changing to 'more cash and care of nature per drop'.

While agricultural activities are performed mostly by private people, food trade is controlled by governments and by a few large international corporations (Allan, 2011). A potential increase in food prices in the near future, mirroring the volatility experienced in 2007 and 2008, raises concerns on how the water sector will be affected. In this regard, it should be noted that global and local water consumption can be altered significantly by issues so diverse as the general increase in living standards in emergent countries (which triggers changes in dietary habits), the use of edible crops as biofuels, the increased price of energy, or the manipulation of prices by governments and international corporations.

Over the years, corporations have evolved towards the ideal of social responsibility and, more recently, to the 'shared value' concept. Social responsibility reports frequently pay attention to the ecological aspects, and deal specifically with water issues (Orr, 2009). This is the rationale behind international initiatives such as the Global Compact, which encourages corporations to contribute to 'sustainable development and to the green economy', or the World Business Council for Sustainable Development. This is largely uncharted territory for Spanish companies, although some of the larger ones are beginning to show an incipient degree of concern about the environment.

In global terms, industrial processes tend to use relatively little water. In fact, most of it comes embodied in agricultural raw materials. The issue with industrial processes is more related to contamination. Since large companies usually have enough economic means to ensure that polluted water effluents do not return untreated to the water cycle, the polluter-pays principle is becoming increasingly widespread in national and international legislation.

All in all, corporate social responsibility is yet to establish itself as a common industrial practice. In this regard, the role of corporations is beneficial to raise awareness as to the relevance of water in maintaining ecosystems. On the other hand – and unless appropriate quality standards are established –, corporate responsibility may end up as a watered-down 'green disguise' to increase the goodwill value of commercial brands.

Conservation groups have attained recognition since the 1960s. Spanish conservation groups currently coexist with the national chapters of international organizations such as the World Wildlife Fund or Greenpeace. Until the implementation of the EU Water Framework Directive, public participation practices were scarce in Spain's water policy. Only those stakeholders with an economic interest in water (electric utilities, urban water supply companies, irrigators) were entitled to an opinion. Hence, conservation groups were often left out of formal negotiations. To make up for this, conservation groups have played an active role in public water debates. They have also been proactive in denouncing illegal water usage and in prompting the authorities to enforce the law in case of abuses. In some cases, these have exposed the weaknesses of water authorities, whose means are often inadequate to cope with ever-increasing responsibilities.

One of the most influential in the field of water resources is the Fundación Nueva Cultura del Agua (FNCA), which originated about fifteen years ago by the initiative of several academics. FNCA comprised a variety of outlooks of the water sector (hydrology, economy, sociology and law) and, due to its presence in a variety of forums, including mass media, they managed to have their unofficial say in water policy. The FNCA played an active role in assisting the socialist party with the cancellation of the Ebro Transfer. It also supported the alternative approach based on desalination plants. While the later have not lived up to their full potential, FNCA is credited for supporting the transition towards state-of-the-art water management practices in the country.

Water policy is currently open to all sectors of society, but the level of public involvement is still low. More recent times have witnessed the birth of public water research centers and private think tanks such as the Water Observatory of the Botín Foundation. Other significant activity has been the survey on the Transparency of the Water Authorities done as a joint venture by Transparency International-Spain and the aforementioned observatory, an ongoing initiative that can be considered a pioneering step at the world scale towards the evaluation of the intangible aspects of IWRM.

4 CONCLUSIONS

Water is prevalent in all aspects of human life, from domestic use to economic activities and ecosystem conservation. IWRM is a welcome addition to the water management community in the sense that it provides a much needed framework to deal with all these aspects in a joint manner. For all its shortcomings, however, IWRM is best described as a useful utopia that prompts water managers and users to think of the broader picture, thus raising awareness about the manifold nature of the resource and making its use more sustainable.

Perhaps the greatest challenge in modern water management consists in the need to integrate tangible realities (volumes of consumptive and non-consumptive water use, pollutant concentrations, economic figures and related jobs along supply-chains) and intangible values (cultural, religious, political, educational and others). The former provide a necessary foothold on objective figures, while the latter are difficult to manage and quantify but usually play a larger role in political decisions. Since all contexts are different, there can be no universal blueprint to underpin water policy. In this sense, tailor-made assessments based on strong public participation processes are advocated as a means to optimize water management at the basin scale.

Another important challenge pertaining water management is the fact that many of those decisions which affect water are made outside the formal planning process. These include key issues such as trade, agriculture, energy, economic or environmental policy, and can be made by actors so diverse as individual farmers, international corporations or supra-national authorities. In this regard, relatively new paradigms like corporate social responsibility or virtual water call for further consideration within the water management process.

REFERENCES

2030 Water Resources Group (2009) *Charting Our Water Future: Economic frameworks to inform decision-making*. McKinsey & Company.

Aldaya, M.M., Allan, J.A. & Hoekstra, A.Y. (2010a) Strategic importance of green water in international crop trade. *Ecological Economics*, 69(4), 887–894.

Aldaya, M.M., Garrido, A., Llamas, M.R., Varelo-Ortega, C., Novo, P. & Casado, R.R. (2010b) Water footprint and virtual water trade in Spain, In: Garrido, A. & Llamas, M.R. (eds.) *Water Policy in Spain*. Leiden, The Netherlands, CRC Press. pp. 49–59.

Aldaya, M.M., García-Novo, F. & Llamas, M.R. (2010c) *Incorporating the Water Footprint and Environmental Water Requirements into Policy: Reflections from the Doñana Region (Spain)*. Papeles de Agua Virtual n.° 5, Fundación Marcelino Botín, Santander.

Allan, J.A. (2006) Virtual water, part of an invisible synergy that ameliorates water scarcity. In: Rogers, Llamas, Martinez-Cortina (eds.) *Water Crisis: Myth or Reality?* Leiden, Balkema Publishers.

Allan, J.A. (2011) *Virtual Water: Tackling the Threat to Our Planet's Most Precious Resource*. London, I.B. Tauris.

Australian Bureau of Meteorology (n.d.) *Australia; Climate of Our Continent*. [Online] Available from: http://www.bom.gov.au/lam/climate/levelthree/ausclim/zones.htm [Accessed November 2012].

Chapagain, A.K., Hoekstra, A.Y. & Savenije, H.H.G. (2006) Water saving through international trade of agricultural products. *Hydrology and Earth System Sciences* 10, 455–468.

Conca, K. (2006) Governing Water: *Contentions Transnational Politics and Global Institution Building*. Cambridge, MIT Press.

De Fraiture, C., Giordano, M. & Yongsong, L. (2007) *Biofuels and Implications for Agricultural Water Use: Blue Impacts of Green Energy*. Colombo, Sri Lanka, International Water Management Institute.

De Stefano, L. & Llamas, M.R. (eds.) (2012) *Water, Agriculture and the Environment in Spain: Can We Square the Circle?* Boca Raton, CRC Press.

Delli Priscoli, J., Dooge, J. & Llamas, M.R. (2004) *Water and Ethics: Overview*. UNESCO. Series on Water and Ethics. Essay number: 1.

Falkenmark, M. & Rockström, J. (2004) *Balancing Water for Humans and Nature: The New Approach in Ecohydrology*. London, Earthscan.

Foster, S. & Ait Kadi, M. (2012) Integrated Water Resources Management (IWRM): How does groundwater fit in? *Hydrogeology Journal*, 20, 415–418.

Fundación Mapfre (2011) *Huella Hídrica, Desarrollo y Sostenibilidad en España*. Madrid, Fundación Mapfre.

Garrido, A., Llamas, M.R., Varela-Ortega, C., Novo, P., Rodríguez-Casado, R. & Aldaya, M.M. (2010) *Water Footprint and Virtual Water Trade in Spain: Policy Implications*. New York, Springer.

Garrick, D., Siebentritt, M.A., Aylward, B., Bauer, D.C.J. & Purkey, A. (2009) Water markets and freshwater services: Policy reform and implementation in the Columbia and Murray-Darling Basins. *Ecological Economics*, 69, 366–379.

Gerbens-Leenes, W., Hoekstra, A.Y. & Van der Meer, T.H. (2009) The water footprint of bioenergy. *Proceedings of the National Academy of Sciences*, 106(25), 10219–10223.

Gleick, P., Allen, L., Cohen, M.J., Cooley, H., Christian-Smith, J., Heberger, M., Morrison, J., Palaniappan, M. & Schulte, P. (2011) *The world's water*. The Biennial Report on Freshwater Resources, Volume 7. Washington D.C., Island Press.

González-Vallvé, J.L. (2012) *El Agua en el Mundo y en España*. La internacionalización del modelo de gestión español. [Online] Club de Roma. Available from: http://www.clubderoma.net [Accessed January 2013].

GWP (2000) *Integrated Water Resources Management*. Global Water Partnership, Technical Advisory Committee. Background Paper number: 4.

GWP (2006) *Setting the scene for change*. [Online] Global Water Partnership Toolbox. Available from: http://www.gwptoolbox.org/images/stories/Docs/gwp_setting%20the%20stage%20for%20change%202006.pdf [Accessed January 2013].

Hoekstra, A.Y. (ed.) (2003) *Virtual water trade: Proceedings of the International Expert Meeting on Virtual Water Trade*. UNESCO-IHE. Value of Water Research Report Series number: 12.

Hoekstra, A.Y., Chapagain, A.K., Aldaya, M.M. & Mekonnen, M.M. (2011) *The water footprint assessment manual: Setting the global standard*. London, UK, Earthscan.

Ioris A. (2008) The limits of integrated water resources management: A case study of Brazils Paraba do Sul River Basin. *Sustainability: Science, Practice, & Policy*, 4(2), 4–11.

Liu, J., Zehnder, A. & Yang H. (2009) Global consumptive use for crop production: The importance of green and virtual water. *Water Resources Research* 45(5), W05428.

Llamas, M.R. (2012) Foreword: The role of the Botín Foundation to support the analysis of issues on Water Ethics. *Water Policy*, 14.

Llamas, M.R. & Martínez-Santos, P. (2005) Intensive groundwater use: Silent revolution and potential source of social conflict. Editorial. *ASCE Journal of Water Resources Planning and Management*, 131(5), 337–341.

Martínez-Santos, P., De Stefano, L., Martínez-Alfaro, P.E. & Llamas, M.R. (2008) Wetland restoration in the Mancha Occidental aquifer, Spain: A critical perspective on water, agricultural and environmental policies. *Restoration Ecology*, 16(3), 511–521.

Mekonnen, M.M. & Hoekstra, A.Y. (2011) *National water footprint accounts: The green, blue and grey water footprint of production and consumption*. UNESCO-IHE. Value of Water Research Report Series number: 50.

Millennium Ecosystem Assessment (2005) *Chapter 7: Fresh Water Ecosystem Services*. Millennium Ecosystem Assessment. Washington D.C., World Resources Institute.

MIMAM (2001) *Libro Blanco del Agua en España*. Madrid, Spain, Ministerio de Medio Ambiente, Secretaría de Estado de Aguas y Costas.

MIMAM (2007) *El uso del agua en la economía española: Situación y perspectivas*. Ministerio de Medio Ambiente.

Molle, F. (2008) Nirvana concepts, narratives and policy models: Insight from the water sector. *Water Alternatives*, 1(1), 131–156.

Mukherji, A. (2006) Is intensive groundwater use a solution to world water crises? In: Rogers, P. & Llamas, M.R. (eds.) *Water Crisis: Myth or Reality?* Proceedings of the Santander Workshop. Harvard University, Complutense University of Madrid, and Marcelino and Botín Foundation. Santander, Spain 14–16 June 2004. Lisse, The Netherlands, Balkema Publishing Co.

Orr, S. & Cartwright, A. (2009) Water scarcity risks: Experience from the private sector. In: Cortina, M., Garrido, A., López-Gunn, E. (eds.) *Re-thinking Water and Food Security*. London, UK, CRC Press. pp. 181–192.

Petrovic, M., Radjenovic, J., Postigo, C., Kuster, M., Farre, M., López de Alda, M. & Barceló, D. (2008) Emerging contaminants in waste waters: Sources and occurrence. In: Hutzinger, O., Barceló D. & Kostianoyv, A. (eds) *The Handbook of Environmental Chemistry*. Water Pollution, Vol. 5, Part S/1. Heidelberg, Springer-Verlag, pp. 1–36.

Rogers, P. & Leal, S. (2010) *Running out of water. The looming crisis and solutions to conserve our most precious resource*. New York, Palgrave-MacMillan.

Salmoral, G., Dumont, A., Aldaya M.M., Rodríguez-Casado R., Garrido A. & Llamas M.R. (2011) *Análisis de la huella hídrica extendida de la cuenca del Guadalquivir. Papeles de Seguridad Hídrica y Alimentaria y Cuidado de la Naturaleza*. Fundación Botín, Observatorio Del Agua. Report number: 1.

Shen, Y., Oki, T., Utsumi, N., Shinjiro, K. & Hanasaki, N. (2008) Projection of future world water resources under SRES scenarios: Water withdrawal. *Hydrological Sciences Journal*, 53(1), 11–33.

Shiklomanov, I.A. (1998) *World water resources: A new appraisal and assessment for the 21st Century*. Milton Keynes, UK, UNESCO.

UN (2009) *Integrated water resources management in action*. United Nations World Water Assessment Program, Dialogue Paper. Paris, UN.

UNESCO, United Nations Educational, Scientific and Cultural Organization (2006) *Water: A Shared Responsibility*. United Nations World Water Development. Report number: 2. Paris, UNESCO.

UNEP (2010) *Clearing the Waters, A focus on water quality solutions*. Nairobi, United Nations Environment Program.

UNEP (2012a) *Global Environment Outlook 5 (GEO5), Environment for the future we want*. Valetta, Malta, United Nations Environment Programme.

UNEP (2012b) *Measuring water use in a green economy, A Report of the Working Group on Water Efficiency to the International Resource Panel*. United Nations Environment Programme.

UNSD (2012) *SEEA-Water: System of Environmental-Economic Accounting for Water*. United Nations. New York, United Nations Statistics Division, Department of Economic and Social Affairs Statistics Division.

Willaarts, B. (2012) Linking land management to water planning: Estimating the water consumption of Spanish forests. In: De Stefano, L. & Llamas, M.R. (eds.) *Water, agriculture and the environment in Spain: Can we square the circle?* Boca Raton, CRC Press.Taylor and Francis.

WSSD (2002) *Johannesburg Plan of Implementation (JPOI)*. [Online] World Summit on Sustainable Development. Available from: http://www.un.org/esa/sustdev/documents/WSSD_POI_PD/English/POIToc.htm [November 2012].

Yang, H., Wang, L., Abbaspour, K.C. & Zehnder, A.J.B. (2006) Virtual water trade: An assessment of water use efficiency in the international food trade. *Hydrology and Earth System Sciences*, 10, 443–454.

Chapter 3

Non-Integrated Water Resources Management

Mark Giordano
Principal Researcher, International Water Management Institute (IWMI), Colombo, Sri Lanka

Tushaar Shah
Senior Fellow, International Water Management Institute (IWMI), Anand, India

ABSTRACT: Integrated Water Resources Management provides a set of reasoned principles that, if followed, would lead us to an improved water future. This promise plus the backing of important international organizations has allowed IWRM ideals to acquire a near monopoly on water management discourse. This is unfortunate because, while the potential benefits of IWRM are large, its implementation comes with its own set of economic, political and time costs, costs which are not always considered in IWRM policy advocacy. Failure to recognize these costs can sometimes result in outcomes counter to the goals of water sector reform. As important, the ubiquity of IWRM in policy discussions means that lower cost and potentially more effective options are sometimes not considered. This chapter highlights these points by first describing the sometimes neglected costs of IWRM implementation, particularly in developing country contexts. It then provides a set of case studies examining solutions to water problems whose methods run counter to IWRM. The case studies include a non-basin scale approach to reduce transboundary conflict in Central Asia, a non-price solution to groundwater overdraft in western India, and non-participatory methods in China to move low-value agricultural water to higher value urban uses without decreasing agricultural production. The overall point of the chapter is not to criticize IWRM. Rather it is to highlight that IWRM principles are simply one option among many for improving the way we use scarce water resources.

Keywords: Integrated Water Management, basin, water pricing, water allocation, groundwater

1 INTRODUCTION

> *'The concept of Integrated Water Resources Management (IWRM) offers solutions to the water crisis in linking water to other vital resources and viewing the whole water cycle together with human interventions as the basis for sustainable water management.' (Advertisement for a Master's Degree course in IWRM).*

The goal of IWRM as stated in the quote above is noble: To achieve Sustainable Water Management. The means for achieving the goal – linking water with other resources and considering the physical and human sides of the entire hydrologic cycle – appear eminently reasonable. It is thus no wonder that IWRM has become the mantra to overcome earlier water management mistakes emanating from narrow, sectoral thinking. We are now awash with calls for IWRM, with global programs to implement it, loans to spur the way, and even, as the quote above shows, formal degree programs to teach us how to do it. The idea of IWRM is so ubiquitous that one would be hard-pressed to find any event in which water management is discussed that does not bring up IWRM as the solution. Good water management has come to be synonymous with IWRM. Put another way, a potential means for achieving good water management, IWRM, has become the goal.

This transformation of a means to an end; of an idea into a 'sanctioned discourse' (Allan, 2003), has shifted our focus away from immediate water management problems and the interventions that could solve them towards idealized, all-encompassing IWRM approaches that we hope will bring future benefit. It is true that short term solutions to immediate problems are usually sub-optimal in an economic or engineering sense. But these sub-optimal solutions can also be of great value when we recognize that water management, at its core, is political and that implementation of IWRM has costs in terms of time, money and political resources. A sub-optimal solution, actually implemented, is better than an 'ideal' that never delivers change on the ground. This fact is perhaps especially important for developing countries where IWRM tends to be driven not by indigenous need but rather by foreign intervention where it arises and is enforced 'through international organizations, loan conditionality, expert consultations, and economic as well as political pressure' (Laube, 2007), often at the expense of alternative solutions and paths.

The goal of this chapter is not to attack the principles of IWRM or its goals as some critiques have done (e.g. Biswas, 2004). Rather it is to remind us that 1) efforts to implement IWRM can have costs, some of them unrecognized, and that 2) because of these costs, we should not forget that there are other options, some of which violate IWRM principles, to solving present water problems. It does this by providing a set of examples highlighting how Non-Integrated Water Resources Management can sometimes bring benefits that in some cases decades of attempts towards IWRM have failed to achieve.

2 WHAT IS IWRM IN PRACTICE?

The GWP has probably produced the most comprehensive guidelines on the concepts of IWRM (GWP, n.d.) As highlighted in that material by the World Bank Institute and others, IWRM attempts to put the Dublin Principles into practice, emphasizing the ideas of integration, decentralization, participation, and economic and financial sustainability with the basin as unit for decision-making. In moving from ideas to actual implementation, Shah & van Koppen (2006) and others have described that application of IWRM has typically meant the establishment of an overall water policy and law which uses the basin as the scale of management, establishes water rights, uses water pricing in the resources' allocation, and includes participation in decision-making.

What has happened when countries have tried to implement this set of practices? The results have not always been as expected as the case of Sri Lanka shows (Samad, 2005). In 1993, the Government began a process of implementing water policy reform under technical assistance of the ADB (Asian Development Bank) in association with the US Agency for International Development. Some 115 stakeholder consultation meetings were held involving government agencies at the national and provincial levels, policymakers, water managers, the private sector, professional bodies, NGOs and all major water users. Working groups were set up involving NGOs to discuss and identify the major problems, suggest policy prescriptions and propose appropriate institutional arrangements to implement the policy. The result was a package of reforms mimicking the IWRM ideal described above. A water policy and water law were established, existing water organizations were to be replaced by river basin organizations, water use rights were established through withdrawal permits, permits were made transferable to encourage water trade towards high valued uses and all water was priced.

Despite following an apparently open process, the reform program was heavily criticized. Sections of the press, non-governmental groups, religious bodies and some farmer organizations argued that the process was in fact top-down and closed despite the apparent efforts for inclusion. Other criticisms included the failure to draft the policy document in the local language, insensitivity to the cultural aspects of water and, importantly, that the whole exercise was to satisfy donor interest rather than cater to national needs.

In the wake of intense agitation by the public and the media against the proposed national water policy, the government first distanced itself and then withdrew the proposals. The result was not simply that the process failed, but also that the opportunity for any reform was greatly reduced. Open discussions of even some of the principles of IWRM, such as cost recovery, have become politically impossible. External organizations trying to help with the water sector are frequently accused of trying to buy up or privatize Sri Lanka's water. Almost 20 years after the reform process started, there has been little change and when a drought hits the country, as is the case now, there is no way to coordinate a response. The process of establishing IWRM sets back progress in tackling Sri Lanka's real water challenges.

In Africa, the push for IWRM has led to another set of problems. Tanzania's 1991 water policy identified water development and provision as key national policy goals and argued for more water storage creation (Shah & van Koppen, 2006). However, Tanzania's budgets were heavily donor dependent and creating new storage and infrastructure went against donor practice at the time. Instead, Tanzania implemented what donors would support: IWRM with state ownership of water resources, water withdrawal permits, water taxes, river basin organizations, and water user associations (WUAs), but no attempt was made to get what Tanzania had defined as its people's needs, more and better-managed infrastructure. Without the right conditions in place, the push for IWRM principles, in particular permit systems to manage water, in Tanzania and other African countries actually disposed the poor from their access to water (van Koppen, 2007) and undermined the goals of better water management.

3 IF WE CAN'T DO IWRM, WHAT CAN WE DO?

The IWRM ideal focuses on what engineers or economists might consider optimal water outcomes. However, the path towards those ideals is long and, as just described, fraught with potentially unrecognized costs. One of the problems with the dominance of the IWRM discourse is that it has caused us to forget that there may be other paths to improving water outcomes, even if those paths may not take us to the ideal or the theoretically optimal. In our effort to find eventual comprehensive solutions, we must not forget that other options are available now. Further, we should not forget IWRM principles do not have a monopoly on our water management options. To highlight this point, we look here at 3 solutions to real water management problems that contradict IWRM principles, but could or already have improved water outcomes.

3.1 Ignore the basin[1]

The basic tenet of IWRM is that the basin is the natural management unit. As we all know though, basin boundaries rarely coincide with the political boundaries in which decisions regarding water use are made. This problem is compounded in many of the world's approximately 300 transboundary basins. Managing an international basin on IWRM principles necessitates that states voluntarily give up some of their sovereign power. For this to happen, each involved state must believe that the power it gives up is offset by the gains from better overall water management. Creating such conditions is especially difficult when uses and negotiating positions are predefined by basin geography, with upstream countries often preferring one set of uses and principles while downstream countries preferring another. An oft cited example of such a conflict is the Nile basin where downstream Sudan and Egypt invoke 'no harm' and 'prior use' principles embodied in the UN Convention on the Law of Non-navigational Uses of the International Water Courses, while upstream Ethiopia favors the 'equitable and reasonable utilization' which concept is found in the same document.

What can be done when states cannot find mutually beneficial solutions? Or, perhaps even worse, what if the apparent opportunities for positive sum outcomes from basin management have been identified but ignored? Such is the case of the Syr Darya River basin shared by the four central Asian republics of Kyrgyzstan, Tajikistan, Uzbekistan and Kazakhstan (Figure 1).

Until 1991, the Syr Darya's management was coordinated by the central government of the Soviet Union in conjunction with its riparian member republics using a benefit-sharing concept. Under this concept, summer irrigation flow was provided to downstream Uzbekistan and Kazakhstan to support agricultural production in exchange for winter energy supplies to upstream Kyrgyzstan to compensate for foregone hydropower production. Even after the fall of the Soviet Union, there was a general consensus that this agreement supported optimized basin water use, and it was formally extended through the 1992 Cooperation in Joint Management, Use and Protection of Interstate Sources of Water Resources Agreement signed by five Ministers of Water Resources of the Central Asian States. However, the experience of almost two decades shows that the implementation of this agreement has been less than perfect as highlighted in Figure 2.

Given that there has been little change for two decades, it may be time to stop thinking about a basin-centered solution. What are the alternatives? Research conducted

1 This section is based on Karimov *et al.* 2011.

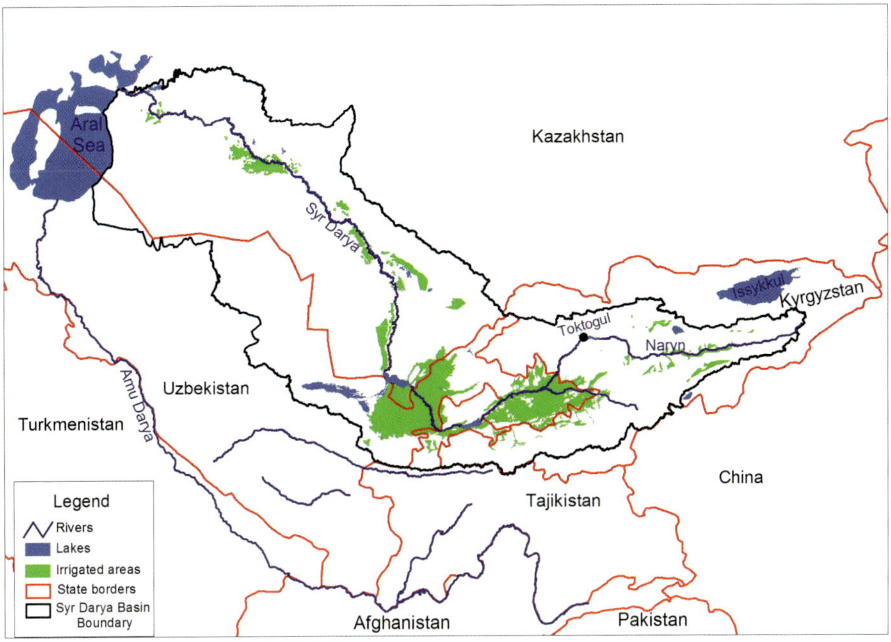

Figure 1 Water sharing in the Syr Darya. The Syr Darya is shared by upstream Kyrgistan and Tajikistan and downstream Uzbekistan and Kazakhstan. Upstream states would prefer to release water from reservoirs such as Toktagul on the Naryn tributary in the winter to produce hydropower. Downstream states would prefer summer releases to support irrigation (Karimov, 2011).

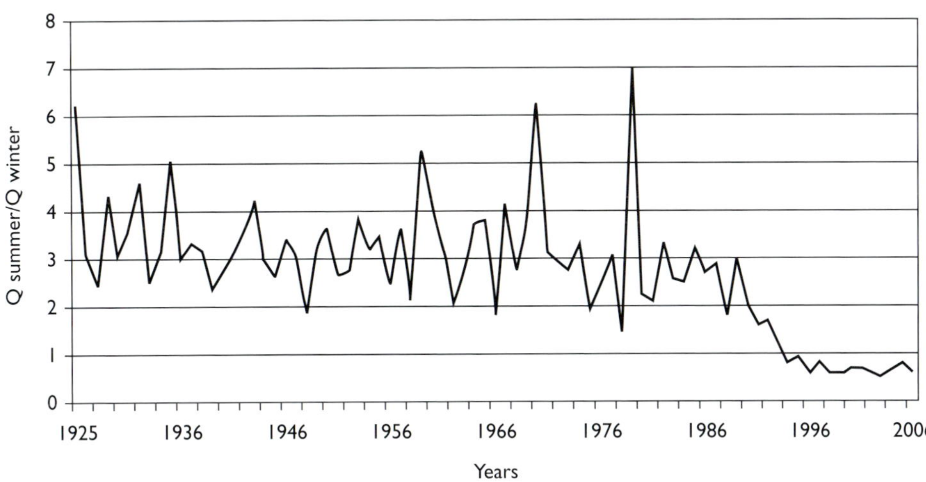

Figure 2 Ratio of the summer to winter flow of the Naryn River. The drop in ratio is caused by upstream Kyrgyzstan changing its release practices after the fall of the Soviet Union. Despite a formal treaty which calls for a change to pre-independence ratios, no change has occurred for 20 years (own elaboration).

from 1965–1974 by 'The Institute of Hydrogeology Uzbekistan' has shown that accumulating the flow of small rivers in the Fergana Valley through underground storage is viable for drinking water supply needs. Following this thinking, it was hypothesized that similar efforts might be feasible for agricultural water storage. Technical feasibility studies were done on the Naryn sub-basin within the Uzbek portion of the Fergana Valley to see if this could be used as a strategy to address at least part of the Syr Darya challenge (Figure 3).

It was found that free capacity of underground aquifers in the canal commands of the Naryn River sub-basin amounted to 769 Mm^3. Additional capacity of 186 Mm^3 could be created by drawing down the aquifers in the summer growing season. The storage from the two sources together would equal nearly half the current shortfall for summer irrigation, caused by the ongoing water dispute.

While careful economic analysis and modeling for each aquifer is required to find out the extent to which groundwater banking is economically viable, field trials under way so far have shown that the idea is in fact technically feasible.

While the long term goal may still be a functioning basin scale agreement, in the shorter term improvement can be made by unilateral action on the part of a single riparian, without focusing on the basin and without causing harm or disadvantage to other farmers.

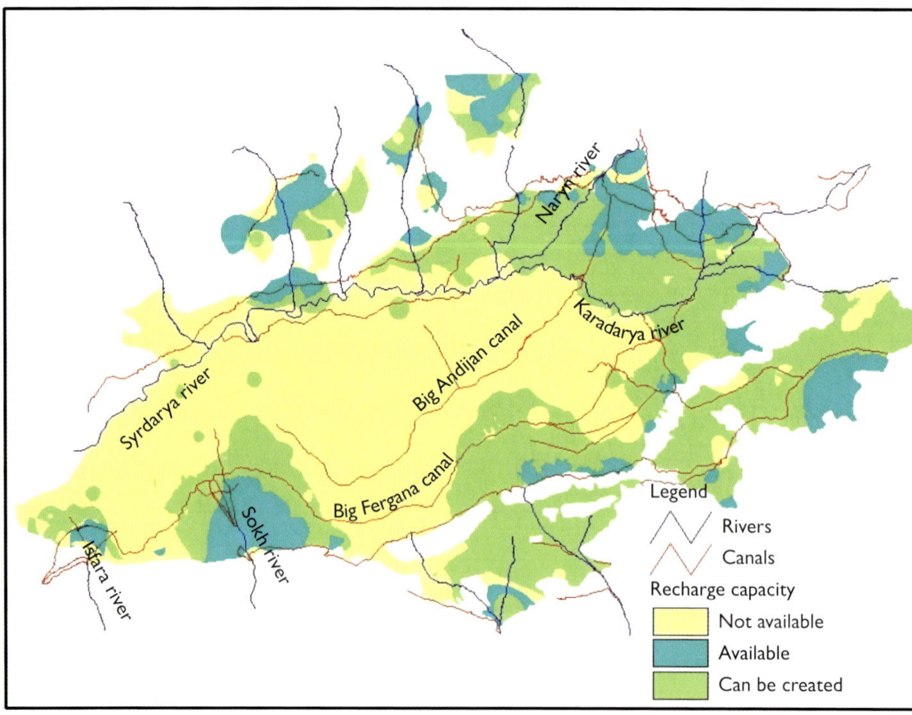

Figure 3 Sokh Aquifer, Naryn River, Uzbekstan. The image shows the areas where water recharge capacity is currently available, as well as where it is not possible, or could be created through planned pumping in the summer growing season to store winter surface water releases from upstream dams (Karimov, 2011).

3.2 Don't charge for water[2]

The issue of groundwater overdraft in India is well known. In the state of Gujarat, free groundwater and the free electricity provided to pump it contributed to severe groundwater overdraft, near bankruptcy of the State Electricity Board, and poor power supply to farmers and other rural residents. The problem has been well known for decades, and the textbook solution simple and following IWRM principles: Price groundwater and electricity to reflect their value. However, those who tried to implement these solutions did not appreciate the political realities of India. Efforts to rationalize pricing were met with great resistance by farmers. Politicians lost their jobs and external funds for modernizing the system were withdrawn. The State Electricity Board continued to generate great losses and was unable to meet the needs of the rapidly growing economy. Farmers had to accept poor quality power supply as the cost of their 'free' supply, and the pressure on aquifers was substantial.

An alternative approach, called the Jyotigram Scheme, diverged from the textbook optimum and embraced subsidies as part of a strategy. But rather than viewing subsidies as a default component of free electricity supply, the Jyotigram Scheme focused on providing rationally managed subsidies where needed, and pricing where possible. Under the program, rural Gujarat has been completely rewired. Villages are given 24-hour, three-phase power supply for domestic uses, in schools, hospitals, and village industries, all at metered rates. Farmers operating tube wells continue to receive free electricity, but for 8 hours, rather than 24 hours and, importantly for the satisfaction of farmers, on a pre-announced schedule designed to meet their peak demands.

The separation of agricultural energy from other uses and the promise of quality supply were sufficient to gain political and social backing for implementation of a new pricing scheme. The Jyotigram scheme has now radically improved the quality of village life, spurred non-farm economic enterprises, and halved the power subsidy to agriculture. While groundwater itself is still free, the program has indirectly raised the price of groundwater supply from tube well owners in the informal market by 30–50 percent, thus providing a signal of scarcity, and reducing groundwater overdraft. The solution may not be perfect, and it is far from the IWRM ideal, but it has proved to be implementable and it has brought substantial improvement in and outside the water sector.

3.3 Is participation necessary?[3]

One rationale for IWRM is that we need fair mechanisms to move water over time to the uses with the highest value. The principles for facilitating this shift include the establishment of use rights which are tradable as well as participation in decision-making on related changes.

The pressures in many parts of China to move water from agriculture to rapidly growing cities are as large as anywhere in the world. The case of the Zhang He irrigation system provides ideas on how this transfer can occur, without reducing agricultural production, water rights or participation. The Zhang He Reservoir in Hubei province was designed to irrigate two rice crops per year. In the 1960s and 70s, the

2 This section draws on Shah *et al.*, 2008 and Shah & Mehta, 2012.

3 This section draws on Molden *et al.*, 2010.

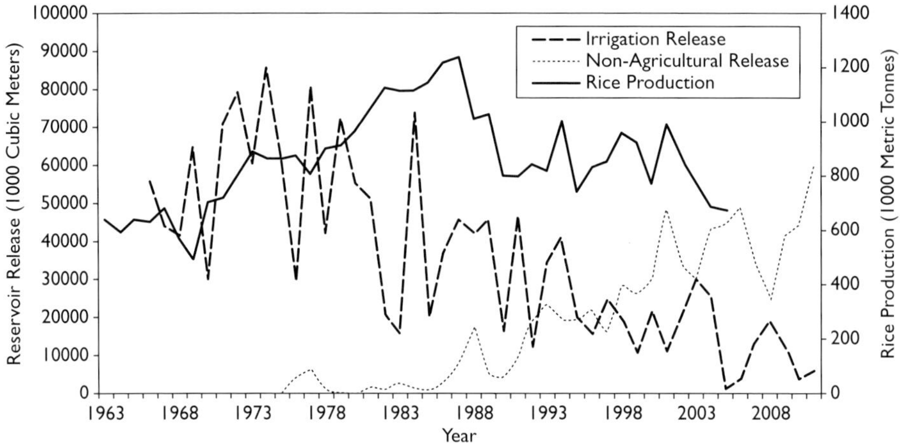

Figure 4 Water allocation and Rice Production, Zhang He Reservoir, Hubei, China (own elaboration from Molden *et al.*, 2007).

reservoir's water was used entirely for agriculture. As the Chinese economy started to liberalize in the 1980s, urban uses expanded and by the 1990s, non-agricultural uses took the majority of the water. Despite the decline in irrigation supplies for the district, agricultural production did not decline (Figure 4).

How did this happen? While Chinese farmers pay for water, it was not the market that provided the incentives for the shift in water use and the large implied use increase in agricultural water productivity. Nor was it a participatory process. Rather, operators have used a top-down approach to simply allocate an increasing amount to cities and less to farmers. Reduced supplies to farmers force a response. Part of the response was the construction of thousands of small reservoirs within the irrigated area to capture runoff generated within the command area, plus to capture return flows from rice cultivation. In addition, research demonstrated that yields would not suffer if rice fields were not left flooded, and alternating wet and dry irrigation could be employed. With the extension of knowledge about alternating wet and dry irrigation of rice, farmers had a technology to help them cope, and the remarkable trend emerged. Crop production remained steady in spite of less water being delivered from the main reservoir to rice cultivators. Productivity of rice has increased. Water productivity gains in the Zhang He irrigation system have skyrocketed.

4 CONCLUSION

We face daunting water management challenges as demand hits the limits of supply, inter-sectoral competition increases, water quality declines and aquatic ecosystems come under threat. The principles of IWRM provide a solid roadmap for us to consider how, as integrated societies, we can best make choices about water allocation and access as well as of the sustainability of water resources and the infrastructure we use to manage those resources. However, moving towards governance or management regimes even approximating IWRM brings its own set of costs and results are far from immediate. Therefore, in trying to improve our water outcomes today, we should not forget that there are alternatives to IWRM.

We don't always have to manage resources, using the basin as the unit. In fact the reality is probably that we rarely do. It can be possible to find unilateral solutions which do not harm other riparians and whose implementation is feasible. In the case of Uzbekistan, formal treaties and 20 years of effort had not brought change. How much longer should the countries wait?

Water pricing and cost recovery are important goals. But in the case of India, it has been politically impossible to implement groundwater pricing despite decades of efforts and national water policies. How much longer could they wait? The imperfect Jyotigram Scheme has already been implemented, and improvements in water and social outcomes are evident even though the water is still 'free'.

All over the world cities suck water away from traditional agricultural uses. It would be nice if this process occurred with the explicit participation and agreement of historic water users and that those users were directly compensated for any loss through an exercise of tradable water rights. But how many examples do we have where this works in practice? In China, administrative approaches, allowing farmers to adapt at the local level and providing new technologies to increase water use efficiencies, allowed water to move to higher value urban uses without impacting food supply-and substantially increased water productivity.

Did any of these solutions follow IWRM guidelines, were they optimal in engineering or economic terms, or did they achieve the goal of long term sustainable water management? No. But they did improve water management outcomes here and now, something that IWRM approaches had not yet been able to do.

We must keep in mind the goals of IWRM, but let us not forget that Non-Integrated Water Resources Management solutions can also take us forward.

REFERENCES

Allan, T. (2003) *IWRM/IWRAM: A new sanctioned discourse?* School of Oriental and African Studies (SOAS) Water Issues Study Group. Occasional Paper number: 50. London, SOAS.

Biswas, A. (2004) Integrated Water Resources Management: A Reassessment. *Water International*, 29, 248–256.

GWP (n.d.) Toolbox. [Online] Global Water Partnership. Available from: http://www.gwptoolbox.org/ [Accessed 12 July 2013].

Karimov, A., Giordano, M., Mukerji, A., Borisov, V. & Djumanov, J. (2011) Of transboundary basins, integrated water resources management (IWRM) and second best solutions: The case of groundwater banking in Central Asia. *Water Policy*, 14(1), 99–111.

Laube, W. (2007) The promise and perils of water reforms: Perspectives from Northern Ghana. *Afrika Spectrum*, 42, 419–437.

Molden, D., Bin, D., Loeve, R., Barker, R. & Tuong, T.P. (2007) Agricultural water productivity and savings: Policy lessons from two diverse sites in China, Water Policy, 9 (Supplement 1), 29–44.

Molden, D., Lautze, J., Shah, T., Bin, D., Giordano, M. & Sanford, L. (2010) Governing to Grow Enough Food without Enough Water—Second-Best Solutions Show the Way. *Water Resources Development*, 26, 249–263.

Samad, M. (2005) Water institutional reforms in Sri Lanka. *Water Policy*, 7, 125–140.
Shah, T., Bhatt, S., Shah, R.K. & Talati, J. (2008) Groundwater governance through electricity supply management: Assessing an innovative intervention in Gujarat, western India. *Agricultural Water Management*, 95, 1233–1242.
Shah, T. & Mehta, M. (2012) *Transformation of Gujarat's electricity utility: Lessons for Revitalising a bureaucratic service delivery agency.* IWMI-Tata Program. Water Policy Research Highlight, number: 6. Anand, India.
Shah, T. & van Koppen, B. (2006) Is India ripe for integrated water resources management?: Fitting water policy to national development context. *Economic and Political Weekly*, 41, 3413–3421.
van Koppen, B. (2007) Dispossession at the Interface of Community-based Water Law and Permit Systems. In: van Koppen, B., Giordano, M. & Butterworth, J. (eds.) *Community-Based Water Law and Water Resources Management Reform in Developing Countries*. Wallingford, Cab International. pp. 46–64.

Chapter 4

Contemporary responses to water management challenges

Xavier Leflaive
OECD Environment Directorate, The Organization for Economic Co-operation and Development (OECD), Paris, France

ABSTRACT: OECD countries face new challenges related to water security. Increasing uncertainty about water demand and availability, combined with deficiencies in past approaches to water management have put the water security of some basins or user groups at risk. Policy responses can be clustered around two main themes. First, water can be more thoroughly managed to drive green growth; this requires i) allocation regimes that reflect governments' priorities for socio-economic development; ii) infrastructures that are scalable to needs and do not generate unrealistic burden on the public purse; and iii) policies that stimulate the development and deployment of innovative technologies and approaches. Second, water resources management would benefit from a financing framework based on a small set of principles. These responses require reforms, which can be very challenging. International experience can help sequence these reforms and design the accompanying measures that facilitate stakeholder engagement. OECD countries and their partners would benefit from sharing experience on these and related issues.

Keywords: water allocation, water security, green growth, financing water management

1 INTRODUCTION

The water challenges OECD countries face call for immediate action. Some of the ways they have been addressed in the past need to be assessed and probably reconsidered. The chapter claims that policy responses can be clustered around three crosscutting themes: Water security, managing water for green growth, and financing water resources management.

This chapter recapitulates what we know about water-related challenges and projected trends. It argues that allocation of water should be reconsidered, as competition between water users intensifies. This text then considers how managing water for green growth can address some of the current and emerging challenges; in such an approach, water allocation regimes should reflect economic development policies; and water and related policies should put emphasis on innovation and investment, particularly in green infrastructures. Financing remains an issue. The article presents a framework which governments may wish to use to assess their financing strategy for water resources management.

Although this chapter extensively relies on OECD analyses and reports, it may not reflect the opinion of the OECD and its member countries. All flaws remain the author's responsibility.

2 WATER CHALLENGES FOR THE FIRST PART OF THE 21ST CENTURY

The OECD Environmental Outlook to 2050 (OECD, 2012c) projects that, without major policy changes and considerable improvements in water management processes and techniques, the situation regarding water resources and water related services is likely to deteriorate by 2050. The situation will result in increased competition for water and increasing uncertainty about water availability. Climate change will increasingly make things more difficult, although demographics and economic development will be more decisive drivers in the coming decades.

Water demand is projected to increase by 55% globally between 2000 and 2050. The increase in demand will come mainly from manufacturing (+400%), electricity (+140%) and domestic use (+130%). In the face of these competing demands, there will be little scope for increasing water availability for irrigation.

The Outlook Baseline scenario projects that by 2050, 3.9 billion people, over 40% of the world's population, are likely to be living in river basins under severe water stress.

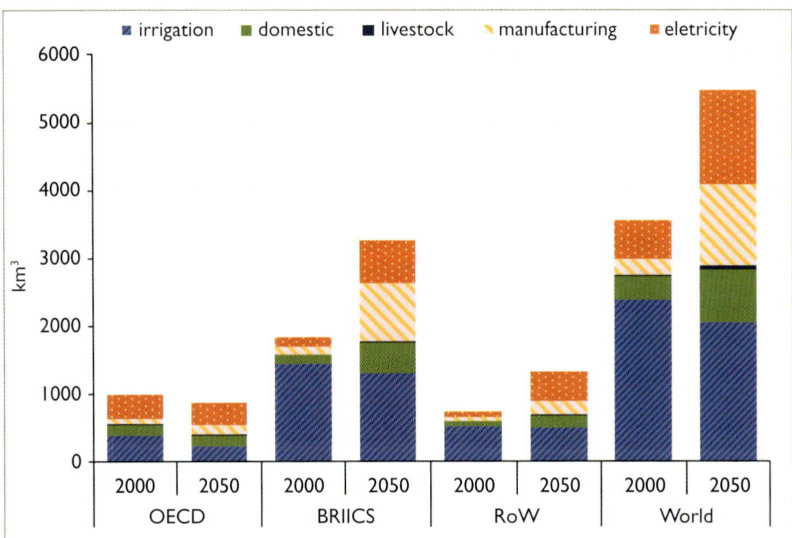

Figure 1 Global water demand: Baseline scenario, 2000 and 2050. This graph only measures 'blue' water demand and does not consider rainfed agriculture. BRIICS stands for Brazil, Russia, India, Indonesia, China, South Africa. RoW stands for Rest of the World (OECD, 2012c; output from IMAGE).

In many regions of the world, groundwater is being exploited faster than it can be replenished and is also becoming increasingly polluted. The rate of groundwater depletion more than doubled between 1960 and 2000, reaching over 280 km^3 per year.

Water quality remains an issue. Continued efficiency improvements in the agricultural sector and investments in wastewater treatment in developed countries are expected to stabilize and restore surface water and groundwater quality in most OECD countries by 2050. However, the quality of surface water outside the OECD is expected to deteriorate in the coming decades, through nutrient flows from agriculture together with poor wastewater treatment. The consequences will be increased eutrophication, biodiversity loss and a more frequent occurrence of diseases. In addition to this, micro-pollutants (medicines, cosmetics, cleaning agents, and biocide residues) are an emerging concern in many countries.

Progress in access to water supply and sanitation in non-OECD countries does not keep pace with demographics and urbanization. Progress has been huge: The number of people with access to an improved water source increased by 1.8 billion between 1990 and 2008, mostly in BRIICS (Brazil, Russia, India, Indonesia, China and South Africa), and especially in China. However, more than 240 million people (most of them in rural areas) are expected to be without access to an improved water source by 2050. The Millennium Development Goal for improved water supply is unlikely to be met in Sub-Saharan Africa. Globally, more city dwellers did not have access to an improved water source in 2008 than in 1990, as urbanization is currently outpacing connections to water infrastructure. The situation is even more daunting given that access to an improved water source does not always mean access to safe water. Almost 1.4 billion people are projected to still be without access to basic sanitation in 2050, mostly in developing countries. The Millennium Development Goal on sanitation will not be met.

These projections have far reaching consequences, on health, on economic development, and on the environment. They can be captured under two categories. One is water security: Without new policies, water users and communities are increasingly vulnerable to water-related risks. The OECD identifies four categories of water-related risks: Risks of water shortage (drought), excess (flood), pollution, as well as the risks of freshwater systems disruption. It is developing a risk-based approach that proposes a framework to 'know' the risks (including risk perceptions), 'target' the risks (reduce risks to acceptable levels) and 'manage' the risks (with a portfolio of policy tools, such as spatial planning, economic and regulatory instruments). This approach could be used by individual countries to review water policies, with a view to strengthen water security at a lower cost for society (OECD forthcoming, 2013a).

Water-related challenges have another, related set of consequences: If not properly managed, water challenges may hinder the growth potential of countries and regions. That water users fail to access the water they need when it is needed can have adverse consequences on growth and investment. These consequences are explored in the following section, with some suggestions on potential policy responses. These are approached from a green growth perspective.

3 MANAGING WATER FOR GREEN GROWTH

The OECD is working to reconcile the demand for continued economic growth and development with the need to ensure that natural assets continue to provide the resources and environmental services on which all human well-being relies. This underpins the concept of 'green growth', which sees sustainable water use as an essential driver, since a lack of water of appropriate quality can significantly hinder growth (OECD, 2011a). Similarly, UNEP (2011) confirms that investments in infrastructure and operation of water-related services can provide high returns for both the economy and the environment. It highlights the need for more private and public investment in green technologies and infrastructure to boost water (and energy) efficiency and sees such investments as critical to building the green economy of the future.

As the OECD report Managing Water for Green Growth (OECD forthcoming, 2013b) claims, managing water for green growth combines securing access to water for those who need it, including ecosystems (losses of biodiversity and ecosystem services come at an increasingly visible economic cost) and allocating water in accordance with the development strategy of a territory (as a pragmatic proxy of social welfare maximisation). While such an approach potentially has universal application, it will necessarily translate differently in different countries given the wide divergence in water security issues and development priorities between countries.

In practice, managing water for green growth does not need to be radically new. Water efficiency and water demand management are essential ingredients for green growth, along with water reuse and recycling. Sound water pricing, as an essential feature of water management, also contributes to managing water for green growth: It generates revenues to finance water-related services and can encourage the efficient allocation and use of water resources. However, several aspects of water management are critical for green growth. OECD (forthcoming 2013b) identifies four such aspects.

The following policy approaches can more systematically harness water management for green growth.

3.1 Make sure water is allocated in line with a development strategy

As seen above, OECD (2012c) argues that competition to access the resource will intensify in a number of basins, including in OECD countries. It follows that allocating water across these competing demands (including the environment) will be increasingly challenging.

Box 1: Economic analysis of the virtual water and water footprint concepts for water policies

Virtual water

The term 'virtual water' began to appear in the water resources literature in the mid-1990s. Professor Tony Allan of London University chose the term to describe the water used to produce crops traded in international markets. During the 15 years since its inception, the virtual water concept has been very helpful in gaining the attention of public officials and policy makers responsible for encouraging wise use of limited water resources.

However, the fundamental shortcoming of the virtual water concept as a policy prescriptive tool is the lack of an underlying conceptual framework. Some researchers have incorrectly described virtual water as analogous to, or consistent with the economic theory of comparative advantage. The virtual water concept is applied most often when discussing or comparing water-short and water-abundant countries. By focusing on the water resource endowment alone, virtual water represents an application of absolute advantage, rather than comparative advantage. For this reason, policy prescriptions that arise from virtual water discussions will not maximize the net benefits of engaging in international trade. Comparative advantage is the pertinent economic concept, and virtual water considers only absolute advantage.

A number of authors have begun describing the important role of non-water factors such as population densities, historical production trends, national food security goals, poverty reduction targets, and the availability of complementary inputs when determining whether to transfer water from one region to another, or to achieve desired outcomes alternatively by transporting or trading agricultural commodities.

Water footprints

The notion of water footprints describes the volume of water required to support production and consumption in selected regions or countries. It is used to assess whether a region or country is consuming resources in a sustainable or unsustainable fashion from a global perspective. However, estimated water footprints are somewhat one-dimensional, as they depict the use of only one resource. In addition, water footprints do not describe the implications of water use. Instead they consider only the amounts of water used in production and consumption activities. Hence, ecological water footprint analysis is not sufficient for determining optimal policy alternatives, as it does not account for the opportunity (scarcity) costs of water resources and the ways in which water is combined with other inputs in production and consumption. Water footprints enable one to compare estimated water use per person or in aggregate across countries, but they are inadequate for evaluating the incremental costs, benefits, or environmental impacts of water use.

Farmers, traders, and public officials must consider many economic and social issues when determining optimal strategies. Virtual water and water footprint concepts will be helpful in policy discussions in many settings, in combination with other environmental, economic, and social indicators. But they will not be sufficient for determining the optimal outcomes of those discussions and establishing economically efficient and environmentally effective policy alternatives.

(OECD, 2010)

Some OECD countries are gaining experience with socially fair and politically acceptable approaches to water allocation. These include water abstraction licences or pricing policies that reflect scarcity and the value of water; market mechanisms, e.g. tradable water rights; and information-based instruments (smart metering). The need to restore environmental flows and to allocate more water to watershed services is already generating initiatives in several countries. The concepts of virtual water and water footprint have pointed to the inherently international dimension of water allocation. However, the policy consequences of these analyses have yet to be deciphered (see the box below on some short comings readily fund in the literature).

In several basins, water is already over-allocated (the needs of ecosystems are not met in practice) and water allocation regimes will have to be reformed to meet future challenges. This difficult policy challenge – diverting water to value-adding activities (including environmental services, see below) – may require reallocation between water users (e.g. from farmers to cities). Experience from OECD and non-OECD countries indicates that building a strong constituency and aligning incentives are two major requisites.

Water allocation remains a bargaining process. More work is needed to fully understand what data is realistically needed to support this process. Valuation (putting a monetised value on water uses) certainly helps, but faces methodological challenges. The OECD is investigating the experience of member countries with the policy instruments that can inform and support that bargaining process.

Water is expected to be allocated in a way that maximizes social welfare. Because such an objective remains abstract and the means to measure it have lacunas, a more practical approach might be to make sure that water allocations reflect economic development strategies. Economic development strategies will translate into land use plans and water needs (by farmers, selected industries, ecosystems, cities, etc.). Water should be allocated in such a way that these needs are met. The greener the development strategy, the greener water allocation will be.

3.2 Invest in ecologically sensitive water storage and water distribution systems

Reliable resources are essential for green growth. However, water storage technologies and infrastructure such as large dams can disturb ecosystem balances. Soft infrastructure (e.g. wetlands, flood plains, groundwater recharge), small-scale dams, rainwater harvesting, or appropriately designed infrastructure are more ecologically sensitive and cost-effective.

Several barriers have to be overcome for green infrastructures to be more widely used. These include inertial effects of existing assets (water infrastructures have a strong technical path dependency), intensity of land use (some green infrastructures may need more land, which may not be readily available in urban or dense areas), the need for emergency action (while green infrastructures may need more time to deliver), fragmented institutions (which make it difficult to have a systemic, cross-cutting approach to investment decisions), functional budgets (which may not account for long term benefits that may accrue to other constituencies), or safety risks (risks related to green infrastructures can be less well understood).

It is essential to review these barriers and to build on the experience of OECD and developing countries in overcoming them.

3.3 Invest in water supply and sanitation infrastructure

In addition to being socially unacceptable, unsafe water and lack of sanitation generate huge health costs and lost opportunities to the economy. Gains are enormous, for every cent invested in improving access to safe water and improved sanitation services. This is particularly true in urban slums.

The trends reported above show that efforts to improve access have been made by local communities, governments and international assistance. More needs to be done to accelerate progress on the ground and to make sure it reaches out to the poorest groups of the population (which has not been the case so far). This may require innovative approaches, to design infrastructures, to organize services (aggregating or bundling in smart ways) and to finance investment, operation and maintenance.

OECD analyses have established that institutions and misaligned incentives (including prevailing business models for water utilities) can hinder the diffusion of innovative technologies or approaches. The efficient allocation of water resources in urban areas is hampered by restrictions in the choice of supply-side options, including by regulatory obstacles to rural-urban water trading or to the use of reclaimed water, where they represent low-cost sources of supply. For instance, under certain circumstances, small-scale distributed systems can lower investment costs, facilitate the use of alternative water sources (such as treated wastewater) and enhance the flexibility of water allocation. Investment in water supply and sanitation systems for green growth requires institutions and regulations which are not driven by a particular technological trajectory, and which do not favour incumbents vis-à-vis newcomers.

3.4 Catalyze investment and innovation

Innovation and investment are two inherent drivers of green growth. Water-related innovation can take many forms, in the fields of agriculture (water-efficient irrigation, less water-intensive crops, less nutrient flows), manufacturing (water-efficient and cleaner processes), water supply (water-efficient appliances), and sanitation (more effective and cost-efficient treatment techniques). There is also room to improve the management of the resource (storage techniques, monitoring river flows and pollution) and the operation of infrastructures (leakage detection, pipe repair, smart water systems).

Water-related innovation will underpin sustained growth and give rise to new economic opportunities. This is the reason why the European Commission (*inter alia*) has set up the Eco-Innovation Partnership on Water (EIP Water), to leverage the benefits of a vivid water industry, which can address environmental issues in Europe, create jobs, and export globally. Several OECD countries explicitly support water-related innovation to both address water challenges and gain a share of fast-growing global markets. USEPA, for instance, considers that investments in water infrastructures generate jobs and spur environmental technology exports. The Japanese METI has set up a national Strategy in water industry. France and Korea have active policies in that domain as well whereas a water innovation hub is being developed in Ontario, Canada.

As mentioned in the section above, these policies should not focus only on the supply side of technologies: They should also strive to stimulate demand and to open markets for innovative solutions. This includes working with developing countries so that they can absorb these innovations, and develop their own capacity to innovate.

Investment in water-related infrastructure and services will follow, when financiers receive clear and stable signals that they will be able to recoup their investment.

This requires carefully crafted strategic financial plans, based on realistic assessments of investment needs (making the best of low-cost options) and financing capacities (based on prices for water services that combine economic, social and financial objectives). This also requires consistency across policy areas (so that initiatives in the agriculture or energy sector do not undermine water policy objectives).

4 A FRAMEWORK TO STRENGTHEN THE FINANCIAL DIMENSION OF WATER RESOURCE MANAGEMENT

Governments around the world struggle to secure financial resources to cover the costs associated with water policies. OECD (2011b) confirms that lack of finance is considered a major governance gap in most OECD countries. The current financial crisis makes the situation even more challenging, as competition to access public finance intensifies.

While the implementation of improved systems for financing water management will necessarily need to reflect local conditions and priorities, there is nevertheless a set of core principles that can be drawn upon.

Two principles have been well-established for a number of years and have formed the cornerstone of both environmental and public policy in many countries.

4.1 The Polluter Pays Principle

'The Polluter Pays principle' creates conditions to make pollution a costly activity and to either influence behaviour (and reduce pollution) or generate revenues to alleviate pollution and compensate for welfare loss. While easy to understand, the Principle is unevenly applied across OECD countries. Implementation can be impaired by property rights, institutional and other barriers (OECD, 2012b). In Spain, for instance (Fuentes, 2011), water prices must cover, but not exceed, the operating and capital costs from the operation of government-funded supply infrastructures (transport, storage and treatment)[1]; they can cover administrative costs as well, to the extent that they are directly related to the operation of these infrastructures. While the recovery of costs that results from the scarcity of water[2] is particularly relevant for a country with a semi-arid climate, scarcity and environmental costs cannot be included in water prices over and above operating and capital costs.

4.2 The Beneficiary Pays Principle

'The Beneficiary Pays principle' allows sharing the financial burden of water resource management. It takes account of the high opportunity cost related to using public funds for the provision of private goods that users can afford. A requisite is that

1 It is worth noting that when capital costs are based on historic (and not replacement) costs, they tend to largely underestimate the financing requirements.

2 The 1999 amendment of the Water Law introduced a factor of 0.5 to 2, to be applied to tariffs reflecting financial costs, depending on whether consumption exceeds or is below reference levels. But these reference levels are likely to be determined with respect to individual concessions and do not reflect scarcity of the resource.

private benefits attached to water resource management are inventoried and measured, beneficiaries are identified, and mechanisms are set to harness them. While drawing a clear distinction between public and private benefits can be challenging, it is consequential from a financing perspective. For instance, if the costs of flood control are readily assumed by the government under a public good rationale (as in Spain), there is a strong incentive for private stakeholders to inflate the estimates of flood control benefits and to reduce their own share of the costs.

Depending on how they are designed and enforced, the Polluter Pays and the Beneficiary Pays principles can contribute to generate funds for water policies and to alleviate the use of the public purse. A fact often gone unnoticed: The two principles can conflict with one another under particular circumstances. Typically, when Payment for Ecosystem Services schemes are not properly designed, they can lead to sharing the cost of pollution. Hanley *et al.* (1998) discuss situations which could be portrayed as 'Pay the Polluter Principle': For instance, farmers who behaved in an ecologically responsible way can be penalised vis-à-vis others, if the less virtuous ones receive a larger incentive to change their behaviour. Similarly, Salzman (2005) highlights the perils of payment for ecosystem services, which, despite their high potential, can create moral hazard, rent-seeking behaviour, free-riding, or perverse incentives.

Payment for ecosystem services can be a very effective tool when the services are clearly defined and properly enhanced. Observers note that this is not always the case, and a number of payments for ecosystem services schemes should be considered as inadequate.

4.3 Equity as a principle for water management financing

A third principle, 'Equity', is also a feature of many policy frameworks for water management. For example, both France and the Netherlands consider equity as a core dimension of water financing. Cross-subsidies across water users have been used if the charges requested from some groups are disproportionate with their capacity to pay. This is one mechanism to address affordability issues, although in many instances it may be more efficient to use well-targeted social policy tools. Equity arguments are also sometimes used when considering the impacts of water policies on revenues (for farmers) or competitiveness (for farmers or industries). Article 4 of the Water Framework Directive (WFD; the anchor of water legislation in Europe) acknowledges that technical feasibility or disproportionate cost issues can legitimately impair the capacities of member states to achieve good ecological status of water bodies.

4.4 Consistency across policies that affect water resources

The OECD recommends adding a fourth principle to this core set: Consistency across policies that affect water resources. Water resources are affected by decisions made in various sectors. Obviously subsidising energy used by farmers to pump ground water may make sense to sustain the revenues for farmers in the short term, but does not lead to water efficient agricultural practices. When they pursue incoherent objectives, incentives generate excessive social costs and adversely affect the outcomes of water resource management. Reforming allocation of public moneys in adjacent sectors (for example, in agriculture, hydropower, energy, urban planning) can be more cost effective than mobilising additional funding in the water sector.

4.5 A set of empirical issues

These four core principles leave a number of practical issues pending. These issues can only be tackled on a case-by-case basis and on empirical ground:

- *Earmarking (parts of) the revenues from water-related taxes* for investments or for the operation of institutions or equipment can secure funding, in particular where there is fierce competition to access the public purse. However, earmarking can undermine overall economic efficiency, if these resources could have been allocated to activities that create more value for the society. Decisions on earmarking will depend on contextual features, including the power of water institutions, at national, basin or local levels;
- *Opportunities to reduce the costs of water management* abound. Initiatives in this domain can at the same time reduce financial needs, and increase the capacity of the sector to raise funds, as willingness to pay usually follows improvements in service levels;
- Water management can *attract private finance* to cover some of the upfront costs related to investment (for example, in relation to storage and distribution infrastructure). This requires robust financial strategies and business models, which secure stable revenue flows with which to repay financiers. The OECD Checklist for Public Action (OECD, 2009) can help governments ensure the policy framework for investment is amenable to private sector participation in water supply and sanitation;
- *Valuation of water-related services* is a requisite to assess benefits and harness beneficiaries. OECD countries gain experience with the combination of multiple methods to more systematically assess the value of these services. This has consequences for the range of policy options that can be considered.

Countries would benefit from considering the four principles and the set of empirical issues listed above to assess and potentially reform the way they finance water resources management.

5 CONCLUDING REMARKS

OECD countries face serious challenges regarding water management. To address them, well-thought and ambitious reform agendas will be required, tailored to national and transboundary circumstances.

The OECD has gained experience in accompanying water policy reforms in member countries (for instance: OECD, 2013), and in countries of Eastern Europe, the Caucasus and Central Asia (OECD, 2011c). Valuable lessons have been learnt from this experience in making water reform happen.

A general lesson is that reform is a process that takes time; planning and sequencing are key. Specific recommendations include:

- Build a broad constituency. Solutions to the water challenges cannot be expected to come from water policies alone. Water authorities need to work with other constituencies, including the agriculture and energy sectors, while taking the environment into account; they also need to work at different levels of government (local, basin, municipal, state and federal levels).
- Explore a mix of policy options and build capacity. A range of policy approaches is available to address water challenges. An optimal policy mix combines a variety of these approaches. Institutions and capabilities have to be adjusted to ensure there is the expertise to make complex technical and non-technical choices.
- Factor in financial sustainability from the start. The financial dimension should be factored in early in the process (to avoid designing a plan that is not financially affordable); cost reduction potentials have to be systematically considered; and financial realism needs to be brought to Water Resource Management plans. Financial incentives from other sectors should be aligned with water policy objectives (e.g. subsidies for energy or agriculture).
- Manage the political process and improve the knowledge base. Hard facts on the economic dimension of water policies can facilitate water policy reforms, demystify taboos and advance debates. This requires information on water demand and availability, and on the economic dimension and distributional impacts of the reform of water policies.

Sharing international experience on water policy reforms can substantiate such a process. The OECD provides a Forum where member countries and partners meet to exchange on their experience and identify good practices.

REFERENCES

Fuentes, A. (2011) *Policies towards a sustainable use of water in Spain.* OECD Economics Department Working Papers, number: 840. OECD Publishing.

Hanley, N., Kirkpatrick H., Simpson I. & Ogelthorpe D. (1998) Principles for the Provision of Public Goods from Agriculture: Modeling Moorland Conservation in Scotland, *Land Economics*, 74, 102–113.

OECD (2009) *Private Sector Participation in Water Infrastructure: OECD Checklist for Public Action.* Practical Guidance document, OECD Publishing.

OECD (2010) *Sustainable Management of Water Resources in Agriculture.* OECD Studies on Water, OECD Publishing.

OECD (2011a) *Towards Green Growth.* OECD Green Growth Studies, OECD Publishing.

OECD (2011b) *Water Governance in OECD Countries: A Multi-level Approach.* OECD Studies on Water, OECD Publishing.

OECD (2011c) *Ten Years of Water Sector Reform in Eastern Europe, Caucasus and Central Asia.* OECD Studies on Water, OECD Publishing.

OECD (2012a) *A Framework for Water Resources Management Financing.* OECD Studies on Water, OECD Publishing.

OECD (2012b) *Water Quality and Agriculture: Meeting the Policy Challenge*. OECD Studies on Water, OECD Publishing.
OECD (2012c), *Environmental Outlook to 2050*. OECD Publishing.
OECD (2013) *Making Water Reform Happen in Mexico*. OECD Studies on Water, OECD Publishing.
OECD (forthcoming, 2013a) *Water Security: A risk-based approach*. OECD Publishing.
OECD (forthcoming, 2013b) *Managing Water for Green Growth*. OECD Publishing.
Salzman, J. (2005) The Promise and Perils of Payment for Ecosystem Services. *International Journal on Innovation and Sustainable Development*, 1, 5–20.
UNEP (2011) *Decoupling natural resource use and environmental impacts from economic growth*. A Report of the Working Group on Decoupling to the International Resource Panel. Fischer-Kowalski, M., Swilling, M., von Weizsäcker, E.U., Ren, Y., Moriguchi, Y., Crane, W., Krausmann, F., Eisenmenger, N., Giljum, S., Hennicke, P., Romero Lankao, P., Siriban Manalang, A. & Sewerin, S. UNEP, Nairobi, Kenya.

Chapter 5

Water policy, agricultural trade and WTO rules

Lee Ann Jackson, Cedric Pene & Marie-Bel Martinez-Hommel
Agriculture and Commodities, Division World Trade Organization, Geneva, Switzerland

Claudia Hofmann & Ludivine Tamiotti
Trade and Environment Division, World Trade Organization, Geneva, Switzerland

ABSTRACT: By impacting national choices that affect relative prices of water, WTO rules may indirectly influence water allocation within and among countries. In addition WTO rules are relevant for national choices regarding behind-the-border-policies, such as labeling, that influence consumer choices. Today, irrigated agriculture accounts for approximately 70 percent of total water withdrawal and trade in agricultural products is sometimes also described as trade in 'virtual water'. This chapter examines in the field of agriculture the rules embedded in the WTO framework that can have direct or indirect impact on the types of policies countries choose to manage water resources. The chapter takes a close look at two broad categories of WTO rules – those focused on subsidies and those focused on consumer information and labeling. This chapter uses economic theory and legal analysis to provide a deeper understanding of the possible implications of WTO institutional framework on outcomes for water policies.

Keywords: trade, water footprint, labeling, WTO, irrigation, subsidies

1 INTRODUCTION

Today, irrigated agriculture accounts for approximately 70 percent of total water withdrawal (International Water Management Institute, 2007). By contributing to the global exchange of agricultural products, international trade can contribute to addressing problems related to the unequal geographical distribution of water. Governments can adopt various policies which will simultaneously affect outcomes in agricultural production, international trade and water resource use. WTO Members have committed to specific disciplines which have implications for the types of policy choices they can make.

This chapter examines the potential impact of WTO rules related to policies affecting agriculture and trade of agricultural products on policies targeting water resource use. The chapter begins by highlighting the reasons for which policy intervention can be justified and describes the analytical concept of virtual water. The third section briefly examines two categories of interventions that can impact the use of water on either the production side (irrigation subsidies) or the consumption side (water footprint labeling). The fourth section focuses on existing WTO rules in the area of subsidies and labeling. The fifth section concludes with suggestions for further work.

2 INTERNATIONAL TRADE AND THE CONCEPT OF VIRTUAL WATER

2.1 Market failures and the (mis)allocation of water

Due to various market failures water tends to be inefficiently allocated and this leads to patterns of use that negatively affect both the quantity and quality of water resources (WTO, 2010). Water does not have the characteristics of classical private goods. Water in many instances is non-excludable, meaning that individuals cannot exclude other users from consuming the good. The consumption of water, in the context of water scarcity, will also generate costs that spillover onto other users. Finally, water markets are characterized by lack of clear information about costs of production and consumption. For these reasons water may not be allocated efficiently either in national or global context. Government intervention may be justified, however not all policies contribute to efficient allocation of water resources. Furthermore, government water policies that favor particular types of agricultural production can influence a country's exports, creating spillover effects in global markets.

In an effort to explicitly recognize the potential impact of the movement of water embodied in traded goods, Professor Tony Allan developed the concept of trade in 'virtual water' (Allan, 2011). This concept is similar to a concept which exists in international trade theory. This general equilibrium trade model (the Heckscher-Ohlin model) states that a country will export the goods which require the use of the country's relatively abundant (and therefore cheap) factors for its production and import the goods which require the use of the country's relatively scarce (and therefore expensive) factors for its production. The idea of virtual water suggests a similar relationship between abundance of water and composition of trade. Water-short countries can reduce the risk of over-exploiting this scarce resource by importing rather than domestically producing water-intensive food products.

While intuitively appealing, the interpretation of trade figures and the policy recommendations that could result from a 'virtual water' analysis are still under debate among experts and must be taken with some caution (Wichelns, 2010). First the price of freshwater used in agriculture generally does not reflect its economic value nor the environmental externalities associated with its use. This distortion can lead to sub-optimal distribution of production and international trade flows from an economic as well as from an environmental point of view.

Secondly, even assuming that costs are accurately reflected in water prices, many other factors such as capital, land availability, farming structure, and technology also affect international trade in agricultural products. The resulting global comparative advantages in the production of different goods by different countries can therefore lead to situations where water-scarce countries export water-intensive agricultural products.

Thirdly, the interpretation of the 'virtual water' content of trade can be misleading when the agricultural good is exported by a water-abundant country where production is less water-efficient than the importing water-scarce country. In that case, the net result in terms of virtual water trade would be negative though the actual global environmental impact could be positive.

Finally, given the high spatial and temporal variability of water productivity and water use environmental externalities aggregate figures of 'virtual water' content of trade should be interpreted carefully.

Recently studies have sought to disaggregate 'virtual water' trade flows, by distinguishing between 'green' water ('The precipitation on land that does not run off or recharge the groundwater but is stored in the soil or temporarily stays on top of the soil or vegetation...') and 'blue' water ('Fresh surface and groundwater, in other words, the water in freshwater lakes, rivers and aquifers') (Hoekstra *et al.*, 2011). This classification assumes that, as a general rule, green water use is characterized by lower opportunity cost and less environmental externalities (the latter include the deterioration of the water quality through the contamination stemming from the use of pesticides and fertilizers), except when replacing high value ecosystems (Aldaya *et al.*, 2010).

Taking due account of the limitations mentioned above, at a global level, the total volume of 'water saved' through trade is estimated at around 5 percent of the global water use in agriculture (Chapagain *et al.*, 2006; Hoekstra, 2010). This figure confirms that the impact of international trade in agricultural products on global water use is significant but should not be overestimated. It is in line with the share of international trade vis-à-vis consumption (international trade represents around 20 percent of the world consumption of wheat, 10 percent in the case of corn and 7 percent in the case of rice). However, the ratio of water saved through imports on water used domestically can be very high for some countries (around 200 percent for Algeria and 70 percent for Morocco and Mexico (Hoekstra, 2010; Mekonnen & Hoekstra, 2011).

2.2 Water market interventions

Market interventions can take many forms. As noted above, public-sector intervention is justified in the context of market externalities and when goods don't behave as classic private goods. The private sector may also have incentives to intervene in the area of water in order to influence the competitive status of their goods. This chapter will now focus on two types of policy interventions where WTO rules may be relevant.

2.2.1 *Production-side interventions: Irrigation subsidies*

The classic public intervention related to water use in agricultural production is the provision of irrigation subsidies. This government support for farmers has been justified based on the notion that the irrigation is a type of public infrastructure. An additional complicating factor in policy making related to water is that certain aspects of water infrastructure, such as irrigation systems, have public good characteristics.

Irrigation subsidies can be broadly divided into two main categories: (i) subsidies to the supply of water for irrigation, which can take different forms like investment and/or operational subsidies to the water provider (later referred to as 'irrigation water subsidies'); and (ii) subsidies to the practice of irrigation (like programs for the development of more water efficient irrigation techniques).

2.2.2 *Consumption-side interventions: Labeling*

Other types of interventions, such as water footprint labeling, are intended to influence the use of water by providing consumers with information about the amount of water that was required to produce particular products. In essence, a water footprint measures direct and indirect water use during production. More specifically, it refers to 'the volume of freshwater used to produce the product, measured over the full supply chain. It is a multidimensional indicator, showing water consumption volumes by source and polluted volumes by type of pollution. All components of a total water footprint are specified geographically and temporally.' (Hoekstra *et al.*, 2011). Water footprint labeling is a tool for increasing consumer awareness as to the environmental impacts of their consumption choices.

3 RELEVANT WTO RULES

Two WTO agreements include disciplines on government subsidies. The 'WTO Agreement on Agriculture' (WTO, 1995a) establishes rights and obligations with respect to domestic support specifically provided to agricultural products. The 'Subsidies and Countervailing Measures Agreement' focuses on subsidies in general. The following two subsections deal with the specifics of these two agreements and the way the disciplines embedded in them relate to each other. The third subsection describes the relevant disciplines in the WTO Agreement on 'Technical Barriers to Trade', which are relevant to labeling.

3.1 WTO Agreement on Agriculture

The 'Agreement on Agriculture' (AoA) establishes disciplines for the so-called 'three pillars': Market access; domestic support; and export competition. In the domestic support pillar, disciplines apply to the support provided by WTO Members either for an agricultural product (product-specific) or as non-product-specific support in favor of agricultural producers in general. It includes not only direct payments but also *inter alia* market price support schemes, low-interest loans to producers or processors, insurance schemes, and input subsidies.

The support is split into several broad categories referred to as 'Green Box', 'Amber Box' 'Blue Box' and 'Development Box'. Amber Box support (Articles 6.1, 6.3 and 7.2, and Annexes 3 and 4) is constituted by all the support which is not classified in any other box. It is deemed to be the most trade-distorting type of support and was subject to reduction commitments under the Uruguay Round Agreement. It is expressed in terms of 'Aggregate Measurement of Support' (AMS) and its bound levels (i.e. maximum permitted support) are listed in Members' schedules of commitments. Amber Box support, either product specific or non-product specific, which remains below a *de minimis* level[1] is not included in the calculation of the current total AMS.

1 *De minimis* level is expressed as 5% of the value of production for developed country Members and 10% as a general rule for developing country Members (Article 6.4).

The Blue Box (Article 6.5) is deemed less trade-distorting than the Amber Box as the support falling into this category must *inter alia* be part of production limiting programs. Under current rules, Blue Box support can be provided without any limitation as long as it meets the above-mentioned requirement[2].

In addition, developing countries benefit from the 'Special and Differential Treatment provisions' of Article 6.2 ('Development Box'), which allow them additional domestic support under development programs, in the form of investment subsidies generally available to agriculture and agricultural input subsidies generally available to low-income or resource-poor producers, and to encourage diversification from growing illicit narcotic crops.

Finally, the Green Box (Article 6.1 and Annex 2) covers the support that has no or at most minimal trade-distorting effect or effects on production. All measures classified under the Green Box must meet Annex 2 general basic criteria, as set out in paragraph 1:

a The support in question shall be provided through a publicly-funded government program (including government revenue foregone) not involving transfers from consumers and;
b The support in question shall not have the effect of providing price support to producers;

Furthermore, such support measures must comply with each policy-specific criterion and condition set out in the different categories (paragraph. 2 to 13) of Annex 2. The support classified under the Green Box can be used without any limitation.

WTO Members notify their domestic support measures to allow the Committee on Agriculture to review the implementation of commitments. Since 1995, 48 Members, out of the current 159, have notified irrigation water subsidies to the WTO, usually as Green Box measures. Irrigation subsidies are in general notified under the Green Box as infrastructural services (paragraph 2 (g)) that are part of the government service programs:

g Infrastructural services, including: Electricity reticulation, roads and other means of transport, market and port facilities, water supply facilities, dams and drainage schemes, and infrastructural works associated with environmental programs. In all cases the expenditure shall be directed to the provision or construction of capital works only, and shall exclude the subsidized provision of on-farm facilities other than for the reticulation of generally available public utilities. It shall not include subsidies to inputs or operating costs, or preferential user charges.

To be eligible under this heading, irrigation subsidies should provide '...services or benefits to agriculture or the rural community' and not involve direct payments to producers (chapeau of paragraph 2).

However, evaluating to which extent irrigation water subsidies can translate or not into 'subsidies to inputs or operating costs, or preferential user charges,' can be a challenging task due to the lack of commonly agreed methodology and definition of

2 It should be noted that only four Members currently use Blue Box support.

what would constitute a non-subsidized input or operating cost or a non-preferential user charge (Charles, 2009). For example, it is not clear amongst WTO Members whether operating costs include irrigation system maintenance costs. If paragraph 2(g) is interpreted as including irrigation system maintenance costs as part of the 'operating costs', then when water charges do not fully cover maintenance costs, the government has to forego certain revenues on that account. Such revenue foregone should not be included in the Green Box but in the calculation of the AMS (according to Annex 3 par. 2 of the AoA) or possibly in another box, if the corresponding criteria are met.

This uncertainty is perceptible in the framework of the Regular Committee on Agriculture (Regular Committee on Agriculture, 1995–2010). Since 1995 WTO Members have been seeking confirmation regarding the compliance of irrigation subsidies notified as 'infrastructural services' (Annex 2, paragraph 2(g)) with Green Box criteria. They raised questions on issues such as whether expenditure for infrastructure services is directed to the provision of generally available public utilities only, and whether these payments do not result in a subsidized provision of on-farm facilities, other than for the reticulation of generally-available utilities. In relation to maintenance costs, some Members consider the 'improvement and maintenance' costs as being part of the support provided under the 'general services' category of the Green Box.

One Member, Mexico, has also notified irrigation water subsidies under two different Green Box categories. Since 1999, Mexico has been consistently notifying its subsidies for infrastructural works in irrigation areas under the general services category of paragraph 2. In 2003, Mexico introduced a new 'water-use acquisition program' (WTO Committee on Agriculture, 2006b), and classified it as 'payments under environmental programs', under paragraph 12 of Annex 2[3]. Mexico justified this classification by the fact that the funds granted to producers were subject to best practices, so as to promote the sustainability of irrigation districts affected by water availability problems, and were thus in conformity with the policy-specific criteria of paragraph 12 (WTO Committee on Agriculture, 2007a; 2007c; 2011a).

Irrigation water subsidies that cannot be classified as Green Box measures could be considered as input subsidies under the AMS, *de minimis* or Article 6.2. If classified as AMS, their value should be measured 'using government budgetary outlays or, where the use of budgetary outlays does not reflect the full extent of the subsidy concerned, the basis for calculating the subsidy shall be the gap between the price of the subsidized good or service and a representative market price for a similar good or service multiplied by the quantity of the good or service' (paragraph 13 of Annex 3). The methodological difficulties mentioned above would also apply here. One Member, India[4], notified first its irrigation water subsidies as non-product specific AMS and then as from 1997/98 as 'other input subsidies'

3 WTO (1995a), Agreement on Agriculture, Annex 2, paragraph 12 states: (a) 'Eligibility for such payments shall be determined as part of a clearly-defined government environmental or conservation program and be dependent on the fulfillment of specific conditions under the government program, including conditions related to production methods or inputs'. (b) 'The amount of payment shall be limited to the extra costs or loss of income involved in complying with the government program.'

4 India shifted its subsidies from non-product specific AMS (reported for marketing year 1996/97) to Development Box (reported for marketing years 1997/98 to 2003/04) (WTO Committee on Agriculture, 1998b and 2011b).

under the category 'Agricultural input subsidies to low income or resource poor producers' of Article 6.2.

3.1.1 SCM Agreement

In addition to the rules embodied in the AoA, the WTO Agreement on Subsidies and Countervailing Measures (SCM Agreement) contains important disciplines on the use of subsidies by WTO Members. Since the 'Peace Clause' (Article 13 of the AoA) lapsed at the end of 2003, the SCM Agreement can be directly applied to agricultural products without the limitations foreseen in the 'Peace Clause'. The SCM Agreement, which prohibits certain subsidies[5], and in addition addresses certain adverse economic and trade effects caused by subsidization, does not address negative environmental impacts of subsidies.

Water subsidies, and in particular subsidies to irrigation, have not been the focus of a specific discussion in the context of the SCM Agreement. In contrast, for example, fossil fuel energy subsidies and dual-pricing in the energy sector have attracted much interest and debate in the recent years (including in the Doha Development Agenda Negotiating Group on Rules).

Such irrigation subsidies, to the extent they are specific as defined by the SCM Agreement (see *infra*), are notifiable to the WTO pursuant to Article XVI:1 of the General Agreement on Tariff and Trade (GATT) 1994 and Article 25 of the SCM Agreement. There is, however, no independent basis on which to determine whether the notifications to the WTO that have been made to date exhaustively catalogue such subsidies (Table 1).

The SCM Agreement defines a 'subsidy' as a financial contribution by a government or any public body within the territory of a Member that confers a benefit. A financial contribution is deemed to exist where: (i) a government practice involves a direct transfer of funds; (ii) government revenue that is otherwise due is foregone; (iii) a government provides goods or services other than general infrastructure; or (iv) a government entrusts or directs a private body to carry out one or more of the types of functions listed in (i) to (iii).

The first step to address when looking at irrigation subsidies under the SCM Agreement would therefore be to assess whether each of the modes by which irrigation subsidies could be provided (examples in Table 1) constitutes a financial contribution by a government, including, where water is provided for irrigation, whether this could be considered to be a government-provided 'good or service other than general infrastructure'. The question of what constitutes 'general infrastructure' in the sense of the SCM Agreement is complex, and subject to a fact-intensive, case-specific analysis.

The second step would be to determine whether a benefit is thereby conferred. The WTO Appellate Body has ruled that a benefit is conferred where a financial contribution is received on terms more favorable than those available to the recipient on the market (Appellate Body Report, Canada – Aircraft). To the extent that an irrigation subsidy took the form of provision of a good or service, Article 14(d) of the

5 Per the terms of the Agreement on Subsidies and Countervailing Measures, those prohibitions apply subject to the provisions of the Agreement on Agriculture. See *infra*.

Table 1 Notifications by WTO Members during the period 2005–2011[6] with reference to 'irrigation'[7].

Country	*Title of the measure*	*Period covered*	*Short description*
Australia	Linking farms and catchments with irrigation modernization initiative	2009/2010 to 2010/2011	Develop the urgently required linkages between modernization and broader catchment management consistent with Government policy and existing programs
Australia	First Farm Grant	2009/2010 to 2010/2011	... including development activities including: Installation of more water efficient irrigation systems
Australia	The water smart farms initiative	2009/2010 to 2010/2011	Supported activities that investigated, planned or implemented improved on-farm irrigation practices, with a focus on ground works programs associated with reconfiguration/ infrastructure programs
Australia	Murray Darling basin irrigation management grants program	2007/2008 to 2010/2011	Assist Murray-Darling Basin irrigators to implement water management strategies to address reduced water allocations
China	Fund for subsidizing transformation of agricultural technology	2005 to 2008	To accelerate the transformation of technological achievements in agriculture, forestry, water conservation and irrigation, and to improve the capacity of agricultural innovation
China	Subsidy for national key construction projects on water and soil conservation	2005 to 2008	To assist small scale farmland irrigation and water and soil conservation projects in rural areas
China	Fund for interest discount of loans for the sake of agricultural water-saving irrigation	2005 to 2008	To support water-saving irrigation technology and the construction of areas using water-saving irrigation
Denmark	Establishment of shelter belts	2005 to 2010	To establish plantings providing shelter and improving the biotope in order to... reduce the need for artificial irrigation in locations threatened by drought
Slovenia	Rural Development Plan of the Republic of Slovenia 2007–2013.	2007–2013	Strengthen the competitiveness of agricultural and forestry sector through infrastructure improvement (rounding up agricultural parcels, upgrading of amelioration systems, irrigation etc.)

SCM Agreement, cited as relevant context for interpreting the term 'benefit', provides that to confer a benefit, goods or services have to be provided for less than adequate remuneration. This in turn is determined in relation to prevailing market conditions

6 Corresponding to WTO document series G/SCM/N/220, G/SCM/N/186, and G/SCM/N/155.

7 This list is not necessarily exhaustive of all the subsidies that may benefit irrigation in the agricultural sector. Notifications with reference to water are in most cases either explicitly not in favour of the agricultural sector or not detailed enough to determine whether the agricultural sector is also covered.

for the good or service in question in the country of provision. In countries where the government is the main provider of the goods or services in question, it may be that the government 'can affect through its own pricing strategy the prices of private providers' (Appellate Body Report, U.S. Softwood Lumber IV). In such cases, it may be necessary to compare prices to a benchmark other than private prices in the country of provision, such as private prices in another market or proxies constructed on the basis of production costs, in order to assess accurately the level of benefit conferred.

Only subsidies that are 'specific' to an enterprise, industry or a group of enterprises or industries are regulated by the SCM Agreement. The third step would therefore be to determine whether the alleged subsidy is specific or not. Some observers consider that any irrigation water subsidy would be considered as specific by the mere fact it benefits the agricultural sector. This view is not universal; however, as others consider that by virtue of the diversity of the agricultural sector, some further limitation would be necessary for such subsidies to be specific. This issue has been discussed, but not resolved, in the Panel Report in 'U.S.-Cotton'.

Under the SCM Agreement, export subsidies and subsidies contingent on the use of domestic goods are prohibited, except as provided in the 'AoA'. The remaining subsidies are considered 'actionable' under the SCM Agreement, which means that they can be challenged if they cause certain specified adverse effects to the interests of another Member. The SCM Agreement defines three such types of adverse effects: (i) Injury to the domestic industry of another Member; (ii) Serious prejudice to the interest of another Member (for example caused by export displacement, or caused by a significant price effect on the market; (iii) Nullification or impairment of benefits accruing under the GATT 1994, in particular where a market access benefit resulting from a tariff bound in a Member's schedule is undercut by subsidization.

A WTO Member that is affected by subsidies granted by another Member or that considers such subsidies to be prohibited can challenge those subsidies in the WTO dispute settlement mechanism to seek the withdrawal of the subsidies, or alternatively in the case of non-prohibited subsidies, the removal of their adverse effects. Another possible recourse is that the affected Member can apply countervailing duties to subsidized imports into its territory if it determines in a properly-conducted investigation that those imports cause or threaten to cause injury to its domestic industry.

Provisions of the SCM Agreement that accorded non-actionable status to certain specific subsidies, including to promote adaptation of existing facilities to new environmental requirements, expired at the end of 1999 due to the absence of a consensus among Members to extend them. As it is the case in other environmental fields, some observers regret the disappearance of this provision that potentially might have explicitly carved out from the outset subsidies designed to favor the transition from water-intensive to water-saving practices.

3.2 Labeling

While environmental labeling schemes can help to correct market failures regarding information on environmental aspects of products, they may also create unnecessary barriers or disguised restrictions to international trade. It is in this context that WTO rules may be relevant.

3.2.1 TBT Agreement

Labeling requirements generally fall within the scope of the WTO Agreement on Technical Barriers to Trade (TBT Agreement)[8]. Since water footprint labeling schemes relate to product characteristics or their related processes and production methods, they may be categorized as mandatory 'technical regulations' or voluntary standards within the meaning of Annex 1.1 and therefore fall under the coverage of the TBT Agreement. In addition, the GATT national treatment obligation in Article III:4, which relates to internal regulations that affect the internal sale, offering for sale, purchase, transportation, distribution or use of products may as well be relevant to water footprint labeling schemes.

The TBT Agreement disciplines the use of technical regulations and standards by applying for instance, the core GATT principle of non-discrimination. In this regard, Members must ensure that technical regulations do not accord less favorable treatment to products imported from other WTO Members as compared to 'like' domestic products[9] of national origin (i.e. national treatment principle) and to like products originating in any other WTO Member (i.e. most-favored nation (MFN) treatment)[10]. However, in some cases, Members may need to make distinctions between like products in order to achieve their regulatory objectives, including objectives of their labeling schemes. The U.S. – Tuna II (Mexico) and U.S. – COOL cases show that any difference in labeling conditions between like products will have to stem from a legitimate regulatory distinction[11].

In addition, the TBT Agreement recognizes Members' right to take regulatory measures to achieve their legitimate objectives, but at the same time stipulates that such measures must not constitute unnecessary barriers to trade[12]. In this context, a measure's contribution to the attainment of the stated objective, the trade restrictiveness of the measure, the nature of risks at issue and the gravity of consequences that would arise from non-fulfillment[13], and the existence of less trade-restrictive alternatives all are taken into consideration in an exercise of 'weighing and balancing'[14].

In this regard, any water footprint labeling scheme would need to be applied to 'like' products in accordance with the non-discrimination obligation. The nature of

8 Annex 1.1 and 1.2 of the TBT Agreement (WTO, 1995b). See also the Preamble of the TBT Agreement. Note however that, in practice, most of such labeling schemes are of voluntary nature.

9 The concept of 'like' products relates to the nature and extent of a competitive relationship between and among two products. The determination of whether these two products are like is made on a case-by-case basis, and may employ four general criteria or characteristics that the products involved might share: (i) physical properties; (ii) same or similar end-uses; (iii) the extent to which consumers perceive and treat the products as alternative means; and (iv) the international classification of the products for tariff purposes. See *EC – Asbestos*, Appellate Body Report, 2001, paras. 99, 101, 102; *U.S. – Clove Cigarettes*, Appellate Body Report. 2012, para. 120.

10 Article 2.1 of the TBT Agreement (WTO, 1995b).

11 *U.S. – Tuna II (Mexico)*, Appellate Body Report, 2012, para. 284; *U.S. – COOL*, Appellate Body Report, 2012, para. 349.

12 Article 2.2 of the TBT Agreement, 1995. Article 2.2 contains a non-exhaustive list of 'legitimate objectives' that includes, *inter alia*, the protection of animal or plant life or health, or the environment.

13 *U.S. – COOL*, Appellate Body Report, 2012, para. 468.

14 U.S. – *Tuna II (Mexico)*, Appellate Body Report, 2012, para. 322; *U.S. – COOL*, Appellate Body Report, paras. 376–378; 471; *US – Tuna II (Mexico)*, Appellate Body Report, paras. 320–321.

a water footprint label would suggest distinctions between products on the basis of non-product related processes and production methods (PPMs), since the use of water during a product's life cycle would not leave a trace in the final product. Recent WTO jurisprudence would indicate that measures relating to non-product related production methods are covered by the TBT Agreement[15]. Turning back to our case of interest, if two products with a different water footprint are found to be 'like products', the extent to which a detrimental impact on imports reflects discrimination or stems from a 'legitimate regulatory distinction' is based on a detailed assessment of the particular circumstances of each case, including the nature, design and application of the label. Finally, an assessment of the necessity of the trade-restrictiveness of water footprint labeling should consider the labeling scheme's objective and efficiency in achieving this, taking into consideration the availability of less trade-restrictive alternative measures to promote the objective pursued[16].

Moreover, the TBT Agreement requires Members to base their technical regulations on 'relevant international standards' where these exist[17]. To the extent that there is an international standard bearing on the matter of water footprint, countries would need to base their measures on the benchmarks established by that international standard[18]. An international standard has been defined as a 'standard that is adopted by an international standardizing/standards organization and made available to the public'[19], 'open' to the relevant bodies of at least all Members[20], and recognized by WTO Members and national standardizing bodies[21]. In addition, the Appellate Body highlighted in the U.S. – Tuna II case that the larger the number of countries that participate in the development of a standard, the more likely it can be said that the respective body's activities in standardization are 'recognized'[22]. An International Water Footprint Standard is currently under development (ISO, n.d.).

3.2.2 Potential trade concerns in relation to water footprint labeling

The implementation of water footprint labeling could raise a number of trade concerns, particularly for developing country Members in relation to their market access. Discussions in the WTO's Committee on Trade and Environment (CTE) on carbon footprint labeling and also recent discussions in the TBT Committee on the issue of water efficiency labeling for household appliances (WTO Committee on Technical Barriers to Trade, 2010), could provide an indication on potential similar concerns with regard to water footprint labeling.

15 See *U.S.—Tuna II (Mexico)*. Appellate Body Report, 2012.
16 An argument has been made with regard to water stewardship as a less trade-restrictive alternative measure to achieve the overall objective of sustainable use, preservation and consumption of water resources.
17 Annex 1.2 and Article 2.4 of the TBT Agreement (WTO, 1995b).
18 Article 2.4 of the TBT Agreement, 1995.
19 *U.S. – Tuna II (Mexico)*, Appellate Body Report, 2012, para. 353.
20 *U.S. – Tuna II (Mexico)*, Appellate Body Report, 2012, paras. 382, 397–398.
21 *U.S. – Tuna II (Mexico)*, Appellate Body Report, 2012, para 389–390.
22 *U.S. – Tuna II (Mexico)*, Appellate Body Report, 2012, para. 390.

Main concerns expressed on carbon footprint labeling refer to methodologies of data collection, costs, and trade impacts (WTO Committee on Trade and Environment, 2010a). Countries have stressed the need for harmonized international standards and guidelines, neutral methodologies, and increased transparency in the collection of data. Concerns with regard to costs include the lack of certification agencies for small and medium-sized enterprises (SMEs) in developing countries (WTO Committee on Trade and Environment, 2010b). Lastly, the proliferation of unilateral voluntary carbon footprint labeling schemes and standards might lead to trade impacts (WTO Committee on Trade and Environment, 2011a and 2011b).

Overall, there seems to be clear need for harmonization and common methodologies with regard to carbon as well as water footprint labeling. For instance, it has been noted that methodologies for water data collection would need to take local factors into account, as well as seasonal and annual variations in water consumption (RPA, 2011). Critics have argued that the 'water label' fails to show the real problem, might lead to consumer confusion, and would need to include information on market distortions (e.g. the price of water does not reflect scarcity). Interestingly, a high water footprint does not necessarily have the same meaning everywhere, since production in water abundant regions compared to water-scarce regions might not be reflected on the label.

In fact, one of the concerns raised in relation to water footprint labeling is that it may not be an effective tool to enhance efficient use of water, particularly given the complexity of the issue. A simple measurement of water consumption seems to fall short of encouraging local communities to improve overall sustainable use, preservation and consumption of water resources. While water footprint assessments should normally include water quantity and quality accounting and impact evaluation, most product labels cannot adequately convey this complex information (WTO Public Forum, 2012).

A few countries have shown an active interest in water footprint labeling and accounting. France, in its Grenelle carbon footprint experiment, has included water footprint as a possible additional environmental indicator for use and net consumption of water (MEDDTL, 2011). Moreover, China, Germany, India, Indonesia, the Netherlands, Spain and the United Kingdom have been active in the development of national water footprint accounting (Water Footprint Network, 2012). As the first government to incorporate water footprint assessment into policy, Spain has made water footprint analysis an obligatory part of the process of developing river basin management plans[23].

Scientific research in Australia, for instance, has focused on the development of water footprint calculation methods (CSIRO, 2009). In the EU, the discussions on water footprint labeling have highlighted some concerns, including the lack of consistency, clarity and transparency in the use of methods, as well as the number of already existing global initiatives that could be built upon in the development of certification standards. Similarly, as can be argued in the case of carbon footprint, a lack of basic understanding or framework of reference on water footprint

23 This has been done as part of the requirements of the Water Framework Directive (Hoekstra, 2011). See also Water Footprint Network (2013).

(including metrics) will limit consumers' decision-making. Today, the water footprint assessment seems to be best viewed as a tool for corporate supply-chain management, while consumer-aimed labeling does not seem sufficiently mature and developed for consumers to relate to (RPA, 2011). Indeed, in the private sector, it is acknowledged as useful for companies to help understand and reduce their water use and associated cost (The Nature Conservancy and the Coca Cola Company, 2010).

4 CONCLUSIONS

This chapter shows that the WTO rules as they apply to agricultural and food products, in areas like subsidies and product labeling may influence water-related policies. More work is needed to clarify key concepts and to enhance transparency in order to have a more comprehensive understanding of the ways in which these rules could alter water resource use and allocation.

WTO texts do reflect an interest in limiting adverse environmental effects. For example, the WTO decision on Trade and Environment (Uruguay Round Trade Negotiations Committee, 1994) refers to the identification of 'the relationship between trade measures and environmental measures, in order to promote sustainable development'. Similarly, the preamble of the AoA calls on WTO Members to 'take into account the need to protect the environment'. Still, neither the SCM nor the Agriculture Agreements take into account specific non-trade concerns such as potential harmful impacts on the environment due to excessive water use. The only attempt until now to introduce explicitly in the WTO rules some notion of 'environmental adverse effect' have been the discussions held on fisheries subsidies under the Doha Development Agenda. These discussions envisaged the prohibition of certain forms of fisheries subsidies that contribute to overcapacity and over-fishing or to provide for a remedy where any Member caused, through the use of any specific subsidy, certain adverse effects in respect of fish stocks in which another Member had an identifiable fishing interest. This negotiation is today far from being concluded and there is no similar discussion taking place in the area of agricultural subsidies. Nevertheless, given the potential economic and environmental impacts of water subsidies on the world market in agriculture, the long-term objective of fair and market-oriented agricultural trading system is consistent and mutually supportive with the objective of 'green trade' (Hoekstra, 2010).

The analysis of impact of subsidies is hampered by the lack of comparable information. More consistent reporting methods would strengthen the robustness of the data on the use and effects of irrigation subsidies. Given its notification system, the WTO could play a role in an effort to collect, for example through its different committees (Committees on Agriculture, Subsidies and Countervailing measures and/or on Trade and Environment), more detailed information and data on water subsidies. In the field of labeling, current WTO rules provide a good basis to allow Member countries to provide consumer information while avoiding 'hidden protectionism'. Here again, as mentioned above, WTO Members have highlighted the need for better information, increased transparency and consistency in the methodologies used, in particular in the TBT and in Trade and Environment Committees. Enhancing the public access to this information, as well as consumers' comprehension thereof, would be a useful contribution to strengthen collective understanding of the complex system of factors affecting global water use.

ACKNOWLEDGEMENTS

The authors are indebted to Ulla Kask, Clarisse Morgan, Jesse Kreier and Devin McDaniels for useful inputs to and comments on this chapter.

REFERENCES

Aldaya, M.M., Allan, J.A. & Hoekstra, A.Y. (2010) Strategic importance of green water in international crop trade. *Ecological Economics*, 69(4), 887–894.

Allan, T. (2011) *Virtual water: Tackling the threat to our Planet's most precious resource*. New York, I.B. Tauris.

Chapagain, A.K., Hoekstra, A.Y. & Savenije, H.G. (2006) Water saving through international trade of agricultural products. *Hydrology and Earth System Sciences*. [Online] 10, 455–468. Available at: http://doc.utwente.nl/58354/1/water_saving_through.pdf [Accessed 15 February 2013].

Charles, C. (ed.) (2009) *The GSI's Method for Quantifying Irrigation Subsidies*. Article Series for the Global Subsidies Initiative (GSI) of the International Institute for Sustainable Development (IISD), Geneva, Switzerland. Available from: http://www.iisd.org/gsi/irrigation-subsidies/irrigation-resources [Accessed 18th October 2012].

CSIRO (2009) *CSIRO partners global Water Footprint Network*. [Online] Available from: http://www.csiro.au/en/Portals/Media/Global-Water-Footprint-Network.aspx [Accessed 30th October 2012].

Hoekstra, A.Y. (2010), The relationship between international trade and freshwater scarcity, WTO Economic Research and Statistics Division, Staff Working Paper Series. Number: ERSD-2010-05.

Hoekstra, A.Y. (2011) *The water footprint: A tool for governments, companies and investors*. [Online] Hét Nationaal Water symposium 19 mei 2011. Available from: http://doc.utwente.nl/78370/1/Hoekstra11water.pdf [Accessed 15 February 2013].

Hoekstra, A.Y., Chapagain, A.K., Aldaya, M.M. & Mekonnen, M.M. (2011) *The water footprint assessment manual: Setting the global standard, Earthscan*, London, UK.

International Water Management Institute (2007) Water for Food, Water for Life: A Comprehensive Assessment of Water Management in Agriculture. London, Earthscan.

ISO (n.d.) *ISO/CD 14046*. [Online] Available from: http://www.iso.org/iso/catalogue_detail?csnumber = 43263 [Accessed 30th October 2012].

MEDDTL, French Ecology, Sustainable Development, Transport and Housing Ministry (2011) *Product environmental footprint policy in France: Current developments*. [Online] Available from: http://www.neaspec.org/documents/carbon_fp/Product%20Env%20footprint%20policy%20 in%20France_Antonin%20Vergez.pdf [Accessed 13th December 2012].

Mekonnen, M.M. & Hoekstra, A.Y. (2011) *National water footprint accounts: The green, blue and grey water footprint of production and consumption*. UNESCO-IHE, Delft. Value of Water Research Report Series. Report number: 50. Available from: http://www.waterfootprint.org/Reports/Report50-NationalWaterFootprints-Vol1.pdf [Accessed 15 February 2013].

RPA, Risk & Policy Analysts Limited (2011) *Assessment of the efficiency of the water footprinting approach and of the agricultural products and foodstuff labeling and certification schemes, ENV.D.4/SER/2010/0051r [Online]*. Report prepared for DG Environment. Executive Summary available from: http://ec.europa.eu/environment/water/quantity/pdf/Executive%20Summary%2027%20Sept_2011%20-%20 revised2.pdf [Accessed 15 December 2012].

The Nature Conservancy and the Coca Cola Company (2010) *Product Water Footprint Assessments*. [Online] Available from: http://www.waterfootprint.org/Reports/CocaCola-TNC-2010-ProductWaterFootprintAssessments.pdf [Accessed 30th October 2012].

Water Footprint Network (2012) *National Water Footprints*. [Online] Available from: http://www.waterfootprint.org/?page = files/NationalWaterFootprint [Accessed 30th October 2012].

Water Footprint Network (2013) *National Water Footprints: Spain*. [Online] Available from: http://www.waterfootprint.org/?page = files/Spain [Accessed 15 February 2013].

Wichelns, D. (2010) An Economic analysis of the virtual water conception in relation to the Agri-Food sector. In: OECD, *Sustainable Management of Water Resources in Agriculture*. OECD Publishing.

WTO (2010) *Trade in natural resources*. World Trade Organization Report.

WTO Public Forum (2012) *WTO Public Forum Session 23, "Water and Agricultural Trade", 24–26 September 2012, Geneva, Swiss.*

AGREEMENTS

GATT (1994) General Agreement on Tariffs and Trade.

Uruguay Round Trade Negotiations Committee (1994) Decision on Trade and Environment adopted by Ministers in Marrakesh on 14 April 1994.

WTO, World Trade Organization (1995a) Agreement on Agriculture (AoA).

WTO, World Trade Organization (1995b) Agreement on Technical Barriers to Trade (TBT).

WTO, World Trade Organization (1995c) Agreement on Subsidies and Countervailing Measures (SCM).

DISPUTES

——DS384, *United States – Certain Country of Origin Labeling (COOL) Requirements (U.S.-COOL)*, Appellate Body Report WT/DS384/AB/R and WT/DS386/AB/R, 29 June 2012.

——DS406, *United States – Measures Affecting the Production and Sale of Clove Cigarettes (U.S. – Clove Cigarettes)*, Appellate Body Report WT/DS406/AB/R, 4 April 2012.

——DS381, *United States – Measures Concerning the Importation, Marketing and Sale of Tuna and Tuna Products (U.S. – Tuna II (Mexico))*, Appellate Body Report WT/DS381/AB/R, 16 May 2012.

——DS267, *United States – Subsidies on Upland Cotton (U.S. – cotton)*, Panel Report WT/DS267/R, 8 September 2004.

——DS257, *United States – Final Countervailing Duty Determination with respect to certain Softwood Lumber from Canada (U.S. Softwood Lumber IV)*, Appellate Body Report WT/DS257/AB/R, 19 January 2004.

——DS135, *European Communities – Measures Affecting Asbestos and Asbestos-Containing Products (EC-Asbestos)*, Appellate Body Report WT/DS134/AB/R, 12 March 2001.

——DS70, *Canada – Measures Affecting the Export of Civilian Aircraft (Canada-Aircraft)*, Appellate Body Report WT/DS70/AB/R, 2 August 1999.

WTO COMMITTEES DOCUMENTS

WTO Committee on Agriculture (1996) *Summary report G/AG/R/7, 26 November 1996*. World Trade Organization.

WTO Committee on Agriculture (1997a) *Summary report G/AG/R/9, 17 January 1997*. World Trade Organization.

WTO Committee on Agriculture (1997b) *Summary report G/AG/R/10, 23 April 1997*. World Trade Organization.

WTO Committee on Agriculture (1997c) *Summary report G/AG/R/11, 28 July 1997*. World Trade Organization.

WTO Committee on Agriculture (1997d) *Summary report G/AG/R/12, 31 October 1997*. World Trade Organization.

WTO Committee on Agriculture (1997e) *Summary report G/AG/R/13, 19 December 1997*. World Trade Organization.

WTO Committee on Agriculture (1998a) *Summary report G/AG/R/14, 30 April 1998*. World Trade Organization.

WTO Committee on Agriculture (1998b) *Notification G/AG/N/IND/2, 17 June 1998*. World Trade Organization.

WTO Committee on Agriculture (1998c) *Summary report G/AG/R/15, 3 August 1998*. World Trade Organization.

WTO Committee on Agriculture (1998d) *Summary report G/AG/R/16, 28 October 1998*. World Trade Organization.

WTO Committee on Agriculture (1999a) *Summary report G/AG/R/17, 25 January 1999*. World Trade Organization.

WTO Committee on Agriculture (1999b) *Summary report G/AG/R/18, 7 May 1999*. World Trade Organization.

WTO Committee on Agriculture (1999c) *Summary report G/AG/R/19, 25 August 1999*. World Trade Organization.

WTO Committee on Agriculture (1999d) *Summary report G/AG/R/20, 22 October 1999*. World Trade Organization.

WTO Committee on Agriculture (2000a) *Summary report G/AG/R/22, 25 April 2000*. World Trade Organization.

WTO Committee on Agriculture (2000b) *Summary report G/AG/R/23, 27 July 2000*. World Trade Organization.

WTO Committee on Agriculture (2000c) *Summary report G/AG/R/24, 31 October 2000*. World Trade Organization.

WTO Committee on Agriculture (2000d) *Summary report G/AG/R/25, 21 December 2000*. World Trade Organization.

WTO Committee on Agriculture (2001a) *Summary report G/AG/R/26, 31 May 2001*. World Trade Organization.

WTO Committee on Agriculture (2001b) *Summary report G/AG/R/27, 14 September 2001*. World Trade Organization.

WTO Committee on Agriculture (2001c) *Summary report G/AG/R/28, 26 October 2001*. World Trade Organization.

WTO Committee on Agriculture (2002a) *Summary report G/AG/R/29, 27 February 2002*. World Trade Organization.

WTO Committee on Agriculture (2002b) *Summary report G/AG/R/31, 27 August 2002*. World Trade Organization.

WTO Committee on Agriculture (2002c) *Summary report G/AG/R/32, 6 November 2002*. World Trade Organization.

WTO Committee on Agriculture (2003) *Summary report G/AG/R/37, 18 December 2003*. World Trade Organization.
WTO Committee on Agriculture (2004) *Summary report G/AG/R/38, 28 May 2004*. World Trade Organization.
WTO Committee on Agriculture (2005) *Summary report G/AG/R/43, 2 September 2005*. World Trade Organization.
WTO Committee on Agriculture (2006a) *Summary report G/AG/R/47, 22 December 2006*. World Trade Organization.
WTO Committee on Agriculture (2006b) *Notification G/AG/N/MEX/13, 2 May 2006*. World Trade Organization.
WTO Committee on Agriculture (2007a) *Notification G/AG/N/MEX/13/Rev.1, 27 March 2007*. World Trade Organization.
WTO Committee on Agriculture (2007b) *Summary report G/AG/R/48, 6 July 2007*. World Trade Organization.
WTO Committee on Agriculture (2007c) *Notification G/AG/N/MEX/13/Rev.2, 25 October 2007*. World Trade Organization.
WTO Committee on Agriculture (2007d) *Summary report G/AG/R/49, 19 November 2007*. World Trade Organization.
WTO Committee on Agriculture (2009) *Summary report G/AG/R/55, 15 September 2009*. World Trade Organization.
WTO Committee on Agriculture (2010) *Summary report G/AG/R/58, 20 April 2010*. World Trade Organization.
WTO Committee on Agriculture (2011a) *Notification G/AG/N/MEX/13/Rev.3, 12 January 2011*. World Trade Organization.
WTO Committee on Agriculture (2011b) *Notification G/AG/N/IND/7, 9 June 2011*. World Trade Organization.
WTO Committee on Trade and Environment (2009) *Document WT/CTE/M/48, Report of the meeting held on 20 November 2009*. World Trade Organization.
WTO Committee on Trade and Environment (2010a) *Document WT/CTE/M/48, Informal Background Note on CTE Debate on Trade and Environment, 29 June 2010*. World Trade Organization.
WTO Committee on Trade and Environment (2010b) *Document WT/CTE/M/50, Report of the meeting held on 2 November 2010*. World Trade Organization.
WTO Committee on Trade and Environment (2011a) *Document WT/CTE/M/51, Report of the meeting held on 6 July 2011*. World Trade Organization.
WTO Committee on Trade and Environment (2011b) *Document WT/CTE/M/52, Report of the meeting held on 14 November 2011*. World Trade Organization.
WTO Committee on Subsidies and Countervailing Measures (n.d.) *Document series G/SCM/N/155, G/SCM/N/186, and G/SCM/N/220*. World Trade Organization.
WTO Committee on Technical Barriers to Trade (2010) *Australia – Joint governments' response to the 2010 Independent Review of the Water Efficiency Labelling and Standards Review Water Efficiency Labelling and Standards (WELS) scheme consultation paper, Notification G/TBT/N/AUS/71.*World Trade Organization.

Section 2

Integrated Water Resources Management: Lessons learnt in Spain

Chapter 6

Virtual water trade, food security and sustainability: Lessons from Latin America and Spain

Daniel Chico
CEIGRAM, Universidad Politécnica de Madrid, Water Observatory of the Fundación Botín, Madrid, Spain

Maite M. Aldaya
UNEP Consultant, Universidad Complutense, Water Observatory of the Fundación Botín, Madrid, Spain

Insa Flachsbarth & Alberto Garrido
CEIGRAM, Universidad Politécnica de Madrid, Water Observatory of the Fundación Botín, Madrid, Spain

ABSTRACT: The relevance of international food trade has been growing in the last decades, as a result of population and income growth, and agricultural and trade policies liberalization. The integration of the world food system has important implications for food and water security. From a food security perspective, trade may provide access to cheaper products, help alleviate food scarcity in importing nations and increase economic returns in exporting nations. Nevertheless, it has distribution effects and may increase the exposure to market forces. This raises the question of whether the welfare gains benefit the most vulnerable. From a water resources perspectives, trade has the potential to alleviate scarcity and increase the efficiency of resource use. It may play a part in ensuring water and water-dependent food security in water short countries. International trade of commodities involves flows of virtual water over large distances, where Virtual Water (VW) should be understood as the volume of water required to produce a commodity However, the use of water for exporting products may deprive some domestic purposes. In addition to this, the social and environmental costs associated with water use are often not taken into account, and remain in the exporting countries. The chapter presents the evolution of the VW trade of six Latin American countries in the period 1996–2008, and reflects on its impacts and potential development. It also presents a study of the VW trade of Spanish feedstuff imports and animal products exports. An econometric granger-causality analysis is performed to provide insights over the determinants of this trade. After these case studies, the chapter discusses the potential role that the World Trade Organization may take in order to regulate water resources. The WTO is one of the main determinants of international trade but its rules do not include food and water security issues. The chapter adds reasons for and against the inclusion of water concerns in the trade regime and the international agreements. Finally, it advocates for the development of institutional or economic solutions to handle market externalities and make the trade regime truly socially-enhancing.

1 INTRODUCTION

International trade of food and feed products has been growing steadily during the last decades to unprecedented levels in the recent world's history. Factors such as population and income growth in relevant developing and emerging countries, agricultural and trade policies liberalization, and intense specialization processes in exporting countries have made world's food systems extremely integrated.

While food trade helps alleviate food and water scarcity in importing countries, it also increases the dependence on market forces and exposure to food price crises. Reliance on international markets will become almost the only option for about half of the world population by 2050 (Falkenmark & Lannersted, 2010). Presently, a renewed craving to obtain access to land and water resources by emerging and wealthy food importing countries has given rise to the process called 'land grab' (Allan, 2012). While many types of direct investment have been included under this denomination, it certainly attests to the world's needs of natural resources to produce more food. There are many governance issues surrounding this phenomenon. In this chapter, we focus only on food trade and its implications on land and water resources.

As explained by classical economic theory (Krugman & Obstfeld, 1995) optimal trade results in a winning situation for all the parts involved. It allows countries to take advantage of their comparative advantages, using resources more efficiently and specialising their productions. Trade offers increased access to products at a cheaper price in the long run, increasing the overall welfare. However, it is not a neutral mechanism, affecting in different ways each actor. The question in relation to food security is whether the welfare gains benefit the poor and enhances food security for the have-nots. In exporting countries food prices might be higher as there is an additional demand from world markets. Importing nations however might suffer from price spikes like the one in 2007/08 (Anderson & Nelgen, 2012). The effects of trade need to be levered in order to counter possible distributive effects, to ensure that improvements in food systems are enhanced instead of threatened. From a water resources perspective, the externalities not included in the production costs may lead to excessive pressure of market forces over the resources, both in quantity and quality.

To this end, this chapter reviews some of the main issues about the relation between trade and natural resources. New analyses focusing on Spain and Latin America will provide insights of various country approaches, some of which recently became massive exporters and some both importers and exporters at the same time. The chapter then reflects on the potential role of the World Trade Organization (WTO) to regulate food trade on the basis of its impact on water resources, Virtual Water (VW) content or domestic pricing.

Section 2 will set the context of the observed increase of trade flows, explaining some of the world trends and factors leading to the augmentation of agricultural production and trade in the last decades. Section 3 will explain the relevance of this growth in trade for global water resources. Section 4 will show some facts about trade in six Latin American countries, and discuss some of the implications for their respective national water resources. Section 5 will focus on the Spanish case of trade in meat and staple products, looking into the explanatory factors of the two way trade (import of staple products and meat exports) and its effects on water resources in the country. Section 6 will discuss some policy options for interdependencies between trade and water. Section 7, will complement section 6 by evaluating the arguments in favor and against the inclusion of water resources considerations in international trade regulations. Finally section 8 will summarize the previous sections and offer some conclusions.

2 WORLD TRENDS OF FOOD TRADE

In the last decades international trade in agricultural products has been increasing. Data from the Food and Agriculture Organization of the United Nations (FAO) and WTO show an increase both in volume traded and value, higher than the increase in production (Figure 1).

Changes in diets and population growth are two key drivers of this increase (Prakash & Gilbert, 2011). High economic development of the middle-income countries has allowed the middle class in these countries to afford largest amounts of food and more expensive diets, leading together with other factors to significant dietary changes (Beddington *et al.,* 2010). Between 1990 and 2005, the number of households with less than $4 per person and day of income went down by 200 million whereas those with more than $4 of daily income increased by 1200 mill. (Chen & Ravallion, 2010). The increase in population is happening mainly in urban areas, also a result of rural-urban migration. It is expected that by 2050, 70% of the world population will live in cities (UN, 2008) (Figure 2). Urbanization is associated with diets that concentrate more calories and proteins on animal products, leading to increases of feed demand.

The mentioned changes in world food systems point to the need of increasing future food production. This increase in food production and thereby trade is projected to be affected by climate change, which may not only shift production areas but

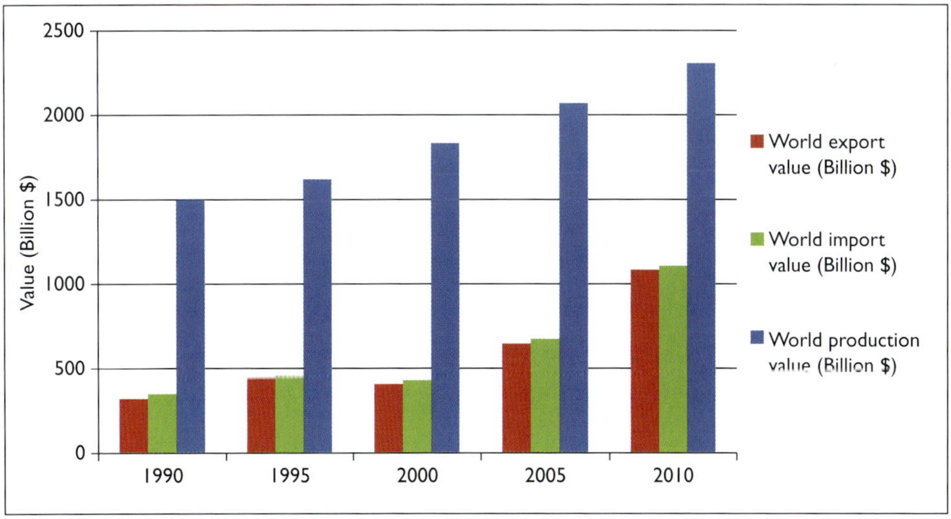

Figure 1 World agricultural production, exports and imports value 1990–2010 in billion $ (FAOSTAT, 2012).

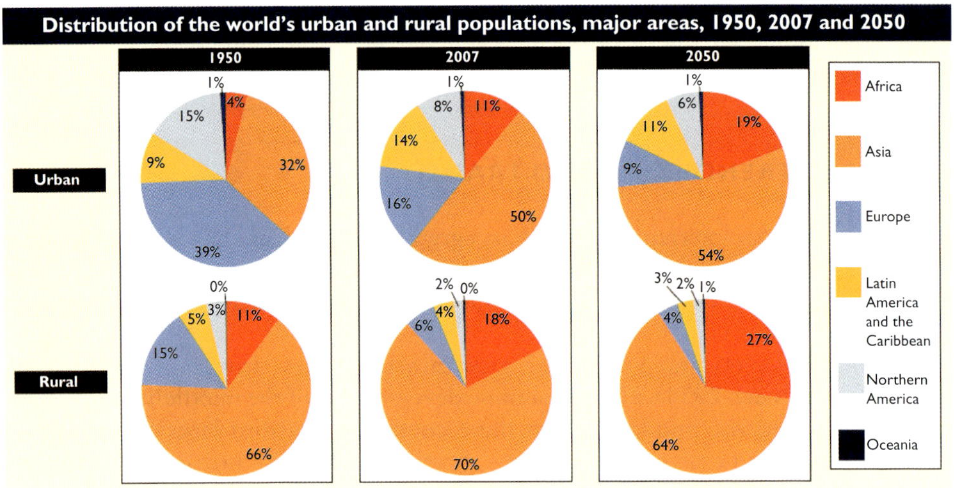

Figure 2 Distribution of the world's urban and rural populations for the relevant world parts in the years 1950, 2007 and 2050 (UN, 2008).

increase the variability of productions as a consequence of a higher number of extreme events (Solomon *et al.*, 2007). Supply shocks are transmitted to food markets, igniting all sorts of reactions from exporters, hoarding and drawing down stocks, a recipe known to cause extreme food price surges (Prakash & Gilbert, 2011). Increased production and exports are concentrated in a small number of countries, with an array of relevant but smaller producing countries complementing the larger ones. Big exporters are increasing their share in the international trade, while total calories exchanged in food markets tripled between 1970 and 2010 (Prakash & Gilbert, 2011).

Other factors, such as crude oil prices, biofuel policies, farm productivity developments and agricultural R&D policies, and market access standard developments (in particular related to genetically modified organisms) will help shaping the food trade in the short and medium term (Anderson, 2010).

3 WHY TRADE MATTERS FOR WATER RESOURCES

According to the World Water Development Report, two thirds of the global population will live in areas of water stress by 2025, even if nothing is done to change present levels of water consumption (UNESCO-WWAP, 2006). In this context, the virtual water trade can efficiently redistribute global water and partially help address the impacts of consumption and production.

International trade of commodities involves flows of virtual water over large distances, where virtual water (VW) should be understood as the volume of water required to produce a commodity (Allan, 2011; Hoekstra, 2010). The global volume of virtual water flows related to international trade in commodities is 1762 Gm^3/yr in the period 1996–2005 (Mekonnen & Hoekstra, 2011) (Table 1).

Table 1 Global Water Footprint (WF) of production for 1996–2005 (Mekonnen & Hoekstra, 2011).

	Agricultural production					
	Crop production	*Pasture*	*Water supply in animal raising*	*Industrial production*	*Domestic water supply*	*Total*
Global WF of production Gm³/yr)						
– Green	5771	913	–	–	–	6684
– Blue	899	–	46	38	42	1025
– Grey	733	–	–	363	282	1378
Total	7404	913	46	400	324	9087
WF for export (Gm³/yr)		1597		165	0	1762
WF for export compared to total (%)		19		41	0	19

An estimated 19 percent of global water use is not for producing domestically consumed products, but rather products for export. As agriculture uses around 92 percent of all water extracted for consumptive purposes, and large amounts of water are embodied in many of the agricultural products traded, the different policy options in this sector deserve careful consideration (UNEP, 2011). About 90 percent of these virtual water flows relate to trade in agricultural products, while the remainder is related to industrial product trade. With the increasing globalization of trade, global water interdependencies and overseas externalities are likely to increase. Dalin *et al.* (2012) report that not only VW trade more than doubled in the period 1986–2008, but also the connections between countries. At the same time, trade liberalization creates opportunities to increase global water use efficiency (Hoekstra, 2010; Dalin *et al.*, 2012).

Nowadays it is widely recognized that international trade in agricultural products can contribute to addressing problems related to the unequal geographical distribution of water (Allan, 2011; Hoekstra, 2010; Hoekstra *et al.*, 2011; Jackson *et al.*, in press; UNEP, 2011). In fact, current global virtual water trade has a moderating impact on the demand for irrigation water, as three of the four major exporters (United States, Brazil and Argentina) produce in highly productive rainfed conditions (Aldaya *et al.*, 2010; de Fraiture *et al.*, 2004; Mekonnen & Hoekstra, 2011) (Figure 3). While major rain-fed exporters produce under rainfed cultivation, most importers would have relied (at least partially) on their blue water resources. Konar *et al.* (2011) estimated the VW traded in agricultural products in 594 Gm³ green water and 78 Gm³ blue water, and argue that in the period 1986–2008 water use efficiency has increased since trade has increased in the water efficient links. The virtual water trade can efficiently redistribute global water and partially help to address the impacts of production. Virtual water trade, thus, can play a role in ensuring water and water dependent food security in water-short countries. Furthermore, the virtual water trade concept makes it clear

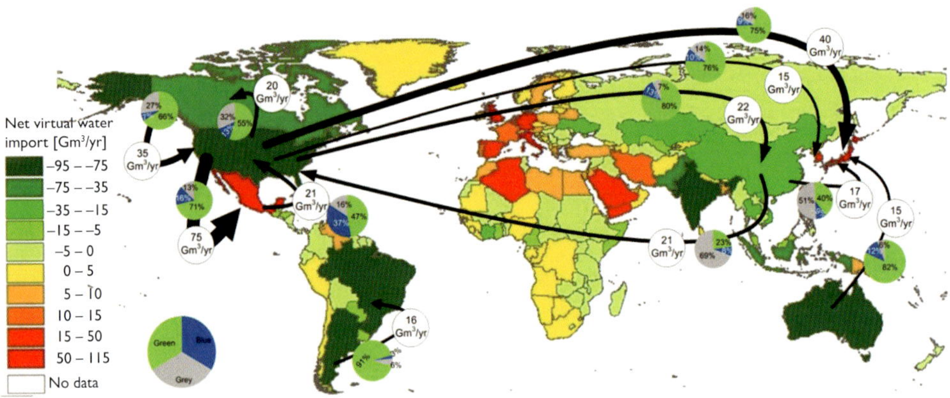

Figure 3 National virtual water balances. Arrows show the gross virtual water flows >15 Gm^3/yr (1 Gm^3 = 1 km^3) (Mekonnen & Hoekstra, 2011).

that, in a reasonably safe, interdependent and prosperous world, a country with limited water resources could depend on imports of agricultural products with high levels of embedded water (e.g. meat, feed) and apply its own water resources to produce other commodities of lower value in terms of water content. Conversely, a country with abundant water resources could benefit from its comparative natural advantage by exporting products that are high in embedded water (UNESCO-WWAP, 2006). Availability of land is also part of the equations (Wichelns, 2010), but agricultural land is also associated with green water availability.

On the other hand, if effective national and international policies and incentives are not properly developed and implemented, virtual water trade can have negative side effects. International trade in water-intensive commodities takes water in the exporting countries, which can no longer be used for other (domestic) purposes. Besides, the social and environmental costs that are often associated with water use remain in the exporting countries. Boelens & Vos (2012) warn against the social impact and effects on local communities of the advocacy for increased efficiency and trade. Agriculture is the main driver of non-point source water pollution, and low or lack thereof of environmental standards in some countries leads to significant impact of agriculture on the quality, and not only quantity, of water resources. These costs are not included in the price paid for the products by the consumers in the importing countries. This explains why some authors contend that the final consumers should be made responsible of the environmental impacts of the products, irrespectively of where they are manufactured (Hoekstra, 2010).

4 IMPACTS OF TRADE IN LATIN AMERICA

In this section we study the VW trade of six Latin American countries, namely Argentina, Brazil, Chile, Colombia, Mexico and Peru, all of which show high relevance in world agricultural commodity markets. Among them are large importing nations as well as exporting nations. Brazil and Argentina are categorised as very large exporting nations supplying the world with increasing amounts of staple food products and feed. Their importance in feeding the world is likely to further increase as large areas have comparative advantages in agricultural production due to rich natural resource endowments (Table 2).

Argentina and Brazil primarily produce for world markets under rainfed conditions, reflecting high uses of green water instead of blue water (see Figure 4). From an environmental point of view, this development poses opportunities and challenges. On the one hand green water is generally associated with lower opportunity costs than blue water (Albersen *et al.*, 2003). Green water cannot be automatically reallocated to uses other than natural vegetation or alternative rainfed crops, whereas blue water can be used for irrigating crops as well as for urban, agricultural and industrial water uses (Garrido *et al.*, 2010). Furthermore, excessive irrigation can cause severe

Table 2 Indicator for comparative advantages in agricultural production due to natural resource endowments in 2009 (own elaboration from Mekonnen & Hoekstra, 2011 and FAOSTAT, 2012).

Country	*Agricultural land (% of total land)*	*Agricultural WF (% of total renewable water)*
Chile	21.17	0.27
Mexico	52.90	3.26
Peru	16.75	0.22

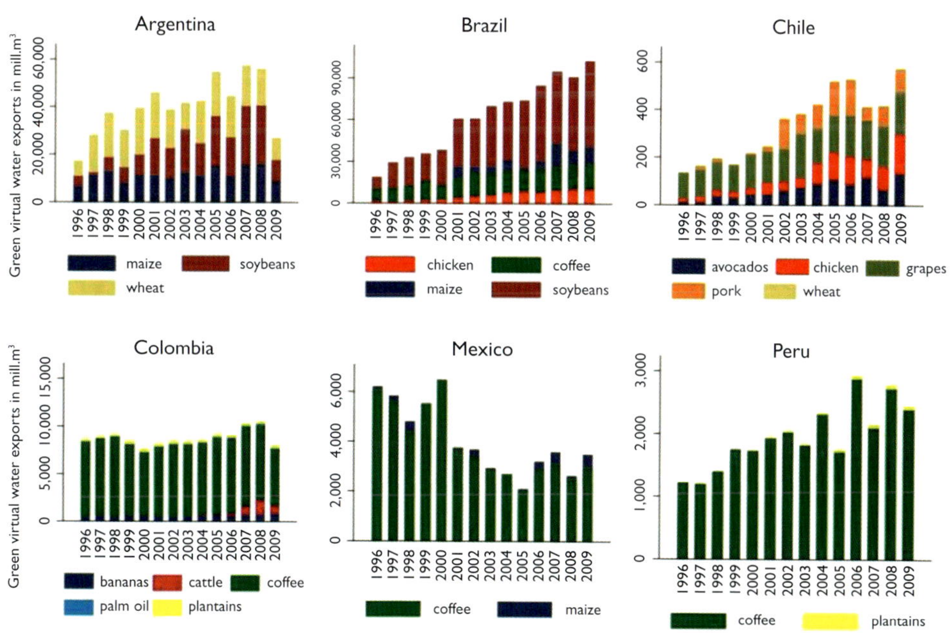

Figure 4 Green virtual water exports per country and products over 1996–2009 (own elaboration from Mekonnen & Hoekstra, 2011; FAOSTAT, 2012).

salinization, water logging and soil degradation (Tilman *et al.*, 2001). Following the notion of opportunity costs, it has been argued that the use of green water in crop production is considered more sustainable than blue water use, except when replacing high value ecosystems (Aldaya *et al.*, 2010; Niemeyer & Garrido, 2011; Yang *et al.*, 2006). On the other hand, expanding rainfed agriculture is often associated with massive land use changes. Especially in Brazil increasing virtual water exports contained in soybeans have led to a threefold land footprint (Figure 5).

Mexico is a large agricultural net importer. In the main agricultural production regions, Mexico must cope with green water constraints and thus highly depends on irrigated agriculture (Hoekstra & Mekonnen, 2011). With growing incomes and population growth, food demand has increased and diets have changed in Mexico. Globalization and liberalized agricultural markets have allowed Mexico to import large amounts of staple food to satisfy this demand, as opposed to the alternative of domestic food production which would have used valuable blue water resources. The substitution of domestic staple food production by imports has led to a shift in agricultural production towards higher value fruits and vegetables as well as livestock production (Figure 4 and 6). Thus globalization has offered Mexico the opportunity to generate more income from agricultural production per drop of water. However, fruits and vegetables are mostly produced under irrigated conditions leading to higher blue water use in relatively water-scarce regions. Furthermore agricultural production has not been stable but has increased substantially due to global market forces. This has resulted in accelerating blue water depletion rates. For example the Rio Grande river basin has already reached or surpassed sustainable extraction rates during some months (Hoekstra & Mekonnen, 2011). The expansion of trade and substitution of production has coincided with alarming prevalence rates of obesity in Mexico (Clark *et al.*, 2012) but causality is difficult to be established.

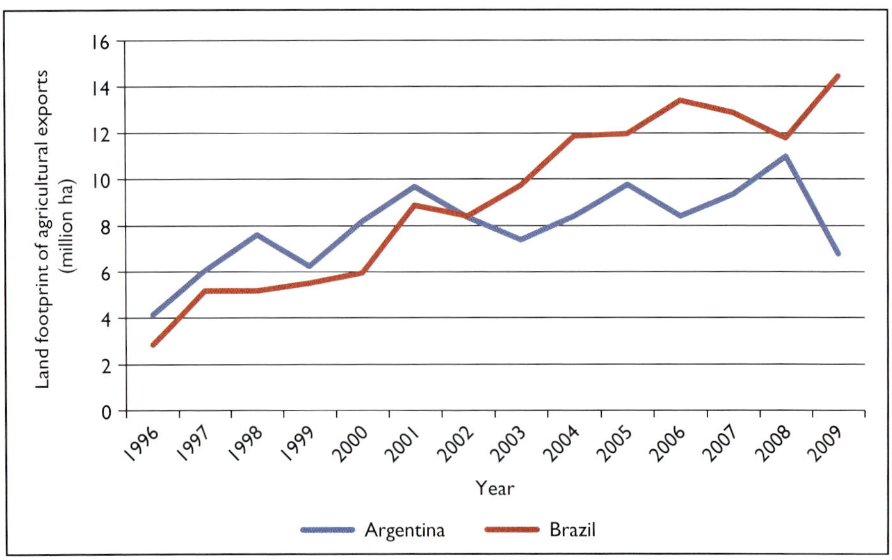

Figure 5 Land footprint of agricultural exports from Argentina and Brazil over 1996–2009 (own elaboration from FAOSTAT, 2012).

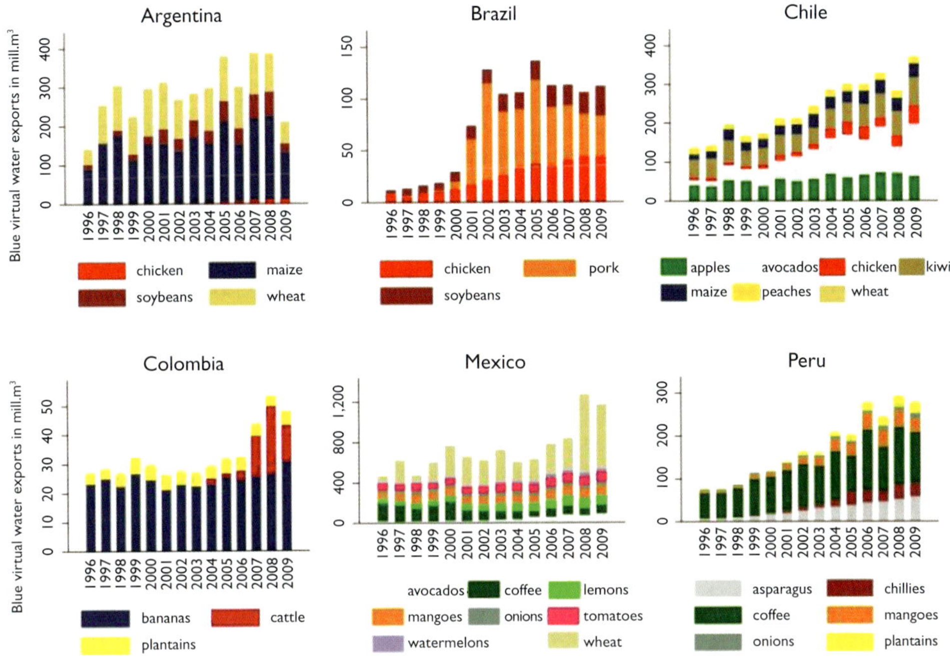

Figure 6 Blue virtual water exports per country and main products (1996–2009) (own elaboration from Mekonnen & Hoekstra, 2011 and FAOSTAT, 2012).

Table 3 Water apparent productivity of blue virtual water exports in Chile over 1996–2009 (own elaboration from Mekonnen & Hoekstra, 2011 and FAOSTAT, 2012).

High water productivity export products	*Lower water productivity import products*
Pork: 3.25 USD/m³	Wheat: 1.47 USD/m³
Apples: 2.21 USD/m³	Maize: 0.43 USD/m³
Chicken: 2.93 USD/m³	

As Figure 4 and 6 illustrate, Chile, Colombia and Peru show similar developments as Mexico, shifting from staple food production to the production of higher value products. However in contrast with Mexico, especially Chile and Peru are becoming net exporters while Mexico is still a net importer. Colombia has always been a large coffee exporter which is mainly produced under rainfed conditions. Table 2 illustrates that Chile and Peru in contrast to Mexico still have sufficient land and water resources on a national level. Using the example of Chile, table 3 illustrates how virtual water trade can favour a development towards more cash per drop. Whether or not trade also leads to more care of nature per drop is still an open question and needs to be evaluated on a case by case basis.

5 EXPLANATORY FACTORS OF TRADE IN SPAIN

The increase in trade in the agricultural sector and the resulting interconnection has also occurred in Spain in just one decade (Garrido *et al.*, 2010). One of the main consequences has been the increase of meat production, linked to an important growth of imports of staple food products, mainly soya and corn. Spain has become one of the main EU meat producers and exporters. Due to land, climate and water constraints, however, the country does not produce enough commodities to meet its feed demand, and relies on international trade to supply it. On average, 25% of the national pork production and 21% of the bovine products are exported to EU countries. Exports to other world regions represent less than 5% of these exports.

In this section we present this two-way virtual water trade and present an analysis of its causality. The interest that this analysis offers is that through the assessment of causality, the explanatory factors of trade may be better understood. Thus, it may help inform policy action. We ask whether the expansion of trade, feed imports and animal product exports happening apparently contemporaneously, is caused by meat imports demand or by feed exports supply. For this purpose, we grouped the selected products in 12 product categories, 5 corresponding to feedstuff products and 7 to animal products.

Figure 7 shows the VW flows, import, production and export, per product category for the periods 1995–1998 and 2006–2008. Note that the scales in both graphics are different. In both periods green water is overwhelmingly predominant, both in production and export. The main VW flows are those associated to the production of pork meat and milk products, together with the production of cereals. In the case of the imports, oil and protein soya meals are the main VW import category, followed by cereals.

Total VW exports are on average 15% of the VW involved in the production. This means that most of the production is consumed inside the country and is not exported. They are mainly linked to the exports of swine and bovine products.

If we compare both graphics, we see how the VW trade has increased. Cereal imports of 4268 Mm^3/yr at the beginning of the period rose to more than 11,000 (Novo *et al.*, 2009). Imports in the form of oil and protein meals rose from 7600 to 12,386 Mm^3. As for the water embedded in national production, it is relevant the change operated towards a higher share of blue water. From 27% it rose to 44% in the analyzed production. VW exports related to these categories are much lower, amounting to a total of 7950 Mm^3/year (average 2006–2008) after growing a 77% in the period. Bovine meat and oil meals lost their preponderance over the period to swine products. This category grew by 330% in the period, compared to just 13% of oilmeals and 43% of bovine products, and the stabilization of cereal VW exports. These four categories make up 94% of the VW exports.

In order to study the causality of this trade, we performed an econometric analysis using a Granger causality test. This test consists in a series of regressions of two variables, an endogenous or 'caused' variable and an explanatory one. First it analyses the endogenous variable against itself, lagged one year. Then it repeats the regressions adding the second variable also one year lagged. If the results from the second analysis increase the explanatory power of the regression, it means that the second

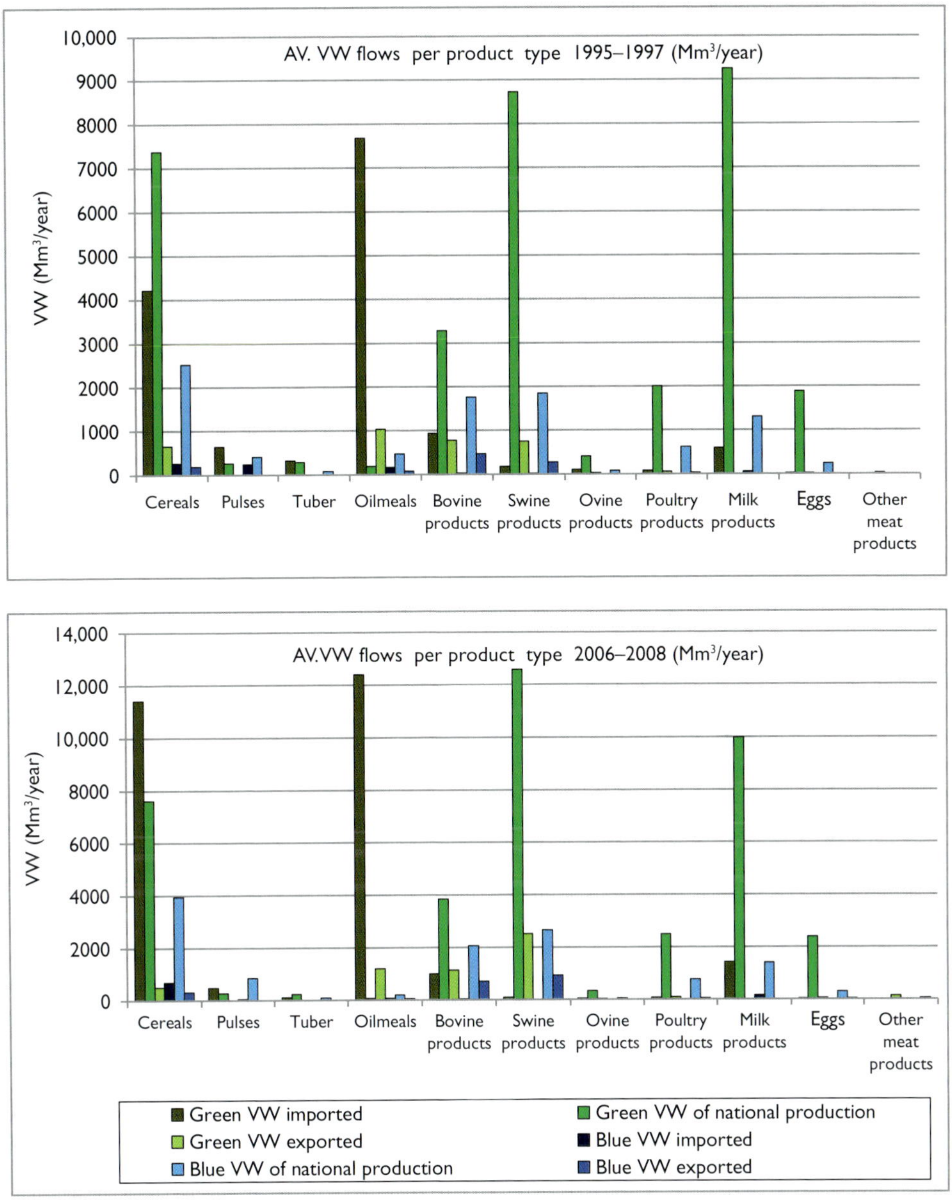

Figure 7 Virtual Water (VW) volumes per product category (Mm³/year) average 1995–1997 and 2006–2008.

variable contains information about the studied variable that its own past did not include. Then, the analysis is carried out swapping the variables, the endogenous variable is used now as the explanatory one and vice versa. We used this analysis between a series of variables related to feedstuffs and to animal production, relate one by one and aggregated.

Figure 8 shows as an example the successive regressions to perform in the case of the Net VW flow (VW of production – VW exported + VW imported) in animal products and the Cereals VW imports.

Table 4 and Table 5 show the results of the analyses performed and the coefficients detailing the significance of the regressions. Test results lower than 0.05 signal that, at statistically significant level, the variable in the upper rows is explained by the variables in the first column, and not the other way round.

In the results obtained, no statistically significant causality relation was found between the net VW flows of animal products and feedstuffs., Only in the relation between egg production and tuber and cereal imports did we achieved significant results. In the same way, VW imports of feedstuff did not significantly conditioned animal products net VW flows. However, the results obtained indicated that at a statistically significant level, VW imports of oilmeals conditioned the VW exports of pork meat, and in general animal products.

$$Animal\ products\ Net\ VW\ flow_t = \gamma_0 + \beta_1 Animal\ products\ Net\ VW\ flow_{t_{t-1}} + \varepsilon_t$$

$$Cereals\ VW\ imports_t = \gamma_0 + \beta_1 Cereals\ VW\ imports_{t-1} + \varepsilon_t$$

$$Animal\ products\ Net\ VW\ flow_t = \gamma_0 + \beta_1 Animal\ products\ Net\ VW\ flow_{t-1} + \beta_2 Cereals\ VW\ imports_{t-1} + \varepsilon_t$$

$$Cereals\ VW\ imports_t = \gamma_0 + \beta_1 Cereals\ VW\ imports_{t-1} + \beta_2 Animal\ products\ Net\ VW\ flow_{t-1} + \varepsilon_t$$

Figure 8 Example regressions of the Granger test between the variables Net VW of animal products and Cereal VW imports.

Table 4 Wald test coefficients of the Granger tests for the causality of the animal products variables explained by the feedstuffs variables. (14 obs.) (own elaboration).

	Variable to be explained (y)→	*Animal products*	*Meat*	*Swine*	*Bovine*	*Poultry*	*Eggs*	*Milk products*
		Net VW Flow						
Net VW flow	Feedstuff	0.981	0.536	0.320	0.398	0.141	0.470	0.536
	Cereals	0.991	0.580	0.378	0.429	0.189	0.876	0.389
	Tuber	0.463	0.645	0.915	0.514	0.381	0.792	0.141
	Oilmeals	0.183	0.587	0.775	0.376	0.297	0.075	0.975
	Forages	0.794	0.709	0.622	0.815	0.235	0.327	0.492
VW imports	Feedstuff	0.687	0.777	0.441	0.602	0.531	0.339	0.056
	Cereals	0.710	0.766	0.438	0.580	0.644	0.679	**0.029***
	Tuber	0.592	0.937	0.605	0.451	0.727	0.603	**0.048***
	Oilmeals	0.170	0.627	0.831	0.577	0.381	0.107	0.764
		VW Exports						
Net VW flow	Feedstuff	**0.025***	**0.025***	0.164	0.097	0.289	0.053	0.873
	Cereals	0.052	0.052	0.287	0.081	0.463	0.056	0.908
	Tuber	0.807	0.793	0.977	0.291	0.277	0.430	0.876
	Oilmeals	**0.031***	**0.032***	**0.033***	0.313	0.138	0.072	0.538
	Forages	0.234	0.232	0.341	0.055	**0.022***	0.079	0.875
VW imports	Feedstuff	0.361	0.366	0.642	0.514	0.474	0.228	0.743
	Cereals	0.478	0.479	0.854	0.533	0.713	0.283	0.859
	Tuber	0.785	0.749	0.949	0.878	0.459	0.857	0.677
	Oilmeals	**0.026***	**0.027***	**0.032***	0.292	0.125	0.071	0.709

Table 5 Wald test coefficients of the Granger tests for the causality of the feedstuff variables explained by the animal products variables (14 obs.).

	Variable to be explained (y)→	*Feedstuff*	*Cereals*	*Tuber*	*Oilmeals*	*Forages*
		Net VW Flow				
Net VW flow	Animal products	0.204	0.184	0.287	0.175	0.505
	Meat	0.504	0.403	0.574	0.570	0.450
	Swine products	0.376	0.309	0.705	0.600	0.149
	Bovine products	0.532	0.641	0.182	0.417	0.976
	Poultry	0.682	0.447	0.782	0.428	0.804
	Eggs	**0.049***	**0.03***	**0.02***	0.201	0.376
	Milk products	0.087	0.096	0.104	0.158	0.562
VW Exports	Animal products	0.509	0.514	0.073	0.126	0.864
	Meat	0.490	0.501	0.073	0.117	0.822
	Swine products	0.168	0.179	**0.031***	0.100	0.689
	Bovine products	0.683	0.618	0.491	0.445	0.721
	Poultry	0.366	0.484	0.284	0.091	0.553
	Eggs	0.5799	0.561	0.496	0.987	0.132
	Milk products	**0.037***	**0.047***	0.241	0.154	0.291

	Variable to be explained (y)→	*Feedstuff*	*Cereals*	*Tuber*	*Oilmeals*
		VW imports			
Net VW flow	Animal products	**0.033***	**0.033***	0.238	0.173
	Meat	0.196	0.171	0.566	0.571
	Swine products	0.259	0.244	0.810	0.449
	Bovine products	0.309	0.274	0.254	0.569
	Poultry	0.449	0.232	0.385	0.399
	Eggs	**0.024***	**0.006***	**0.006***	0.146
	Milk products	**0.002****	**0.002****	**0.045***	0.127
VW Exports	Animal products	0.056	0.065	0.058	0.232
	Meat	0.055	0.066	0.065	0.213
	Swine products	**0.013***	**0.012***	**0.018***	0.155
	Bovine products	0.539	0.613	0.524	0.700
	Poultry	0.060	0.073	0.139	0.118
	Eggs	0.693	0.613	0.126	0.704
	Milk products	**0.004****	**0.003****	0.079	0.083

Note: * Significant at a 95% confidence level, ** Significant at a 99% confidence level.

These results indicate that it is the import of oilmeals and cereals which 'granger – cause' the VW exports of meat products, mainly pork products, and egg production. This means that it is the availability of these imports, a supply effect originating in exporting countries, the one that has allowed the development of the meat exports, instead of the process being demand-driven. Based on our results, the possibility of Spanish livestock farms of procuring massively from foreign supplies their feed needs is at the root of the expansion of production and exports of meat products and fattened live animals. For the country's water resources this has allowed for freeing blue water resources for other crops and uses, including environmental uses. On the other hand, the heavy concentration of intensive livestock production has led to serious water pollution problems in specific areas.

6 THE FRAMEWORK FOR TRADE REGULATION: PROS AND CONS OF REGULATING TRADE

As mentioned in section 3, agricultural products represent about 90% of all the water virtually traded in the world. WTO rules for agricultural commodities affect trade flows, prices and production, thus influencing the use of land and water resources around the world. WTO rules have direct and indirect impacts, because of substitution effects and secondary impacts, on scarce land and water allocation. This opens the question of including water issues as part of the international trade agreements. Presently, WTO disciplines focus on the production side, by means of irrigation and other type of subsidies, and on the demand side on technical barriers of trade based on footprint labeling and consumers information.

Any WTO acceptable subsidy must fall in the so-called green box. This kind of subsidies must not distort trade, or at most cause minimal distortion. They have to be government-funded (not by charging consumers higher prices) and must not involve price support. They tend to be programs that are not targeted at particular products, and include direct income supports for farmers that are not related to (are 'decoupled' from) current production levels or prices. They also include environmental protection and regional development programs, and infrastructural services (water supply facilities). Annex 2, paragraph 2 (g) of the Agreement on Agriculture includes under the denomination of infrastructural services '[expenditures that] shall be directed to the provision or construction of capital works only, and shall exclude the subsidized provision of on-farm facilities other than for the reticulation of generally available public utilities. It shall not include subsidies to inputs or operating costs, or preferential user charges' (WTO, 2012). Furthermore, as per paragraph 12, payments under environmental programs would be acceptable provided that (a) Eligibility for such payments shall be determined as part of a clearly-defined government environmental or conservation program and be dependent on the fulfillment of specific conditions under the government program, including conditions related to production methods or inputs, or (b) The amount of payment shall be limited to the extra costs or loss of income involved in complying with the government program.

Adverse subsidies are especially scrutinized when they cause injuries to industries in another Member, but WTO rules do not address the environmental impacts of subsidies. Finally, in the Agreement of Technical Barriers to Trade (TBT) there are some definitions about specific product 'characteristics' whose compliance would be mandatory. However, standards have to be approved by a Recognized Body, address products' characteristics and production methods and compliance is voluntary. Article 12 'Special and Differential Treatment of Developing Country Members' recommends that Members shall provide differential and more favorable treatment to developing country Members to this Agreement, through the following provisions as well as through the relevant provisions of other articles of this Agreement.

The key principles enshrined in the TBT are non-discrimination; avoidance of unnecessary trade barriers and harmonization. However, discrimination is permitted if a Member claims that some imports may have detrimental impact on competitive opportunities that stem from legitimate objectives (Art. 2.2) including, among others, protection of human health and safety, and of the environment. Harmonization is a stated preference for international standards, one of which could be an 'International Water Footprint Standard'. However, this option seems to be premature.

There are a number of reasons providing support for considering water in trade regulations:

- If trade is becoming so important (Carr *et al.*, 2012), the trade regime has to handle market externalities. Based on standard trade theory, markets are only efficient if, among other conditions, all parties are responsible of their private and social costs. In the absence of regulation, prices would be distorted if some parties do not take into account their external costs and some other do.
- In the absence of environmental regulations, a 'race-to-the-bottom' might ensue, with parties competing for the lowest possible standards to reduce export costs and gain competitive advantage. It is assumed that the weakest countries may suffer the worst consequences. A contrary hypothesis is the Porter Hypothesis, which is discussed below.
- Consumers of importing countries should be made responsible of the environmental consequences of the products they consume, and nudged to consume less of those most harmful by higher prices or simply by not having available.
- Finally, trade and globalization would exacerbate world inequities in water access, and provide further incentives for pursuing land-grabs in developing countries.

There are a number of reasons providing support for NOT considering water in trade regulations:

- The Porter Hypothesis[1] may be true: national environmental protection increases the competitive edge of exporters, improving their terms of exchange.
- Regulatory mechanisms are not feasible, making room for more serious unintended consequences.
- It creates and reinforces dependencies between countries, reducing the likelihood of war (Pinker, 2011).
- The pressure by NGOs, consumer groups and countries on corporations, and the certifying agencies that are created to assess impacts, will be more powerful to improve sustainable production systems.
- It would be extremely difficult for the WTO to set and enforce rules that provide for obligatory labeling, monitor adverse water subsidies or establish TBTs based on environmental or resource impacts.

1 Porter Hypothesis: States that strict Environmental regulation can induce efficiency and encourage innovations that help improve commercial competitiveness (Porter & van der Linde, 1995).

7 POLICY IMPLICATIONS

The need for water governance at the global scale results from growing concerns over, first, water security in many parts of the world, and secondly, whether the existing commodity market system can deliver security as well as the necessary stewardship of water resources (Allan, 2012; Sojamo *et al.*, 2012). Even if international trade presently involves a significant part of products for which production is water-intensive, and virtual water flows are mainly subordinated to world trade rules, the policy linkages between international trade and impacts on freshwater have rarely been analyzed. It is therefore worthwhile further exploring how to improve global water governance through trade.

It is well-known that water is seldom the dominant factor determining trade in water-intensive commodities. Many factors influence virtual water trade, such as direct or indirect subsidies, availability of land, labor, technology, level of socio-economic development, national food policies and international trade agreements (Aldaya *et al.*, 2010; Rogers & Ramirez-Vallejo, 2011). Currently, virtual water flows are mainly subordinated to world trade rules (Hoekstra *et al.*, 2011). The European Single Market and WTO frameworks are potentially suited to address the link between international trade and sustainable water use. Hoekstra (2010) identifies several mechanisms to better ensure that trade and sustainable water use go hand in hand such as product transparency or an international water pricing protocol. Trade will never contribute to optimal production and trade outcomes from a water-perspective as long as water remains so much underpriced (Hoekstra, 2010). There is a need to arrive at a global agreement on water pricing structures that cover the full cost of water use, including investment costs, operational and maintenance costs, a water scarcity rent and the cost of negative externalities of water use. Without an international treaty on proper water pricing it is unlikely that a globally efficient pattern of water use will ever be achieved. However, finding an harmonized water pricing mechanism may be so elusive that second best solutions may be more feasible. Improving trade and water governance also requires an effort from the water community, applying integrated water resource management, considering both surface and groundwater governance, which are very different and still challenging, particularly for the latter.

More recently, the WTO has started looking at the WTO interventions that can influence water-related policies on either the production side (irrigation subsidies) or the consumption side (water footprint labeling) (Jackson *et al.*, in press). More work is needed to clarify key concepts and to enhance transparency in order to have a more comprehensive understanding of the ways in which these rules alter water resource outcomes.

Even if it is not yet widely recognized, the private sector has also a vital role to play in food-water securitization. Food supply chains operate beneath a complex pact between the state and the market. The agents in these food supply chains – mainly farmers – determine whether food-water is managed sustainably and securely (Allan, 2012; Sojamo *et al.*, 2012).

At the national level, virtual water trade may enable economies to efficiently redistribute water resources over time and space (UNEP, 2011). Disadvantages, however, may include increased water scarcity and pollution, if water demand and pollution control are not carefully managed. Countries with poor or weakly enforced environmental regulations which allow pollution to flow across international boundaries are of great concern (UNESCO-WWAP, 2006). Since virtual water trade is mainly a consequence of agricultural (crop and livestock) policies, decisions in this sector deserve careful consideration. The implementation of national food policies (through subsidies, taxes, tariffs, food aid and others) can distort markets and marginalize the rural poor; and inadequately organized and non-pro-poor international trade liberalization can exacerbate this. Because of the difficulties poor families face in accumulating any surpluses – food or financial – they find it difficult to maintain consumption when their incomes are interrupted or their crops fail (UNESCO-WWAP, 2006).

Water and food security is today much more related to economic capacity and trade, than to physical water scarcity. Knowledge about the virtual water flows entering and leaving a country can cast a completely new light on the actual water scarcity of a country. This shift in perception means that it forces a re-consideration of what are the main problems of food security, away from pure physical scarcity and technological fixes. The main issues that have to be addressed globally in relation to food security are: first, the hidden monopolies that currently exist in the WTO, food sovereignty, the potential threat of political embargoes and the need for domestic social changes to be fulfilled.

8 CONCLUSIONS

International trade of food products is likely to increase in the next decades as a consequence of global population and economic trends. From a water resources perspective, trade poses a threat to water resources while offering a solution for water and food security. On the one hand it allows countries to feed themselves and develop economical activities without pressing too hard on their natural resources, as in the case of Mexico or Spain, or Brazil and Argentina. On the other hand it involves a reorganization of the benefits of water use and the impacts of water consumption. Spain has, in a way, relocated its water consumption for feedstuff production in order to develop its swine and poultry sector, based on imports, and expanded crops, like olives and grapevines, for which it has a comparative advantage. Still, this has come at the cost of serious water quality problems in specific areas, which could be expected to not having appeared without the pressure of international demand. Colombia, Mexico, Chile and Peru are increasing the economic productivity of water consumption by substituting productions and relying on trade for supply of water-intensive, low-value commodities. Importing nations may free water resources for other uses, including environmental ones, but increase their external dependency for food security, and may affect the economic capacity of rural areas.

Exporting nations may benefit from trade, if adequate incentives and mechanisms are set in place to avoid environmental pressure. The increase in the land footprint of Brazilian export agriculture is an example of this.

At a global level, trade can cause an increase in global water use efficiency. It may also increase the capacity of nations to fight water scarcities and food shortages through imports. While trade is concentrating production in countries that benefit from comparative advantages, smaller new actors are appearing (Prakash & Gilbert, 2011) that may introduce flexibility in the global food trade system.

Therefore, to make it truly socially-enhancing the trade regime has to handle market externalities. A relevant question is whether this has to be done through their inclusion in international trade agreements or leave it to national policies. Even though there are reasons for and against regulating trade to enhance water resources management, the relevant impact of international trade for water and food security calls for the inclusion of water and food security concerns in trade organization, and the development of institutional or economic solutions to this relation.

REFERENCES

Arrami, M. & Garner, H. (2008) A tale of two citations. *Nature*. [Online] 451 (7177), 397–399. Available from: http://www.nature.com/nature/journal/v451/n7177/full/451397a.html [Accessed 20th January 2008].

Albersen, P.J., Houba, H.E.D. & Keyzer, M.A. (2003) Pricing a raindrop in a process-based model: General methodology and a case study of the upper-zambezi. *Physics and Chemistry of the Earth*, 28, 183–192.

Aldaya, M.M., Allan, J.A. & Hoekstra, A.Y. (2010) Strategic importance of green water in international crop trade. *Ecological Economics*, 69(4), 887–894.

Allan, J.A. (2011) *Virtual Water: Tackling the Threat to Our Planet's Most Precious Resource*. I.B. Tauris.

Allan, J.A. (2012) Food-water security: Beyond hydrology and the water sector. In Lankford B., Bakker K.,. Zeitoun M & Conway D. (eds.). *Water Security*. London, Earthscan.

Anderson, K. (2010) Globalization's effects on world agricultural trade, 1960–2050 *Phil. Trans. R. Soc.*, B 365, 3007–3021.

Anderson, K. & Nelgen, S. (2012) Agricultural trade distortions during the global financial crisis, *Oxford Review of Economic Policy*, 28(2), 235–260.

Beddington C.J., Crute, I.R., Haddad, L., Lawrence, D., Muir, J.M., Pretty, J., Robinson, S., Thomas, S.M. & Toulmin, C. (2010) Food Security: The Challenge of Feeding 9 Billion People. *Science*, 327(5967), 812–818.

Boelens, R. & Vos, J. (2012) The danger of naturalizing water policy concepts: Water productivity and efficiency discourses from field irrigation to virtual water trade Agricultural Water Management, 15, 16–26.

Carr, J.A., D'Odorico, P., Laio, F. & Ridolfi, L. (2012) On the temporal variability of the virtual water network. *Geophysical Research Letters*, 39.

Chen, S. & Ravallion, M. (2010) The Developing World is poorer than we thought, but no less successful in the fight against poverty. *The Quarterly Journal of Economics*, 125(4), 1577–1625.

Clark, S.E., Hawke, C., Murphy, S.M.E., Hansen- Kuhn, K.A. & Wallinga, D. (2012) Exporting obesity: US farm and trade policy and the transformation of the Mexican consumer food environment. *International Journal of Occupational and Environmental Health*, 18(1), 53–65.

De Fraiture, C., Cai, X., Amarasinghe, U., Rosegrant, M. & Molden, D. (2004) *Does International Cereal Trade Save Water? The Impact of Virtual Water Trade on Global Water Use*. Report number: 4. Comprehensive Assessment Secretariat. Colombo, Sri Lanka.

Dalin, C., Konar, M., Hanasaki, N., Rinaldo, A. & Rodriguez-Iturbe, I. (2012) Evolution of the global virtual water trade network *Proceedings of the national academy of science of the United States of America*, 109(16), 5989–5994.

FAOSTAT (2012) *Production and trade statistics*. [Online] Available from: http://faostat.fao.org/site/291/default.aspx [Accessed December 2012].

Falkenmark, M. & Lannerstad, M. (2010) Food Security in water-short countries—Coping with carrying capacity overshoot. In Martinez Cortina, L., Garrido, A. & Lopez-Gunn, E. (eds.) *Re-thinking Water and Food Security: Fourth Marcelino Botín Foundation Water Workshop*. Liden, The Netherlands, Taylor & Francis. pp. 3–22.

Garrido, A., Novo, P., Rodríguez-Casado, R., Varela-Ortega, C. & Aldaya, M.M. (2010) Economic aspects of virtual water trade: Lessons from the Spanish case. In Martinez Cortina, L., Garrido, A. & Lopez-Gunn, E. (eds.) *Re-thinking Water and Food Security: Fourth Marcelino Botín Foundation Water Workshop*. Liden, The Netherlands, Taylor & Francis. pp. 145–159.

Hoekstra, A.Y. (2010) *The relation between international trade and freshwater scarcity. Working Paper ERSD-2010-05*, January 2010, World Trade Organization, Geneva, Switzerland.

Hoekstra, A.Y., Aldaya, M.M. & Avril, B. (eds.) (2011) *Proceedings of the ESF Strategic Workshop on accounting for water scarcity and pollution in the rules of international trade.* Value of Water Research Report Series, Report number: 54. UNESCO-IHE.

Hoekstra, A.Y. & Mekonnen, M.M. (2011) Global water scarcity: monthly blue water footprint compared to blue water availability for the world's major river basins. Value of Water Research Report Series. Report number: 53. UNESCO-IHE.

Jackson, L.A., Pene, C., Martinez-Hommel, M.B., Hofmann, C. & Tamiotti, L. (in press) Water Policy, Agricultural Trade and WTO Rules. Martinez-Santos, P., Aldaya, M.M. & Llamas, M.R. (in press) *Integrated Water Resource management.* Taylor & Francis, London.

Konar, M., Dalin, C., Suweis, S., Hanasaki, N., Rinaldo, A. & Rodriguez-Iturbe, I. (2011) Water for food: The global virtual water trade network. *Water Resources Research,* 47.

Krugman, P.R. & Obstfeld M. (1995) *International Economics: Theory and Policy.* 3rd Edition. New York, Harper Collins.

Mekonnen, M.M. & Hoekstra, A.Y. (2011) *National water footprint accounts: The green, blue and grey water footprint of production and consumption.* Value of Water Research Report Series. Report number: 50. UNESCO-IHE.

Niemeyer, I. & Garrido, A. (2011) International Farm Trade in Latin America: Does it Favour Sustainable Water Use Globally? In: A. Y. Hoekstra, M. M. Aldaya, & B. Avril (eds.) *Proceedings of the ESF Strategic Workshop on accounting for water scarcity and pollution in the rules of international trade.* Value of Water Research Report Series, Report number: 54. UNESCO-IHE.

Pinker, S. (2011) *The better angels of our nature. Why Violence Has Declined.* Allen Lane, Penguin Books.

Porter, M. & van der Linde, C. (1995) Toward a New Conception of the Environment-Competitiveness Relationship, *Journal of Economic Perspectives,* 9(4), 97–118.

Prakash, A. & Gilbert C.L. (2011) Rising vulnerability in the global food system: beyond market fundamentals. In Prakash, A. (ed.) (2011) *Safeguarding food security in volatile global markets.* FAO, Rome, Italy.

Rogers, P. & Ramirez-Vallejo, J. (2011) Failure of the virtual water argument. Hoekstra, A.Y., Aldaya, M.M. & Avril, B. (eds.) (2011) *Proceedings of the ESF Strategic Workshop on accounting for water scarcity and pollution in the rules of international trade* Value of Water Research Report Series, Report number 54. UNESCO-IHE.

Sojamo, S., Keulertz, M., Warner, J. & Allan, T. (2012) Virtual water hegemony: the role of agribusiness in global water governance. *Water International,* 37(2), 169–182.

Solomon, S., Qin, D., Manning, M., Chen, Z., Marquis, M., Averyt, K.B., Tignor, M. & Miller, H.L. (eds.) (2007) *IPCC Contribution of Working Group I to the Fourth Assessment Report of the Intergovernmental Panel on Climate Change.* Assessment Report number: 4. Cambridge, United Kingdom.

Tilman, D., Fargione, J., Wolff, B., D'Antonio, C., Dobson, A., Howarth, R. & Schindler, D. (2001) Forecasting agriculturally driven global environmental change. *Science,* 292(5515), 281–284.

UN (2008) *World Urbanization Prospects: The 2007 Revision. Department of Economic and Social Affairs*. [Online] Economic & Social affairs: Population Division, New York, US. Available from: http://www.un.org/esa/population/publications/wup2007/2007 WUP_Highlights_web.pdf [Accessed January 2013].

UNEP (2011) *Water footprint and corporate water accounting for resource efficiency*. [Online] United Nations Environment Programme, Paris. Available from: http://www.waterfootprint.org/Reports/ UNEP–2011.pdf [Accessed January 2013].

UNESCO-WWAP (2006) *Water- a shared responsibility*. [Online] The UN World Water Development Report. Report number: 2. http://unesdoc.unesco.org/images/0014/001444/144409E.pdf [Accessed January 2013].

Yang, H., Wang, L., Abbaspour, K.C. & Zehnder, A.J.B. (2006) Virtual water trade: An assessment of water use efficiency in the international food trade. *Hydrology and Earth System Science,* 10, 443–454.

Wichelns, D. (2010) Virtual Water: A Helpful Perspective, but not a Sufficient Policy Criterion. *Water Resources Management*, 24, 2203–2219.

Chapter 7

Ten years of the Water Framework Directive in Spain: An overview of the ecological and chemical status of surface water bodies

Bárbara A. Willaarts
Water Observatory, Botín Foundation, Madrid, Spain
Research Center for the Management of Agricultural and Environmental Risks (CEIGRAM), Technical University Madrid, Madrid, Spain

Mario Ballesteros
Research Center for the Management of Agricultural and Environmental Risks (CEIGRAM), Technical University Madrid, Madrid, Spain

Nuria Hernández-Mora
Research Center for the Management of Agricultural and Environmental Risks (CEIGRAM), Technical University Madrid, Madrid, Spain
Founding Member, Fundación Nueva Cultura del Agua (FNCA), Zaragoza, Spain

ABSTRACT: The Water Framework Directive (WFD) represents a significant departure from prior water management and planning practices in Spain. The traditional water policy, oriented towards supply augmentation to meet increasing water demands, is required to shift focus and prioritize the protection of the aquatic environment and ecological health. Despite the long hydrological planning history in Spain, little information was available on the ecological status of surface and ground waters prior to the implementation of the WFD. The requirements of the Directive have implied a significant effort for Spanish authorities to develop all biological, hydromorphological and chemical data required to determine the status of water bodies. This study provides a first national overview of the ecological, chemical and overall status of Spanish surface water bodies (rivers, lakes, transitional and coastal), identifies the most problematic areas and discusses the consistency and adequateness of the methods and indicators used across basins. To carry out the analysis, we use the status information contained in the draft or approved River Basin Management Plans (RBMPs) available at this point. Our results show that almost 50% of all surface waters in Spain are in poor status, mainly due to their poor ecological status. 43% of surface water bodies have not yet been evaluated so the information presented in the study is necessarily partial and incomplete. Available data shows that rivers and transitional waters are the water bodies in poorest status. The Southern basins draining into the Atlantic Ocean show the worst rank, with over 53% of surface waters in poor status. The lack of reference conditions to measure some key indicators such as fish fauna and most hydromorphological conditions, impedes making robust comparisons across basin districts at this point.

Keywords: water quality, WFD implementation, ecological indicators, heavily modified water bodies, ecological potential

1 INTRODUCTION

The Water Framework Directive (WFD 2000/60/EC) responded to the need for a common, coherent and integrated policy framework for European Union (EU) Member States (MS) in order to effectively tackle the problems of water quality deterioration, loss of aquatic ecosystem functionality and increasing water scarcity pervasive throughout Europe. The WFD incorporates into a legally binding instrument the key principles of integrated river basin management (EC, 2012), bringing together economic and ecological considerations, incorporating stakeholder perspectives into policy making, and accounting for the interrelationships between water management and other sectoral policies.

The WFD establishes a clear program and calendar for MS to develop River Basin Management Plans (RBMPs) by December 2009 in order to help meet its goals. It calls for the determination of the current health of Europe's aquatic ecosystems, the identification of significant pressures and impacts, and for the development of effective programs of measures that tackle these problems and allows us to reach the good status[1] of all European waters by 2015. The Directive recognizes that only by improving the health of Europe's aquatic ecosystems we will be able to guarantee access to sufficient water resources to meet our sustainable water needs in the mid and long term.

In many European countries, Spain among them, the WFD represents a significant departure from prior water management and planning practices. Water management at the river basin scale has been a trademark of Spanish water policy since the 1920s, and the development of river basin management plans has been a legal requirement since the approval of the 1985 Water Act. However, the transposition of the WFD into Spanish law in 2003 has shifted the emphasis of water policy from supply augmentation to meet increasing water demands toward the protection of the aquatic environment and ecological health. As a result, data and monitoring requirements and water planning processes have also changed. These requirements have implied a significant effort to develop the biological, hydrological and morphological data and reference conditions necessary to assess the status of water bodies thus helping us improve our understanding of the health of our surface and groundwater.

The WFD establishes iterative 6 year-planning cycles that start with the characterization of the river basin district, the monitoring and the assessment of status, the establishment of objectives, and finally the development of the program of measures and its implementation (EC, 2012). After approval of the plan, monitoring and evaluation of the effectiveness of measures provide the basic information for the next

1 Good ecological and chemical status for surface waters and good quantitative and chemical status for groundwater.

planning cycle. In the WFD's calendar, the first RBMPs had to be approved in December 2009 for the 2009–2015 planning cycle. While most countries have complied with this calendar, four MS – Belgium, Greece, Portugal and Spain – had not submitted all RBMPs to the European Commission (EC) as of December 2012.

In compliance with WFD requirements (art. 18), in November 2012 the EC issued its first WFD implementation report, an assessment of the RBMPs submitted by MS. This report is an integral part of the Blueprint to Safeguard Europe's Water Resources (COM (2012) 673), a policy document that identifies basic obstacles for action to protect Europe's water resources and proposals to overcome them, based on a comprehensive evaluation of existing European water policy.

The Commission's implementation report offers a mixed assessment of the first WFD planning cycle (EC, 2012). While some progress has been made, according to information provided by MS in their RBMPs only 41% of Europe's waters were in good status in 2009 and 51% will reach good status by 2015. The main pressures on water bodies that will hinder the attainment of WFD goals in 2015 are related to hydromorphological pressures, pollution and over-abstraction. Significant is also the lack of information on the ecological status of 15% and on the chemical status of 40% of Europe's surface waters. In some member states this lack of basic information is even more acute, with over 50% of their surface waters in unknown status.

Given the delay in approving Spain's RBMPs the Commission's assessment does not include Spanish river basins. As the Commission points out in its report, these delays will have significant impacts in the second implementation cycle (2015–2021) both within the MS that have not approved their plans as for others they share catchments with (EC, 2012), as is the case of the shared river basins between Portugal and Spain. During 2012, RBMPs have been approved for eight river basin districts in Spain, and the remaining plans are scheduled to be approved in 2013. However, according to WFD's calendar, work in the second implementation cycle is scheduled to begin in early 2013. In order to build on past achievements and set the groundwork to overcome existing obstacles and limitations, it is essential to start from an evaluation of progress made, however preliminary. The work presented in this chapter aims to fill this gap, by offering a comprehensive assessment of the information provided by the draft RBMPs that have been published and submitted to public consultation in most river basin districts in Spain, even if they have not yet been approved.

This chapter is divided into five sections. After this introduction we outline the main objectives of this work. We then go on to discuss the sources of information used and describe the indicators and methodological approach that have been applied to determine the status of surface water bodies in Spain. In the fourth section we discuss the results whereas in sections five and six we discuss the main results and present the most important conclusions.

2 OBJECTIVES

This chapter has two main goals: 1) the assessment of the ecological and chemical status of surface waters (surface water bodies in the WFD terminology), identifying the primary challenges to the achievement of the good status goal, and the regions where the most important problems are concentrated; and 2) an evaluation of the methodologies used to perform the assessments in order to identify whether all Spanish RBDs have had the same level of ambition in achieving the good status of surface water bodies. This second objective is in line with the findings of the Commission in its assessment of the first cycle of RBMP (EC, 2012) which identifies important deficiencies in the methodologies employed by member states to evaluate ecological and chemical status of surface waters. Although our work also evaluates the status of groundwater bodies, chapter 7 of this book deals with this aspect so we have not included it here.

3 DATA AND METHODS

3.1 Data sources

The information on the ecological and chemical status was obtained from the assessments included in seventeen River Basin Management Plans that have been published so far (Table 1). When plans have not been approved, we've used the assessments contained in the RBMPs' draft or in the Significant Water Management Issues document (SWMI). In these cases we also held consultations with the River Basin Management Agencies (RBA) responsible for developing these plans and requested updated information. The status information recorded for every Surface Water Body (SWB) was integrated into a geodatabase. The cartographic sources on SWB' types were obtained from the Spanish Integrated Water Information System (Sistema Integrado de Información del Agua, SIA) (MAGRAMA, 2012a).

In order to facilitate the interpretation of the results presented in this chapter for readers not familiar with Spanish river basins, Figure 1 illustrates the delimitation of river basin districts in Spain following WFD criteria.

3.2 Determination of the ecological, chemical and overall status

According to the WFD, the status of a SWB is defined by both its ecological and chemical status. The ecological status provides information on the health of ecosystem's composition and functioning, being determined by the quality of the biological community (phytoplankton, phytobenthos, benthic fauna, macrophytes and fish fauna), the hydromorphological conditions and the physico-chemical quality (e.g. oxygenation, acidification or presence of non-priority substances like sewage effluents or pharmaceuticals products) (EC, 2005). The chemical status informs on the degree of pollution for priority substances including metals, pesticides and various industrial chemicals (EC, 2005). See box 1 for WFD definitions on good ecological and chemical status.

The ecological and chemical statuses are assessed taking into account the different nature of SWBs. Hence, prior to the status assessment, water bodies have been classified into different categories (rivers, lakes/wetlands, transitional or coastal waters) and typologies within each category (for instance 33 types of rivers such as coastal Mediterranean rivers or high mountain rivers) based on the water characteristics. The degree of alteration of each category also allows classifying them as natural, heavily modified or artificial (EC, 2003). Next, the ecological status has been recorded on the scale of high, good, moderate, poor or bad. 'High' denotes largely undisturbed conditions, whereas the remaining classes represent increasing deviation from this natural or 'reference'

Table 1 Data sources used in this analysis.

River Basin District (RBD)	*Management level*	*River Basin Organization (RBO)*	*Document used for the assessment*[1]	*Source*	*Period*[2]
Cantabrico (West)	National	Confederación Hidrográfica Cantábrico (CHC)	RBMP	CHC (2011a)	2009
Cantabrico (East)		Confederación Hidrográfica Cantábrico (CHC)	RBMP	CHC (2011b)	2009
Douro		Confederación Hidrográfica Duero (CHD)	RBMPd	CHD (2010)	2009
Ebro		Confederación Hidrográfica Ebro (CHE)	RBMPd	CHE (2012)	2008
Guadalquivir		Confederación Hidrográfica Guadalquivir (CHG)	RBMP	CHG (2012)	2009
Guadiana		Confederación Hidrográfica Guadiana (CHGn)	RBMPd	CHGn (2011)	2009
Júcar		Confederación Hidrográfica Júcar (CHJ)	SWMI	CHJ (2009)	2007
Miño-sil		Confederación Hidrográfica Miño-Sil (CHMS)	RBMPd	CHMS (2010)	2009
Segura		Confederación Hidrográfica Segura (CHS)	SWMI	CHS (2008)	2007
Tagus		Confederación Hidrográfica Tajo (CHT)	RBMPd	CHT (2012)	2007–2010
Balearic islands	Regional	Govern de les Illes Baleares (GIB)	RBMPd	GIB (2011)	2005–2008
Cantabrico		Agencia Vasca del Agua (AVA)	RBMPd	AVA (2012)	2004–2008
Andalusia's mediterranean river basins		Junta de Andalucía (JA)	RBMP	JA (2012a)	2009
Catalan river basin district		Agencia Catalana del Agua (ACA)	RBMP	ACA (2011)	2008
Galicia-coast		Augas de Galicia	RBMP	AG (2012)	2009
Guadalete and Barbate		Junta de Andalucía (JA)	RBMP	JA (2012b)	2009
Tinto, Odiel and Piedras		Junta de Andalucía (JA)	RBMP	JA (2012c)	2009
Canary islands		Gobierno de Canarias (GC)	–	–	–

1 Approved River Basin Management Plan (RBMP); River Basin Management Plan draft (RBMPd); Special Water Management Issues (SWMI).
2 Indicates the year in which the data was last updated or the period in which the data was obtained.

Figure 1 Geographic location of Spanish River Basin Districts ('RBD') (MAGRAMA, 2012b).

Box 1: Definitions for surface water bodies in good status (WFD, 2000)

High or good ecological status is defined when the values of the biological quality elements for the surface water body show low levels of distortion resulting from human activity, but deviate only slightly from those normally associated with the surface water body type under undisturbed conditions. If these levels of distortion and deviation become more important, the ecological status will fall to moderate, poor or bad, depending on the degree of deviation from undisturbed conditions.

Good chemical status is recorded when a water body achieves compliance with all the environmental quality standards established in Article 16 (*Strategies against pollution of water*) and Annex IX (*Emission limit values and environmental quality standards*) of the Water Framework Directive, and in other relevant EU Community legislation setting environmental quality standards. If not, the water body shall be recorded as failing to achieve good chemical status and thus classified as in poor chemical status.

Good overall status is achieved when a surface water body reaches a high or a good ecological and chemical status.

condition. Reference conditions of SWBs are type and region specific, implying that they vary from rivers to lakes or coastal waters and also across different ecological regions.

For those SWBs with a high degree of alteration (heavily modified or artificial), ecological potential is defined instead of ecological status. Maximum ecological potential as the reference condition for these modified SWBs is intended to describe the best approximation to a similar natural aquatic ecosystem, but taking into account the constraints imposed by the social and/or economic uses that altered them, leading to heavily modified or artificial water bodies designation (EC, 2003).

The determination of the ecological status/potential is determined by the 'one out, all out' principle (EC, 2005) (Figure 2). Accordingly, the ecological status of a SWB is determined by the worst scoring quality element among the three complementary indicator types (biological, hydromorphological and physico-chemical). Likewise, the chemical status of a SWB needs to comply with all of the quality standards, otherwise it will fall to poor status (EC, 2005). The overall status is determined as 'good' when both the ecological and the chemical status are high/good and good, respectively. Otherwise, the SWB is classified as in 'poor' status. According to the WFD, for those SWBs that do not reach the 'good' status now, specific objectives and programs of measures need to be developed within the RBMPs to ensure compliance with the goal of achieving the good status objective by 2015, and exceptionally by 2021 or 2027.

The status of a SWB is largely influenced by the type of indicators used as well as by the criteria used to define the different class boundaries (Irvine, 2004). To ensure that

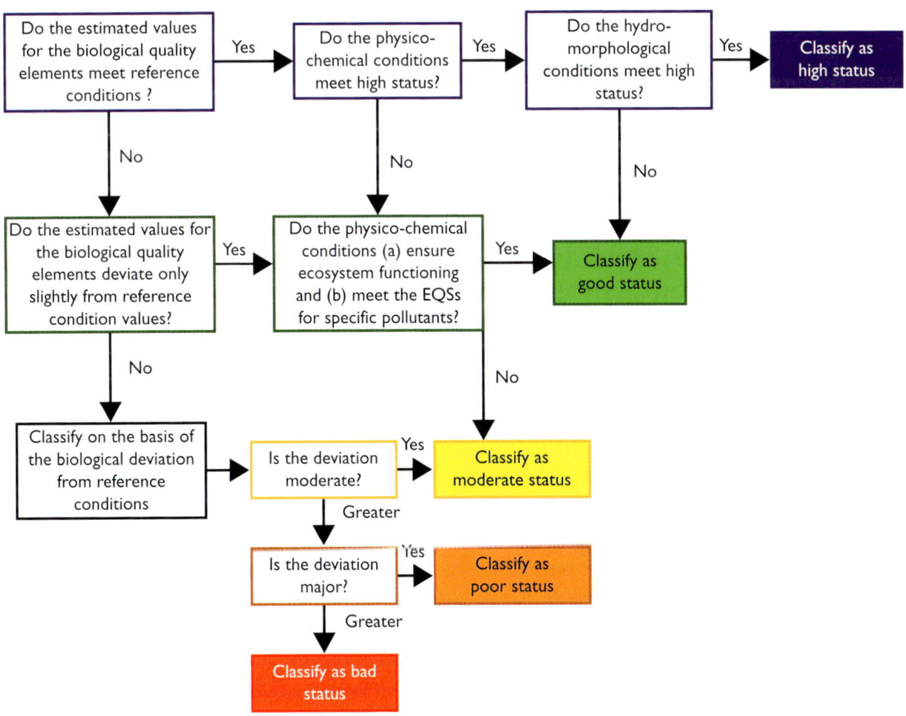

Figure 2 Steps in the determination of the ecological status of a SWB according to the WFD (EC, 2005).

all countries and RBDs have an equal level of ambition in achieving the good surface water status, the WFD required every Member State to develop standardized methods for monitoring the different ecological quality elements. Also, the WFD enforced the establishment of an inter-calibration process so that class boundaries for each quality element were consistent across boundaries and WFD compliant (EC, 2005).

The Spanish legal instrument supporting the implementation of the WDF, the so-called Hydrological Planning Technical Guidance (Instrucción de Planificación Hidrológica or IPH in Spanish), was developed with the aim of providing technical guidance on how to develop RBMPs. The IPH detailed the type of indicators to be used in the SWBs' assessment, although reference conditions were not always provided due to the lack of data (for instance, for fish fauna or most hydromorphological indicators). This data gap has hampered to a large extent the implementation of common and standardized procedure to assess the ecological status in this first planning cycle. In some RBDs, expert criteria and literature sources have been used to qualitatively determine the hydromorphological quality of SWBs in the absence of reference conditions. The partial assessment plus the potential lack of robustness of some of the evaluation outputs, hinders the comparison of SWB assessment across RBDs. Table 2 summarizes the indicators recommended by the IPH and most widely used by RBDs to determine the ecological status of SWBs.

4 RESULTS

4.1 Classification of surface water bodies in Spain

Table 3 summarizes the spatial distribution and representativeness of surface water bodies (SWBs) in Spain, differentiating between those classified as natural and those classified as heavily modified or artificial. Rivers are the most abundant water bodies and have a combined length of over 70,000 km.

Lakes and wetlands are spatially less significant water bodies, although they have great ecological and socioeconomic importance in certain areas of Spain, such as the Daimiel and Doñana wetlands. Most of the lakes and wetlands classified as heavily modified are also located in the Southern part of the country in the Guadalquivir basin, namely the Doñana region. Transitional waters mostly refer to deltas and estuaries and despite their limited extension, they exhibit a high degree of human alteration with 52% their total area classified as heavily modified. Coastal waters refer to the waters lying one mile from the coastline; they occupy around 18,000 km^2 and their degree of human modification is small. No information was available on surface water bodies for Canary Islands at this point.

Almost 9% of Spanish rivers have been declared as being heavily modified because of hydromorphological or hydrological[2] alterations. This denomination implies that

2 Hydrologically modified SWB are those located downstream from reservoirs used primarily for irrigation water supply and that, as a result, have an inverted hydrological regime because of increased water releases during the irrigation season (spring and summer) and lower water flows in the winter months (IPH, 2008).

Table 2 Most frequent indicators used in the assessment of the ecological status of SWBs in Spanish River basin Management Plans (own elaboration).

Quality element		*Indicators*			
		Rivers	*Lakes*	*Transitional*	*Coastal*
BIOLOGICAL	Phytobenthos	Specific Pollution Sensitivity Metric (IPS) Multi-metric Diatom Index (MDIAT)	–	–	–
	Benthic fauna	Iberian Bio-monitoring Working Party (IBMWP) Multi-metric-type Index	Shannon Index Taxon diversity	–	Multivariate-AZTI's Marine Biotic Index (MAMBI)
	Fish fauna	Rate of autochthonous species	Rate of autochthonous species	–	–
	Phytoplankton	–	Chlorophyll a Bio-volume Algae Group Index (AGI)	Chlorophyll a	Chlorophyll a Number of cells per taxon
	Macrophytes	–	Presence of introduced macrophytes % of natural vegetation	–	–
	Macro Algae	–	–	Cover value per taxon	Quality of Rocky Shores (CFR) CARLIT Index (Cartography of littoral and upper-sub littoral rocky-shore communities) Cover Value per taxon
	Angiosperms	–	–	Cover value per taxon	*Posidonia oceanica* multivariate index (POMI)

(*Continued*)

Table 2 (Continued).

	Quality element	*Indicators*			
		Rivers	*Lakes*	*Transitional*	*Coastal*
HYDROMORPHOLOGICAL	Hydrological regime	Environmental flows Index of hydrological alteration	Environmental flows Average inflow	–	–
	Fluvial continuity	Length of rivers free of artificial barriers Typology of barriers	–	–	–
	Morphological conditions	Riparian Forest Quality: QBR index Fluvial Habitat Index (IHF)	Depth-average change Riparian Forest Quality: QBR index	Depth-average change Percentage soft substrate surface Area of Intertidal surface	Max and Min Depth (BMVE) Average slope and granulometry (D50) Area of Intertidal surface
	Tidal regime	–	–	Environmental flows Water residence time Degree of wave exposure	Degree of wave exposure Speed and direction of dominant sea currents
PHYSICO-CHEMICAL	Water transparency	–	Depth (Secchi disk)	Depth (Secchi disk) Suspended Solids and Turbidity Index (NTU)	Depth (Secchi disk) Suspended Solids and Turbidity Index (NTU)
	Thermal	Mean average water temperature	Mean average water temperature	Mean average water temperature	Mean average water temperature
	Oxygenation	Dissolved oxygen Biological Oxygen demand 5 days (DBO_5)	Dissolved oxygen Tasa de saturación del oxígeno	Dissolved oxygen Rate of dissolved oxygen saturation	Dissolved oxygen Rate of dissolved oxygen saturation
	Salinity	Electric conductivity at 20°C	Electric conductivity at 20°C	Salinity (PSU) Electric conductivity at 20°C	Salinity (PSU)
	Acidifcation	pH	pH, Alkalinity	–	–
	Nutrients	Ammonia, Nitrates, Phosphates	Ammonia, Nitrates, Phosphates	Ammonia, Nitrates, Phosphates	Ammonia, Nitrates, Phosphates
	Non-synthetic pollutants	Non- synthetic contaminants listed in Annex II of the Spanish Royal Decree 1290/2012 & non-synthetic pollutants from List II of preferential hazardous substances listed in Annex IV of the Spanish Royal Decree 907/2007		Non-synthetic pollutants listed under Annex II of the Spanish Royal Decree 1290/2012 & non-synthetic substances listed under the Spanish Coastal Law 22/1988	
	Synthetic pollutants	Synthetic contaminants listed in Annex II of the Spanish Royal Decree 1290/2012 & synthetic pollutants from List II of preferential hazardous substances listed in Annex IV of the Spanish Royal Decree 907/2007		Synthetic pollutants listed under Annex II of the Spanish Royal Decree 1290/2012 and synthetic substances listed under the Spanish Coastal Law 22/1988	

Table 3 Categories of Surface Water Bodies (SWBs) in Spain (own elaboration).

Surface water bodies	*Total*	*Heavily modified (HMWB)*	*Artificial (AWB)*
Rivers (km)	70,648	6154 (8.7%)	253 (<1%)
Lakes (km²)	1010	161 (15.9%)	31 (<1%)
Transitional (km²)	632	329 (52%)	–
Coastal (km²)	17,276	435 (2.5%)	–

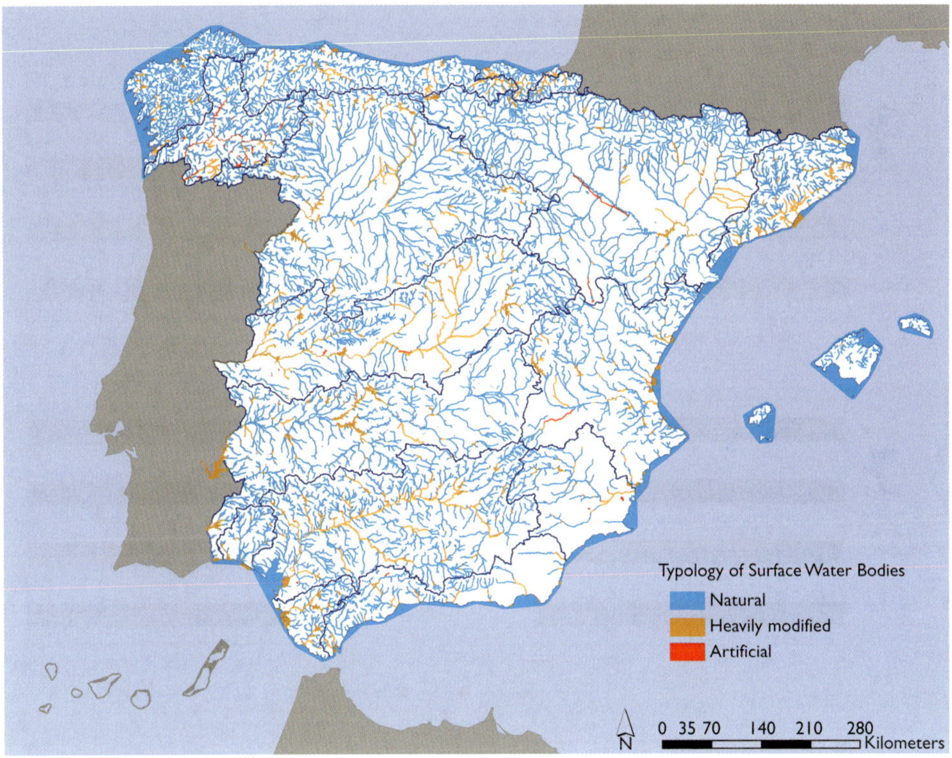

Figure 3 Distribution of natural, heavily modified and artificial surface water in Spain (own elaboration).

they do not have to reach good ecological status (GES), only good ecological potential (GEP), a more modest goal.

As Figure 3 indicates, the largest number of rivers classified as heavily modified are found in the Southern half of Spain, particularly in the Tagus (1100 km) and in the Guadalquivir river basins (1075 km). This trend seems to indicate the high pressure rivers suffer in this part of the country. Nevertheless, it is remarkable the difference in the amount of SWBs declared as heavily modified among RBDs. At this stage, it is early to determine whether the classification of such a large number of SWBs as heavily modified in some RBDs responds to the existence of conditions that are difficult to alter, or on the contrary the goal is to limit the ambition of the planning objectives.

4.2 National overview of the ecological, chemical and overall status

Figure 4 summarizes the results of the evaluation of the ecological, chemical and overall status of the SWBs in Spain in the first river basin planning phase (2009–2015). Overall natural lakes are the most problematic from the perspective of ecological status, with almost 60% of the evaluated lakes with moderate to poor status (Figure 4a). It is however significant that over 50% of lakes have not been evaluated because of lack of data. Rivers hold the second ranking in water bodies with the poorest ecological status, with only 51% in good or very good status. Coastal waters are the water bodies suffering less pressure, with 82% of them in good status.

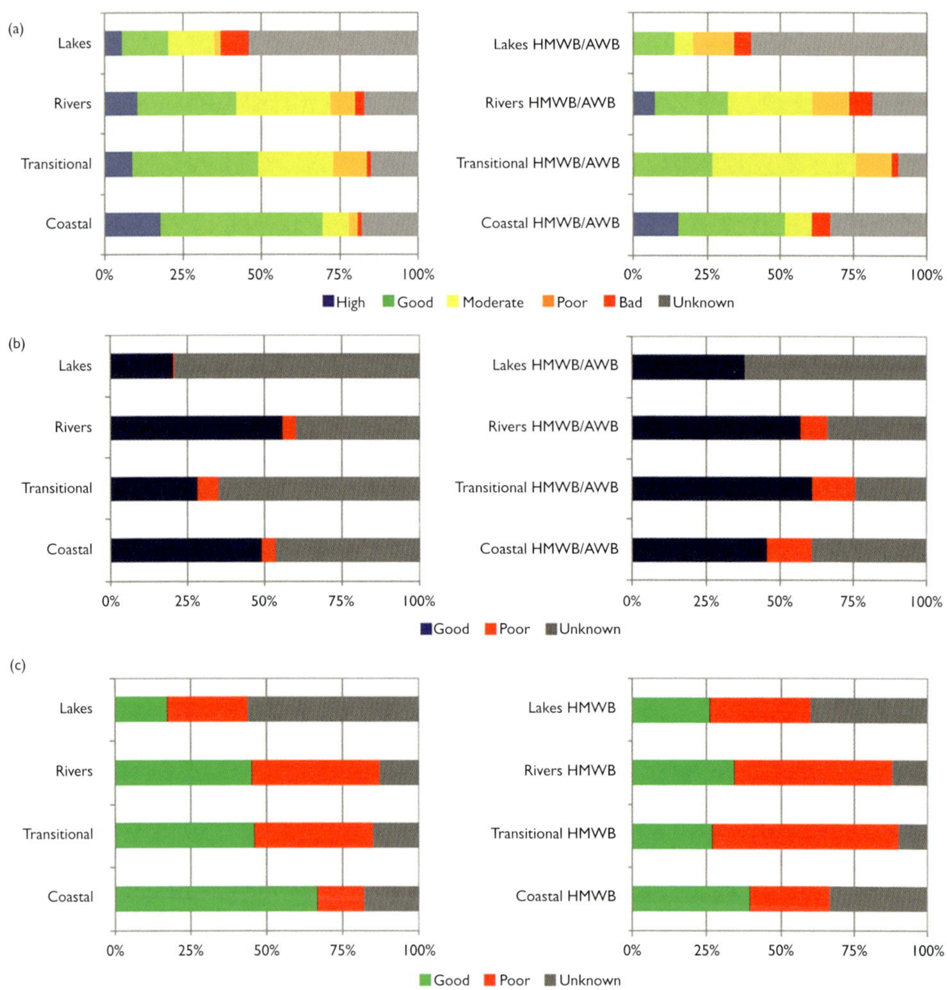

Figure 4 Ecological (a), chemical (b) and overall status (c) of natural, heavily modified (HMWB) and artificial (AWB) Surface Water Bodies in Spain (own elaboration).

Among the heavily modified water bodies (HMWB) and artificial water bodies (AWB), transitional waters hold the poorest ecological status (70%), followed by the lakes (65%) and rivers (61%). On the contrary, 75% of the coastal areas reach the 'good ecological potential' according to the information gathered. The results of this assessment show that there is still an important knowledge gap, particularly for lakes and wetlands with over half pending evaluation because of lack of data.

Regarding their chemical status (Figure 4b), the vast majority of the evaluated water bodies exhibit a good status but there is a significant lack of data. Over 50% of all natural rivers and coastal waters have a good chemical status. However, these results might change significantly since the remaining 50% of these water bodies have not been evaluated. The chemical status from HMWB and AWB shows a similar pattern, although the fraction of water bodies not reaching the good status is slightly higher (9 to 14%).

When looking at the overall status (Figure 4c), natural lakes are the SWB showing the worst state with only 17% reaching the good status. Rivers and transitional waters exhibit a better situation with nearly 50% in good status. Coastal waters seem to be, at this point, the SWB in better condition, with almost 75% reaching good status. The main driver determining the poor status of SWBs is the poor ecological status of a large number of SWBs over any major chemical problem. The high number of non-evaluated water bodies in all categories and particularly for chemical status, evidences the information needs still existing in Spain.

The overall results show that less than 50% of SWBs are in good ecological status (with rivers emerging as one of the water body types in poorest status) and a large proportion of them do not have information for the chemical status. A recent report from the European Environment Agency showed very similar results for SWBs across Europe (EEA, 2012).

Figure 5 illustrates the ecological and chemical status of all SWBs across Spain. The figure shows how the driving factor behind the poor or moderate status of SWB in all RBD is the poor ecological status. According to the information available, the most important ecological problems in Spain are located in the Douro, Guadiana and the Catalan RBD. The northern basins of Cantábrico, Galicia-Coast and Miño-Sil show the best status, with a large number of rivers in good and very good status.

As was mentioned in section 3, not all indicators have been used to determine the ecological status of SWBs. In most RBDs only macro-invertebrates and diatoms have been used to determine the biological quality, and hydromorphological indicators have not been used for the most part. It is therefore likely that when these indicators are incorporated in the second planning cycle, the overall picture of the ecological status of SWB in Spain may worsen. Furthermore, the aggregate fashion in which the information is presented in most RBMPs does not allow us to distinguish what indicators have been used to assess the different quality elements nor their score. This circumstance limits our ability to identify the primary pressures on SWBs.

Regarding the chemical status, most basins have the majority of their water bodies in good chemical status, although there is a major information gap in the Ebro and Júcar basins in Eastern Spain. Specific chemical problems are found in rivers downstream of big cities, like for instance the upper Tagus downstream from the Madrid metropolitan region, the middle Guadalquivir basin, or the Catalonian RBD around the Barcelona

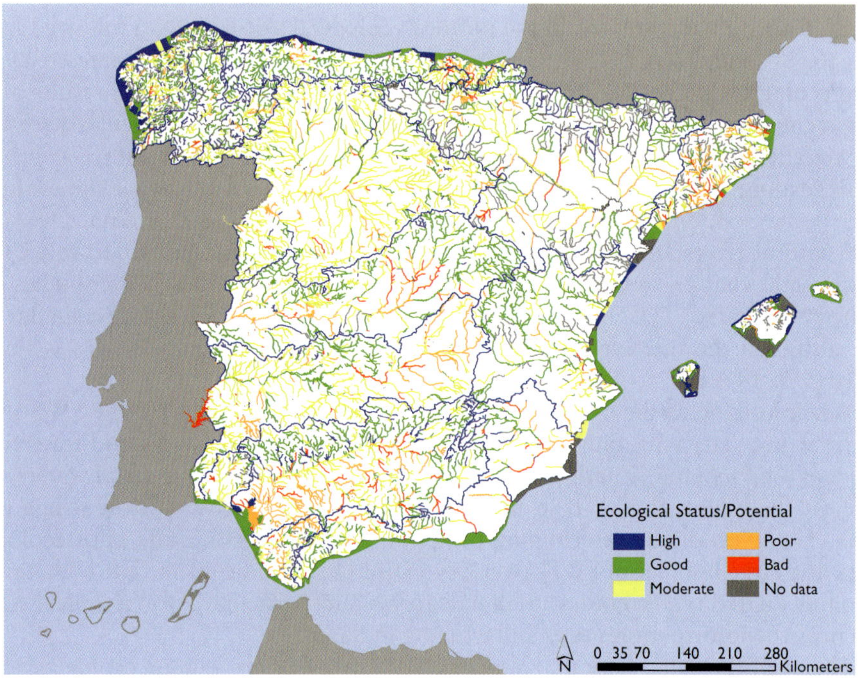

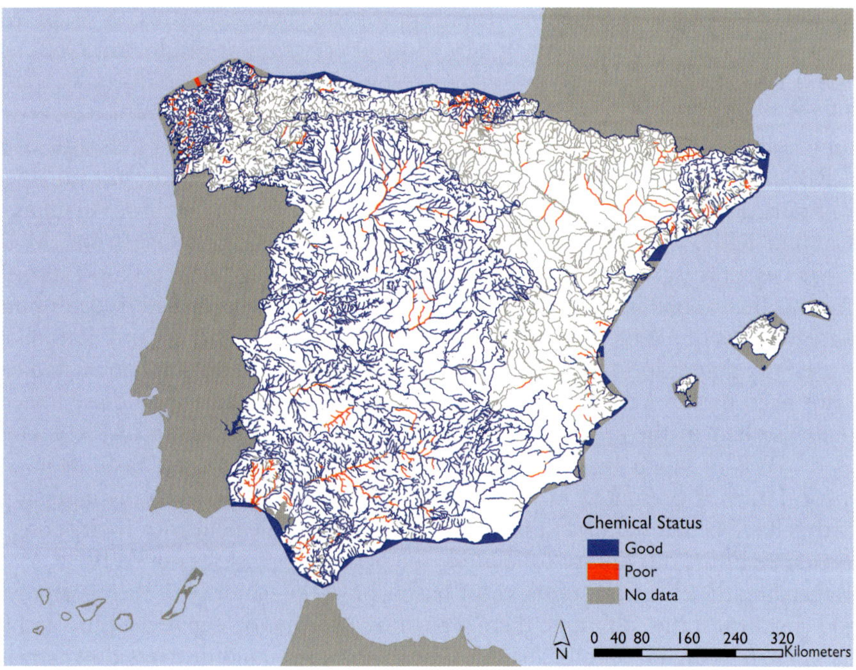

Figure 5 Ecological (top) and chemical status (below) of Spanish rivers, lakes, transitional and coastal areas (own elaboration).

Metropolitan area. Acute problems of chemical pollution are also found around important agricultural areas, as in the (Duoro, Guadiana and Guadalquivir river basins.

Table 4 summarizes the information on the overall status of SWBs. The information in the table shows that, according to the information available, the RBD in poorest status is the Douro basin, with 79% of its total SWBs in poor status. When looking at the extension of SWBs in poor status in this basin the situation gets worse: 82% of its 12,930 km of rivers do not reach the good status and 67% of the total area of lakes and HMWB rivers (dams) are in poor status too. The RBD where the SWB are in best conditions is Western Cantabria, where around 73% of the total number of water bodies are in good conditions, this is to say 77% of the total length of rivers and 97% of coastal, lakes and transitional waters. Miño-Sil and Galicia Coast RBD also have a large fraction of their SWB in good status (≈70%).

Overall, the most important hotspots are found in those RBDs located in the Southern half of the country and flowing into the Atlantic Ocean. In these basins around 53% of their SWBs are in poor ecological conditions, with 67% of its river's length and 41% of the coastal, transitional and lakes area not reaching the good status. The situation is only slightly better in the rest of the territory, particularly in the Southern basins draining to the Mediterranean Sea. These results might change in the next planning cycle since the Segura basin for instance has only assessed the status of 6% of its total coastal and transitional water bodies. The Ebro RBD is also lacking the assessment of 100% of its coastal, transitional and lakes and almost 30% of its rivers.

Nevertheless, these results need to be interpreted with caution since the assessment process has not been homogenous across all RBDs in Spain. As was mentioned in Section 3.2, the IPH did not provide reference conditions for key quality elements, namely fish fauna and hydromorphological elements like fluvial continuity or hydrological regime. Among the seventeen RBMPs here assessed, only two (Catalan RBD or the Inland River Basins of the Basque Country) have developed metrics and reference conditions to fill this gap in the SWB assessment. The majority of RBDs have omitted those elements in the actual assessment; or their evaluation has been based on expert judgment (Table 5). Additionally, the designation of a large portion of problematic SWBs as heavily modified in some river basin districts such as the Tagus or the Guadalquivir, results in a lower percentage of water bodies in poor status. Other basins, such as the Douro, have maybe reported a more realistic picture of the status of the river basin.

5 DISCUSSION

The results of this assessment show that the main underlying driver for not achieving the good status in the majority of SWBs in Spain is related with their poor ecological status. The chemical status of the vast majority of inland and coastal waters appears to be good, although precaution needs to be taken when interpreting these results. Foremost, because an important number of the priority-substances and priority hazardous substances included in the chemical assessment have been set out late in time, in the 2008/105/EC Directive, limiting the options for RBDs to consider them all. As a result, some basins have chosen to perform their assessment without considering all those substances (mainly due to the lack of monitoring capacity), while others

Table 4 Number and extension of Surface Water Bodies (SWB) in poor status across Spanish River Basin Management Districts (own elaboration).

Geographical distribution	*River Basin Management Districts (RBD)*	*Total number of SWB*	*Number of evaluated SWB*	*Evaluated SWB with 'Poor Status' (%)*	*km SWB*	*km SWB evaluated*	*Evaluated km in 'Poor Status' (%)*	*Total km^2 SWB*	*Evaluated km^2 SWB*	*Evaluated km^2 in 'Poor Status' (%)*
North Atlantic	Douro	710	695	79	12,929	12,929	82	362	341	67
	Inland RB Basque Country	66	66	65	584	584	69	624	624	9
	Eastern Cantabria	72	69	49	952	936	61	3	3	0
	Galicia-Coast	464	460	30	4229	4198	32	3362	3360	14
	Miño-Sil	278	274	31	3954	3954	26	166	134	40
	Western Cantabria	293	289	27	3694	3677	23	1664	1663	3
	Sub-total			**47**			**49**			**22**
North Mediterranean	Catalan River Basin District	346	251	70	3835	2482	77	1635	1608	34
	Balearic Islands	172	108	32	578	261	63	3806	1922	4
	Ebro	821	386	42	12,303	8690	47	910	0	–
	Sub-total			**48**			**62**			**19**

South Atlantic	Guadalete and Barbate	97	78	55	997	927	80	769	761	20
	Guadiana	313	302	71	7157	7157	79	985	964	47
	Tinto, Odiel and Piedras	68	55	55	783	692	78	360	358	50
	Guadalquivir	443	443	43	9126	9126	57	1917	1917	31
	Canary Islands RB	–	–	–	–	–	–	–	–	–
	Tagus	324	308	44	7360	7095	40	545	545	57
	Sub-total			**53**			**67**			**41**
South Mediterranean	Andalusia Medit RB	175	173	47	1999	1984	60	2156	2156	2
	Segura	114	90	49	1435	1394	41	1320	84	53
	Jucar	349	256	39	5141	3978	36	2361	2177	20
	Sub-total			**45**			**45**			**25**
TOTAL		5105	4303		77,055	70,067		22,945	18,617	
National Average		–	–	**49**	–	–	**56**	–	–	**28**

Table 5 Quality elements included by the different River Basin Districts in evaluation of surface water bodies. This evaluation is based on the results of 17 River Basin Management Plans drafts or approved. (own elaboration).

	Biological quality elements		*Hydromorphological quality elements*		
	Only diatoms and macroinvertebrates	Diatoms, macro-invertebrates and fish fauna	None	Only morphological conditions	Expert criteria
Number of RBDs	15	2	8	4	5

have delayed the assessment of the chemical status until monitoring options are available. Consequently, the current picture is still partial and incomplete.

The prevailing poor ecological status found indicates that the hydrological functioning of these water systems is altered, probably as a result of multiple factors depending on the geographical context. At this point, is not possible to discern what is the ecological quality elements mostly affected, because the majority of RBMPs only provide the overall ecological status without detailing the individual element scores. This lack of information impedes for now establishing a solid relationship between main pressures (e.g. pollution from urban or industrial areas, highly regulated water bodies, over-extraction of water or diffuse pollution) and resulting impacts in SWBs. Obtaining such information might be helpful to know if the program of measures elaborated in each RBMP, regarding the achievement of the 'good status' of all water bodies by 2015, fits to actual pressures and impacts occurring in their regions, helping this way, effectively, to improve the status of water bodies where needed.

Another remaining challenge in most Spanish basins for the next planning cycle refers to the need of moving forward in the establishment of an standardized evaluation program across all RBDs. Prior to the implementation of the WFD, the monitoring of surface water bodies in Spain was limited to control just the quality aspects of waters (e.g. SAICA system, which includes the ALERTA and ICA network). No monitoring network was in place to track the ecological status of surface water bodies, in contrast for instance to the extensive research done in other Member States like UK (Staddon, 2010). The only available information on biological conditions prior to the WFD implementation in Spain was limited to academic work on certain river basins. With the enactment of the WFD, the surveillance program for surface water bodies developed in Spain has allowed the creation of a network of 2201 river monitoring stations, 434 in dams and 169 in lakes and wetlands (MAGRAMA, 2012c). This network has meant a significant step forward, although important monitoring gaps still exist and the current national economy does not favor its completion for the next planning cycle, starting now in 2013. In fact, budgetary restrictions at both the regional and national levels are restricting the existing network both in the frequency as well as the extent of monitoring efforts.

Due to the lack of monitoring infrastructure, ecological knowledge and reference conditions for some key indicators, many RBDs have chosen not to perform a full and complete surface water status assessment as outlined in Section 4. This issue has two major implications. First, the assessments now available do not portray the real status of the majority of surface water bodies and therefore do not ensure the same level of ambition in achieving the good status across RBD. Second, when those missing assessment indicators are included in the next planning cycle, the overall picture will likely deteriorate and we may find that the measures in place do not address the real pressures on SWBs. In this context, the chances of achieving the good status of all water bodies in the short to mid-term will decrease.

Munné *et al.* (2012) provide a clear example on the different outputs that can be obtained when using different indicators to evaluate the ecological status of SWBs. When determining the biological quality of SWB in the Catalonian RBD, they found that depending on the criteria used (diatoms, macro-invertebrates and/or fish fauna) both the number and the location of SWBs in good status vary. For instance, when only considering one of these three criteria separately, between 56–62% of the evaluated water bodies do reach the good biological quality. However, when considering all three criteria together, the fraction of SWBs in good status only reaches 36%. Furthermore, the location of these SWBs failing to meet the good status, changes in space as Figure 6 shows. The areas inside the circles show an example of SWBs that do meet the criteria for some biological quality elements but not for others. The most

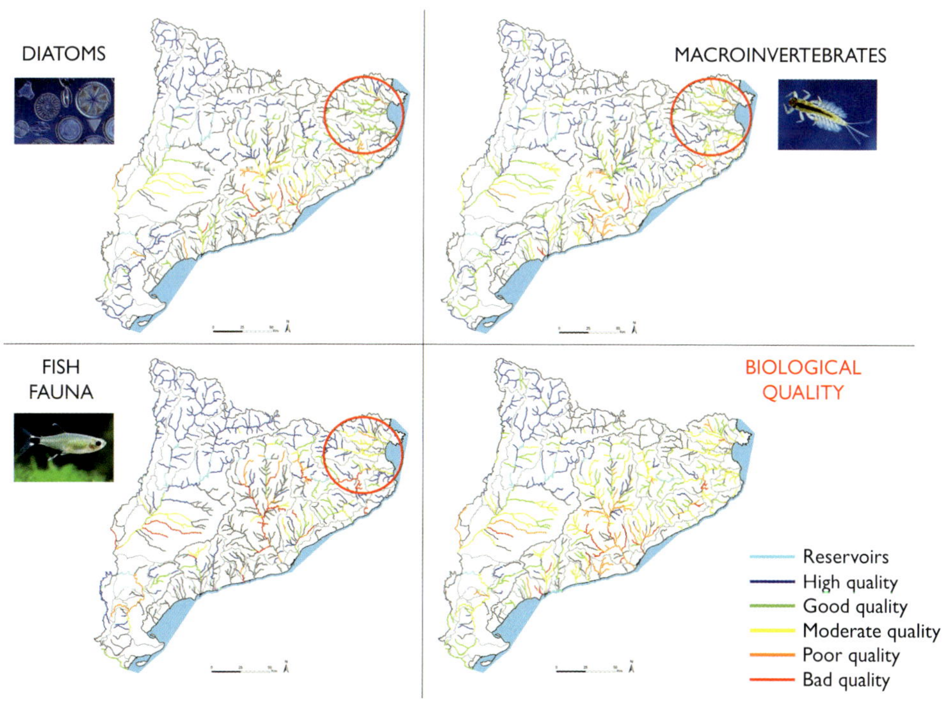

Figure 6 Biological quality of surface water bodies in Catalonian River Basin District. Water bodies with no data are shown in grey. Reservoirs are shown in light blue. (own elaboration from Munné *et al.*, 2012).

important reason is that biological elements like fish fauna are affected by different pressures compared to diatoms or macro-invertebrates, so the SWBs that are defined as being in poor status vary depending on the type of indicators used.

6 CONCLUSIONS AND OUTLOOK

The current planning cycle has meant an important step forward in improving our understanding of the ecological and chemical status of SWBs in Spain. However, the information available is still partial and incomplete, and therefore it is difficult to have a comprehensive understanding on the true situation of surface water bodies and on the pressures affecting them. As we have seen throughout this chapter, several are the reasons that explain this limitation.

In terms of the definition of the ecological status of surface water bodies, most RBDs have only used macroinvertebrates and diatoms as biological indicators. Fish and hydromorphological indicators have for the most part not been used. As a result, it is very likely that the current determination of ecological status of surface water bodies while currently problematic (overall less than 50% of SWBs are in good or very good ecological status) will likely deteriorate when all indicators are considered. This is particularly true in Spain, where rivers are heavily regulated and therefore hydromorphological alterations are widespread.

In what pertains to the chemical status of surface water bodies, available information is also limited and incomplete. On one hand the assessment approach used by different RBDs has varied. On the other hand, monitoring networks do not yet have the capacity to evaluate all substances included in the recent 2008/105/EC Directive. Finally, the chemical status of over 50% of SWBs has not been assessed. These are important limitations to have a comprehensive understanding of the chemical status of Spanish SWBs.

An additional problem is the lack of information available to assess the appropriateness of the designation of SWBs as heavily modified. The significant differences in the percentage of SWBs designated as heavily modified in different river basins seem to indicate that the criteria used is not comparable. Heavily modified water bodies do not need to reach the more demanding good ecological status, but rather, only the good ecological potential. It is unclear whether RBDs that have heavily relied on the designation of HMWBs have done so in response to objective criteria or because of a decision to lower the planning objectives for these problematic water bodies.

Finally, the aggregate form in which the information is presented in the planning documents for the different RBDs makes it difficult to establish causal relationships between the status of SWBs and the potential pressures and impacts affecting them. Furthermore, the inclusion of new ecological status indicators such as fishes or hydromorphological alterations in the next planning cycle and the evaluation of the chemical status of more water bodies will probably alter the overall assessment of the situation of Spanish waters. It will therefore be important to closely review the effectiveness of the programs of measures in dealing with existing problems and challenges and, more importantly, to adapt these programs as new information and assessments become available.

REFERENCES

ACA, Agència Catalana del Agua (2011) *Plan de gestión del distrito de cuenca fluvial de Catalunya*. Generalitat de Catalunya.

AG, Augas de Galicia (2012) *Plan Hidrológico de la Demarcación Hidrográfica de Galicia Costa*. Xunta de Galicia.

AVA, Agència Vasca del Agua (2012) *Proyecto de Plan Hidrológico de la Demarcación Hidrográfica del Cantábrico Oriental en el Ámbito de las Cuencas Internas del País Vasco*. Gobierno Vasco.

CHC, Confederación Hidrográfica del Cantábrico (2011a) *Propuesta de Proyecto de Plan Hidrológico de la Demarcación Hidrográfica del Cantábrico Occidental*. Ministerio de Agricultura, MAGRAMA, Alimentación y Medio Ambiente.

CHC, Confederación Hidrográfica del Cantábrico (2011b) *Propuesta de Proyecto de Plan Hidrológico de la Demarcación Hidrográfica del Cantábrico Oriental*. MAGRAMA.

CHD, Confederación Hidrográfica del Duero (2010) *Propuesta de Proyecto del Plan Hidrológico de la parte española de la Demarcación Hidrográfica del Duero*. MAGRAMA.

CHE, Confederación Hidrográfica del Ebro (2012) *Propuesta de Proyecto de Plan Hidrológico de la Cuenca del Ebro*. MAGRAMA.

CHG, Confederación Hidrográfica del Guadalquivir (2010) *Propuesta de Proyecto de Plan Hidrológico de la Demarcación Hidrográfica del Guadalquivir*. MAGRAMA.

CHGn, Confederación Hidrográfica del Guadiana (2011) *Propuesta de Proyecto de Plan Hidrológico de cuenca en la Demarcación Hidrográfica del Guadiana* (parte española). MAGRAMA.

CHJ, Confederación Hidrográfica del Júcar (2009) *Esquema Provisional de Temas Importantes (Demarcación Hidrográfica del Júcar)*. MAGRAMA.

CHMS, Confederación Hidrográfica del Miño-Sil (2010) *Propuesta de Proyecto de Plan Hidrológico de la Demarcación Hidrográfica del Miño-Sil*. MAGRAMA.

CHS, Confederación Hidrográfica del Segura (2008) *Esquema Provisional de Temas Importantes (Demarcación Hidrográfica del Segura)*. MAGRAMA.

CHT, Confederación Hidrográfica del Tajo (2012) *Propuesta de Proyecto de Plan Hidrológico de Cuenca de la parte española de la Demarcación Hidrográfica del Tajo*. MAGRAMA.

EC, European Commission (2003) *Common Implementation Strategy for the Water Framework Directive: Identification of Water Bodies*. [Online] Guidance Document Number: 2. Office for Official Publications of the European Communities, Luxemburg. [Online] Available from: https://circabc.europa.eu/faces/jsp/extension/wai/navigation/container.jsp [Accessed September 2012].

EC, European Commission (2005) *Common Implementation Strategy for the Water Framework Directive: Overall Approach to the Classification of Ecological Status and Ecological Potential*. [Online] Guidance Document Number: 13. Available from: https://circabc.europa.eu/faces/jsp/extension/ wai/navigation/container.jsp [Accessed September 2012].

EC, European Commission (2012) *River Basin Management Plans: Report on the Implementation of the Water Framework Directive (2000/60/EC)*. [Online] Available from: http://ec.europa.eu/environment/water/water-framework/pdf/COM-2012–670_EN.pdf [Accessed November 2012].

EEA, European Environmental Agency (2012) *European waters – assessment of status and pressures (EEA Report No 8/2012)*. [Online] Available from: http://www.eea.europa.eu/publications/european waters-assessment-2012 [Accessed November 2012].

GIB, Govern de les Illes Baleares (2011) *Propuesta del Plan Hidrológico de la Demarcación de Baleares* (Versión 2.0). Conselleria de Medi Ambient i Mobilitat. Direcció General de Recursos Hídrics.

Irvine, K. (2004) Classifying ecological status under the European Water Framework Directive: The need for monitoring to account for natural variability. Aquat. Conservat. Mar. Freshwat. Ecosyst., 14(2), 107–112.

IPH, Instrucción de Planificación Hidrológica (2008). *ORDEN ARM/2656/2008, de 10 de septiembre, por la que se aprueba la instrucción de planificación hidrológica.* BOE 229, 38472–38582.

JA, Junta de Andalucia (2012a) *Plan Hidrológico de la Demarcación Hidrográfica de las Cuencas Mediterráneas Andaluzas.* Consejería de Agricultura, Pesca y Medio Ambiente.

JA, Junta de Andalucía (2012b) *Plan Hidrológico de la Demarcación Hidrográfica del Guadalete y Barbate.* Consejería de Agricultura, Pesca y Medio Ambiente.

JA, Junta de Andalucía (2012c) *Plan Hidrológico de la Demarcación Hidrográfica del Tinto, Odiel y Piedras.* Consejería de Agricultura, Pesca y Medio Ambiente.

MAGRAMA, Ministerio de Agricultura, Alimentación y Medio Ambiente (2012a) *Servicio de cartografía del Sistema Integral de Información del Agua.* [Online] Available from: http://servicios2.magrama.es/sia/visualizacion/descargas/documentos.jsp [Accessed September 2012].

MAGRAMA, Ministerio de Agricultura, Alimentación y Medio Ambiente (2012b) *Spanish River Basins.* [Online] Available from: http://www.magrama.gob.es/en/agua/temas/planificacion-hidrologica/marco-del-agua/Demarcaciones_hidrograficas.aspx [Accessed November 2012].

MAGRAMA, Ministerio de Agricultura, Alimentación y Medio Ambiente (2012c) *Agricultura, Alimentación y Medio Ambiente en España 2011* [Online] Available from: http://www.magrama.gob.es/es/ministerio/servicios/publicaciones/memoria2011_cap.aspx [Accessed December 2012].

Munné, A., Tirapu, T., Sola, C., Olivella, L., Vilanova, M., Ginebreda, A. & Prat, N. (2012) Comparing Chemical and Ecological Status in Catalan Rivers: Analysis of River Quality Status Following the Water Framework Directive. In: H. Guasch *et al.* (eds.) (2012) Emerging and Priority Pollutants in Rivers. *Environmental Chemistry,* 19, 243–266.

Staddon, C. (2010) *Managing Europe's Water Resources: 21st Century Challenges.* Farhnam, Ashgate Books.

WFD, Water Framework Directive (2000) *Directive 2000/60/EC of the European Parliament and of the Council of 23 October 2000 establishing a framework for Community action in the field of water policy.* [Online] Available from: http://eur-lex.europa.eu/LexUriServ/LexUriServ.do?uri = CELEX:32000 L0060:EN:NOT [Accessed September 2012].

Chapter 8

Intensive groundwater use in agriculture and IWRM: An impossible marriage?

Lucia De Stefano, Elena López-Gunn & Pedro Martínez-Santos
Water Observatory, Botín Foundation; Universidad Complutense de Madrid, Madrid, Spain

ABSTRACT: Groundwater is widely used in many countries worldwide. Particularly in semiarid regions, such as Spain, aquifers provide a series of practical advantages which make them preferential sources of supply for many uses. In this country, groundwater accounts for one third of the total irrigated surface and one fourth of the total water consumption, as well as for one third of the economic value of Spain's irrigated crop production. This chapter examines groundwater-based agricultural development in Spain from the perspective of the European Union Water Framework Directive. Based on the example of different case studies, it is concluded that its principles and requirements remain distant from what is actually happening on the ground, where economic, political and social reasons prevail on legal considerations set by national and supranational regulations. In this Chapter greater emphasis is put on irrigation because this sector is by far the main user of groundwater.

Keywords: Groundwater, Spain, Water Framework Directive, intensive use

1 INTRODUCTION

The importance of groundwater use in many regions of the world is widely recognized in the literature (Giordano, 2009), and so is the difficulty to control it (Shah, 2009; De Stefano & López-Gunn, 2012) due to its intrinsic nature and the benefits generated by its use. Similarly, the implications of intensive use on the water table levels and the quality of groundwater are well documented (Custodio, 2002). Spain, with a semi-arid climate in most of its territory, is no exception to this, and is an interesting case to study groundwater development trends and the different policy experiments to deal with the consequences ensued from intensive groundwater use.

The 2000 Water Framework Directive (WFD; 2000/60/CE) is often portrayed as the crystallization of several principles of Integrated Water Resources Management (IRWM) (e.g. policy integration, cost recovery, participation, environmental sustainability, basin level management) into a legal document of mandatory application to all the member states of the European Union (EU). The WFD constitutes an environmentally oriented umbrella under which EU countries should reformulate their water resources planning and achieve good status of all their waters (groundwater, surface, coastal and transitional waters) by 2015 or 2027 at the very latest. Thus, the Directive sets similar objectives for all types of waters, including groundwater, and underpins the unity of the hydrological cycle. This objective suggests that the WFD strives at re-establishing the balance between the three tiers of 'sustainable development' by acknowledging the economic and social dimensions of water management and emphasizing the environmental one, which traditionally has been underrepresented in water management, as proven by the poor quality of many aquatic ecosystems. Thus, the benchmark used to measure whether EU Member States have succeeded in the implementation of the WFD is the environmental quality of their water resources (good status of all waters). This means that even if countries strictly meet the milestones and formal requirements of the Directive and set up a 'model' water management system, actually they will be judged by the outcome of their water planning, i.e. ensuring that present and future water uses are compatible with good status of the resources they rely on.

Since the WFD is heralded – mostly with reason – as a IWRM-oriented policy, the chapter looks at groundwater use in Spain through the lens of the Directive and argues that its principles and requirements seem to be still very distant from what is actually happening on the ground, where economic, political and social reasons prevail on legal considerations set by national and supranational regulations. The chapter shows that since groundwater became a key water resource for socioeconomic development in many areas in Spain, a wide array of measures has been applied to deal with the negative consequences of intensive use on the physical system. However, it argues that three issues have been constantly eluded: a) the recovery by groundwater users of the cost of measures to increase water availability in intensively used aquifers; b) the poor transparency and accountability in relation to groundwater uses and the cost of groundwater overdraft for the public budget; and c) the poor connection between land use and water policies, which creates a tension between attempts to limit further groundwater exploitation and the support to socio-economic development highly dependent on water. While there is no significant progress on those three aspects it is unlikely that the objective of good status of all waters will be achieved. The next section (2) presents an overview of groundwater resources in Spain according to the WFD baseline requirements, pointing to the main challenges that intensive groundwater development entails in terms of quantitative and chemical degradation of the resource. In Spain, two main instruments are envisaged to deal with the negative consequences of groundwater use: a strong – at least on paper – Water Act and the self-management of water users. Section 3 outlines Spain's legal framework for groundwater management and its main pitfalls, and provides an overview of the emergence of collective action by groundwater users, either for the initiative of the Water Administration or as a spontaneous bottom-up phenomenon. Section 4 presents four cases of intensive water use for agriculture in the southern half of Spain and highlights common trends in the evolution of public and user-driven strategies to deal with intensive water use. Section 5 concludes summarizing the main challenges ahead if the goals of the WFD have to be achieved.

2 GROUNDWATER RESOURCES AND THE WATER FRAMEWORK DIRECTIVE[1]

According to the classification defined by the WFD, in Spain there are 744 groundwater bodies[2] (GWBs) (Figure 1), covering almost two thirds of the country (about 355,000 km² of Spain's 505,992 km² total area). The estimated overall groundwater demand[3] is now 7000 hm³/year, which represents about 22% of Spain's total water demand (31,500 hm³/year), and an increase of about 27% relative to one decade ago (5500 hm³/year) (MMA, 2000). Over 80% of groundwater demand occurs in 7 of the 25 existing River Basin Districts (RBDs), most of which correspond to the semi-arid part of the country. In absolute terms, the RBD with the highest groundwater demand is Jucar (1600 hm³/year), followed by Douro and Guadalquivir with close to 1000 hm³/year, while estimated extractions are around 500 hm³/year in Segura, Guadiana, the Internal Basins of Catalonia and the Andalusian Mediterranean Basins. The highest shares (over 40%) of groundwater demand (relative to the RBD total) can be found in the Canary and Balearic Islands, Jucar and the Internal Basins of Catalonia and the Andalusian Mediterranean Basins.

Overall, agriculture is the main groundwater user in Spain (73%), followed by domestic water and industrial uses connected to the urban water supply network that represent 21% of the estimated groundwater demand. However, these figures range widely at the regional level. Groundwater plays a key role in the urban supply in some RBDs such as the Canary Islands, Balearic Islands, Jucar, the Internal Basins of Catalonia (where industrial use is especially important), and RBDs in northern Spain. Conversely, groundwater-based irrigation is small in the northern basins.

Using the available official data, De Stefano *et al.* (2012) estimated that the consumption of groundwater for irrigation is approximately 3200 hm³/year over an area of 10,000 km², which represents about one third of the total irrigated surface (33,000 km²) and one fourth of the total water demand (or approx. 12,000 hm³/year). The economic value of agricultural production using groundwater is about 4700 M€/year or 30% of the total value of Spain's irrigated crop production. While the reverse relationship – less economically productive and labour-intensive crops are associated to less degradation – still has to be explored, it can be said that the production of high value crops usually is associated to poor groundwater status. As it will be shown in Section 4, this has direct consequences on the viability of strategies aimed at managing intensive groundwater use.

The WFD requires that all the GWBs be in (at least) good status by 2015, although it is possible to request time extensions to the following planning horizons 2021 and 2027, or to set less stringent objectives for those GWBs where good status cannot be

1 This section draws from De Stefano *et al.* (2012; 2013).

2 A groundwater body includes one or several aquifers (or portions of them) whose waters have common characteristics and are confronted with similar challenges – either qualitative or quantitative.

3 We present data of *demand* and not actual *withdrawal* as no official data are available for the latter. Demand figures are estimated by the Water Authorties using a mix of different sources and methods (withdrawals estimated from crops requirements, registered water rights, etc.) For more on this see De Stefano *et al.* (2013).

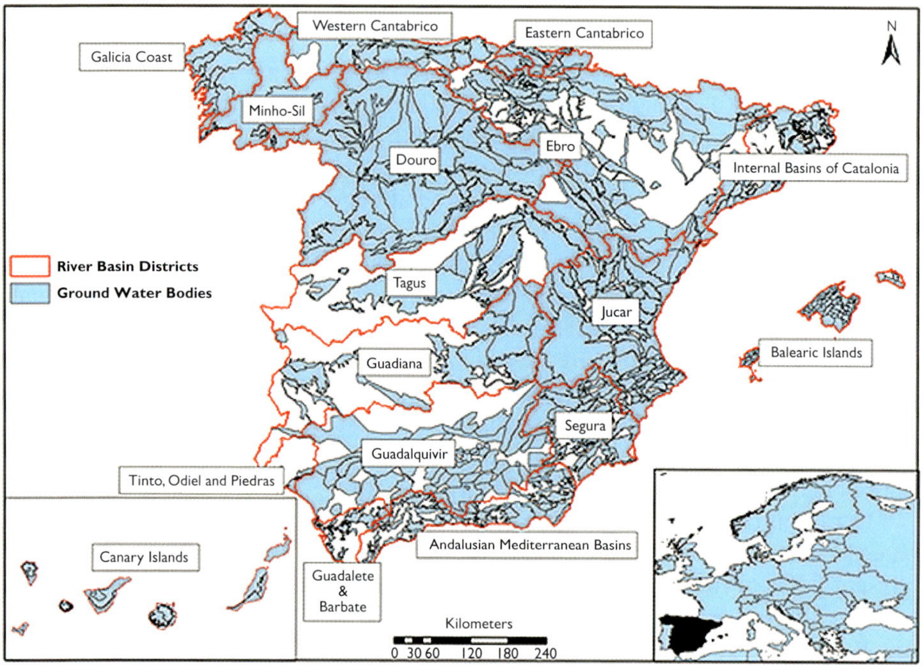

Figure 1 Groundwater bodies in Spain (Modified from De Stefano *et al.*, 2012).

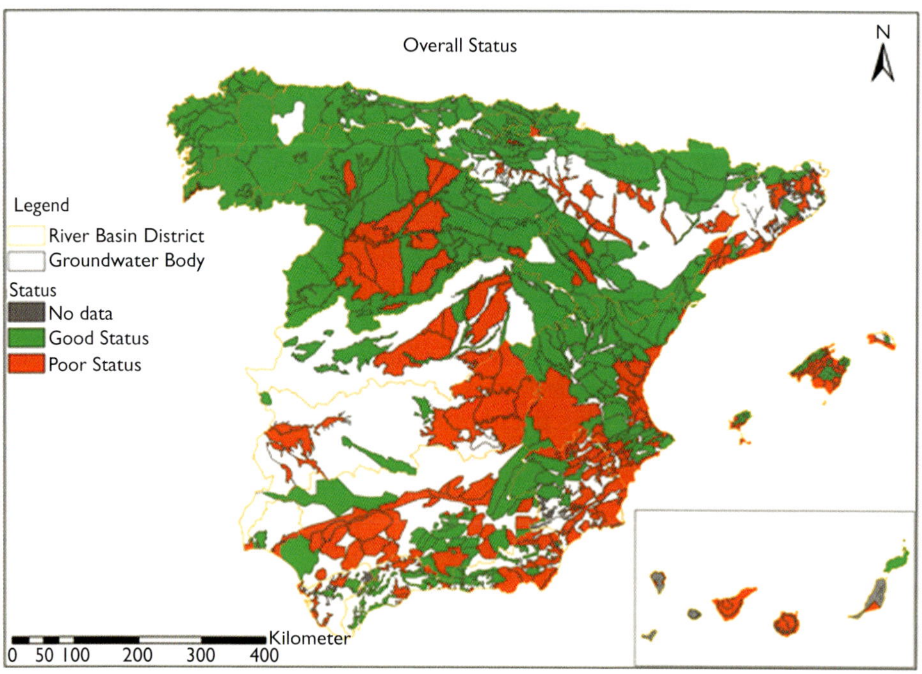

Figure 2 Current status of groundwater bodies in Spain (Modified from De Stefano *et al.*, 2013).

met without incurring disproportionate social or economic costs. Figure 2 shows a preliminary overview of the status of groundwater bodies according to the WFD documents made available in fall 2012, which state that 413 of the Spanish 744 groundwater bodies are currently in good status (55%), while 22 GWBs are still under study (3%).

Based on the baseline water status, the River Basin Management Plans (RBMPs) set the framework for implementing a Programme of Measures (PoM) to improve or maintain (in case of water bodies that are already in good status) the status of water bodies. According to the projections of the effects of the PoM, 590 groundwater bodies will meet the objective of good status by 2027 (80% of the GWBs), while less stringent objectives have been set in 33 GWBs (4%) (Figure 3). In the remaining 121 GWBs (16%) there are insufficient data to predict the achievement of good status by 2027. Pollution, mainly by nitrate, is the main cause of non-compliance with the objectives of good status: out of 309 GWBs currently in poor status, 273 (or 88%) do not comply with the required quality standards. Similarly, qualitative problems are the main reason for establishing less stringent objectives for 2027 in 33 GWBs.

These figures provide a first approximation to the challenge faced to meet the WFD objectives, and should be interpreted with caution for a number of reasons. First of all, some RBDs data is provisional because the final RBMPs were still in process of approval at the termination of this Chapter. Secondly, the number of water bodies in a certain status is an indicator with limited significance of the severity of the problem,

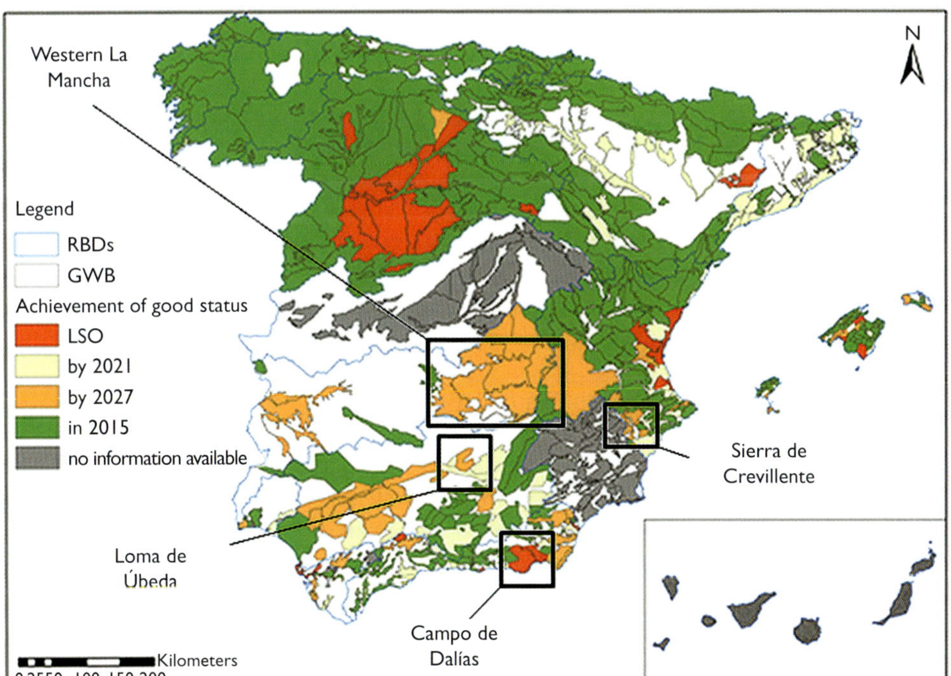

Figure 3 Expected year of achievement of good status by GWB. In Segura, Tagus and the Canary Islands data about the achievement were not available at the end of this study (own elaboration based on De Stefano *et al.*, 2013).

as there is no lower or upper limit to the size of the water body. Thus, if one uses the number of water bodies in poor status as an indicator of the challenge ahead, the level of fragmentation or aggregation of the water bodies may actually give a biased view of the magnitude of the problem. In practice, the obstacles to be overcome will be determined by the severity of the pollution (in terms of magnitude or nature); by how entrenched the problem of overdraft is; and by how water quality and quantity problems interact. Having in mind that the ultimate objective of this status classification is informing groundwater management decisions, its most important contribution is the spatial localization and the characterization of groundwater bodies, as that baseline will determine the battery of measures that should be planned and implemented within the River Basin Management Plans (RBMPs).

3 THE REALITY OF GROUNDWATER RESOURCES MANAGEMENT IN SPAIN

As explained in the Introduction, this section provides an overview of the legal context of groundwater development and its pitfalls, and of the emergence of collective action by users, which mostly represents a reaction of water users to the negative consequences of intensive groundwater use on the resource.

3.1 The administrative situation of groundwater

Spain's groundwater management issues are deeply rooted in the past. Since the first national-scale water law was passed in 1866, most surface waters have been considered public domain. This is a corollary of the idea that neither the market nor individuals can guarantee an adequate sharing of limited, first-need resources. Thus, the government is traditionally responsible for assigning, ordering and making water uses compatible (Del Saz *et al.*, 2002). These principles were not considered applicable to groundwater, which was excluded from public ownership from the very outset. This is largely a consequence of the relatively low importance of groundwater resources use at the time. Besides, quantitative hydrogeological methods were yet to become widespread and most groundwater-related issues were surrounded by a halo of mystery and uncertainty.

The 1866 and 1879 Spanish Water Laws, the latter valid until 1985, allowed for a dual ownership regime. Legal constraints on private ownership of groundwater were generally small and arose as a means to avoid damages to third parties. These included a minimal distance between wells, as well as other limitations designed to avoid interference with public surface waters and to guarantee the safety of buildings, railways and roads (Aragón, 2003).

In contrast, the 1985 Water Law placed groundwater under public ownership. This reform also created the Registry and the Catalogue as legal instruments to underpin groundwater management. Both legal instruments were designed as inventories to keep track of the ownership and characteristics of every well when the Water Law entered into force. All users were compelled to join one or the other within three years. Wells included in the Registry acquired the legal status of temporary private wells. In practice, this meant that private ownership would be respected for fifty years. After that time, ownership would be restored to the public domain. Owners would then be enabled to acquire a concession to keep using the water. Alternatively, well owners wishing to maintain private ownership could choose to apply for inclusion in the Catalogue of private waters. Finally, new groundwater uses established after January 1st 1986 had to be registered in the Registry as concessions for the use of water, the associated water resource being fully public. The hybrid water right system was devised to circumvent the potential compensation to existing water rights under the Spanish Constitution (which forbids expropriation of rights without compensation) and in practice it became a management challenge because of the co-existence of private and public rights (Lopez-Gunn *et al.*, 2012a).

The given three-year deadline and the legal advantages of the Registry had a two-fold aim. Firstly, to ensure diligence among applicants so that an inventory of wells could be compiled as soon as possible; and secondly, to encourage applicants to join the Registry instead of the Catalogue so that, ideally, all groundwater would be public domain within fifty years. Unfortunately, the system featured several loopholes. On one side, those failing to join the Registry within a three year period would still be in full possession of their rights, because inclusion in the Catalogue could not be established as a prerequisite for ownership. As a result, the desired effect was not achieved. Moreu (2002) estimates that only 10–20% of well owners joined the Registry in time. In contrast, the overwhelming majority maintained their private ownership for an indefinite period without even declaring their wells. This is a clear indicator that the reforms were poorly implemented in practice, and that users either did not really understand the importance of the new rules or deliberately decided to ignore them. In some cases, thousands of illegal wells were drilled overnight simply because new users ignored the fact that they had to ask for permission. In other cases, illegal drilling became rampant as soon as prospective users realized that water authorities lacked the expertise and means to exert control over aquifers (Moreu, 2002; Díaz-Mora, 2002; Yagüe *et al.*, 2003). The 1991–1995 drought catalyzed the drilling frenzy, while the EU Common Agricultural Policy fuelled intensive irrigation and thus proved counterproductive for the purpose of bringing illegal users under control (Martínez-Santos *et al.*, 2008).

Further reforms of the water law took place in 1999 and 2001, but these failed to tackle the groundwater problem. Thus, by the time the Water Framework Directive was transposed into national legislation in 2003, groundwater management was severely hampered by the existence of thousands illegal wells. In fact, Spain's White Book of Water described the situation to be 'very discouraging' in most regions (MMA, 2000). Prosecution attempts on the part of water authorities have found stiff opposition among influential farmer lobbies over the years, to the point that thousands of lawsuits are yet to be resolved across the country.

Overall, the legal situation of groundwater in Spain remains uncertain. The total number of wells is unknown, though it is known that new unlicensed boreholes are drilled every day. The path towards an effective management of groundwater goes through bringing water books up to date. On one hand this refers to registering existing rights, where it is estimated that in the case of groundwater only 30% of wells are registered. On the other hand it refers to the much trickier issue of unlicensed or informal water use, and how to find solutions to incorporate these into existing registries through negotiation, or if necessary have the institutional capacity (and support), to close all relevant abstractions to ensure that State capacity and legitimation is not undermined.

3.2 Users collective management as an emergent reality

Groundwater is considered as a classic example of a common pool resource and thus is historically linked with the relevance of collective action by users as a potentially valuable management model. Under this model users play a key role in the use of groundwater. Collective management is the most basic level of decentralized management, which places special emphasis on co-management between users and the authorities, and it is one of the basic measures to manage, regulate and control the use of groundwater resources. This is partly because one of the most important problems groundwater management is facing is related to asymmetric information between users and the regulatory agency. Often the latter – e.g. the River Basin Authority – faces high costs to obtain a reliable and updated information on the use of groundwater resources. Although this is changing, in part due to the use of remote sensing, it is estimated that in Spain the authorities have good information on a relatively small number of existing wells. Therefore, there is a problem of lack of information and a difficulty of controlling the actions of thousands of users and increasing evidence that collaborative and co-management approaches are necessary for common pool resource management such as groundwater (Blomquist, 1992; Berkes, 2009; López-Gunn *et al.*, 2012a; Plummer, 2009; Shah, 2009). However, these co-management approaches could not be sufficient unless they are supported by robust and clear legal frameworks and a strong implementation and monitoring regime on the part of responsible authorities.

A first step to understand and untangle the groundwater mismanagement is to analyze the emergence of collective action by groundwater users and the key factors for their success and failure in their efforts to manage groundwater resources. This is particularly useful to understand when it evolves beyond individual gain towards collective (and long-term) benefits. Although intuitively it is assumed that groundwater users would spontaneously organize once a common interest is detected, the theories of collective action (Olson, 1965) state that this is not necessarily the case. Spain, along with countries like Mexico and India, has gained valuable experience in a number of regulatory initiatives led by users (Shah, 2009). Well known worldwide for its ancient tradition of irrigation communities of surface irrigation (like the case of Herdedades in the Canary Islands), nearly 60% of Spain's irrigated land is in the hands of irrigation communities or other collective groups (Valero de Palma, 2011). In the case of groundwater, there is now a 40 years accumulated stock of experience from the oldest associations of groundwater users and new lessons from the recent emergence of organizations to manage non-conventional water resources, often closely correlated with groundwater-intensive use (Rica *et al.*, 2011). In general, the emergence and evolution of collective institutions have experienced three waves. The first wave refers to the long and well-documented history of surface water communities. The second wave refers to the institutions of groundwater users under different organizational types to face water quantity and quality issues. The third wave marks the emergence of user groups related to the use and operation of new water resources like desalinated, recycled, or recharged water made possible by advances in technology and knowledge.

The analysis of the second wave in collective groundwater institutions shows that these have been developed mainly through the initiative of users across a wide spectrum of organizational forms, both in the public and private sphere. The pioneering groundwater user organization was formed in 1976 in the Delta del Llobregat, when groundwater was still fully in the private domain, and since then there have been at least 19 other collective institutions. Many of these have emerged bottom-up, to defend their water rights either before or after the declaration of aquifer overexploitation under the Spanish 1985 water law. The declaration of aquifer overexploitation in theory introduces some punitive measures like closure of the groundwater resource for other users and potential restrictions on their water rights in the form of restrictions on their allocated water quotas. Yet the main effect was that the threat of declaration over-use mobilized farmers to protect their rights. In some cases like the Llobregat in Catalonia emergence of a water group however happened spontaneously and it is a clear example of strong collaborative management between users and the administration.

Fieldwork conducted by the Water Observatory for the period 2010–2012 confirms that most organizations emerged spontaneously due to the collective interest from users (Rica *et al.*, 2012), as a reaction to drought or, in most cases, to defend their private water rights when facing a potential declaration of overexploitation. Only a minority (3) were created under the direction of the State as a result of a declaration of overexploitation, which forces the constitution of a groundwater use group for aquifers declared overexploited. In terms of evolution however there are some relevant trends to observe for the coming years, including those related to changes coming from the implementation of the EU WFD, and the new requirement to establish 'groundwater bodies community groups' (CUMAS in Spanish).

Since about 50% of groundwater bodies in Spain have been declared as in poor status this potentially means the top-down creation of an equivalent number of groundwater body community groups (around 300 CUMAS). This raises some pertinent questions on a) the capacity of the State to support their creation, and then the operational viability of these CUMAS; b) what useful lessons can be learnt from the experience gathered so far from the handful of existing communities, which as explained earlier, have mainly developed bottom-up; and c) a final issue relates to what new experiences are being documented in recent years that could provide useful information for realizing the full potential of a successful co-management model.

In this context the most relevant signals refer to the potential to use other policies, like energy or rural development or trade (e.g. export cooperatives, etc.) which in synergy with water resources management, can provide a more robust basis for collective action. This idea comes from fieldwork, where the most obvious collective action was related to energy, with farmers working collectively to negotiate more favourable energy rates (the cases of Almeria or Crevillente discussed below highlight this). Moreover, there is an increasing blurring in terms of water 'sources', ranging from traditional surface water, groundwater and more recently non-conventional water like recharged, desalinated or re-used. This is now part and parcel of modern farm management with a portfolio of water sources designed to lower risk and which are blended to adjust the right water quality at the right time for specific crops.

4 THE REALITY OF GROUNDWATER RESOURCES MANAGEMENT IN SPAIN

This section outlines four case studies where intensive use of groundwater for irrigation has resulted in important socio-economic development but also in the drop of groundwater levels and environmental degradation. They showcase the difficulties of a command-and-control approach solely based on the enforcement of the 1985 Water Act and how solutions envisaged so far by water users and the Water Authorities do contribute to improve the water supply guarantee for irrigation but not to revert groundwater degradation trends. In all these cases, the issue of how to reduce abstractions while ensuring the socioeconomic viability of the existing system which is heavily dependent on groundwater is a still unsolved challenge.

4.1 Sierra de Crevillente

The Sierra de Crevillente aquifer is a small (100 km^2) calcareous unit located in the Jucar basin. The Jucar is Spain's main groundwater basin (43,000 km^2) (see its location in Figure 3) and it is home to 4.5 million people, peaking seasonally at nearly 6 million due to tourism. Irrigation accounts for 80% of the total water uses, with a total irrigated surface of 357,000 ha. Out of these, 200,000 ha rely exclusively and 26,000 ha partially on groundwater (JRBA, 2004). Many examples of intensive groundwater use may be found within the Jucar basin. These are particularly frequent towards the south, around the Vinalopó valley. Dry climatic conditions, together with a geological context with a predominance of permeable terrain and a topography that does not allow the construction of reservoirs, have led water users to rely almost entirely on aquifers. Table 1 shows the degree of exploitation (extraction/aquifer recharge) of some of the main aquifers in the Vinalopó valley.

In the Sierra de Crevillente aquifer, intensive groundwater agriculture, supplied exclusively through private wells, has taken place since the 1960s. Abstractions

Table 1 Groundwater extraction as a percentage of recharge in selected aquifers of the Vinalopó area (Aguilera & Murillo 2009).

Aquifer	*Extraction as % of recharge*
Serral-Salinas	571
Madara	110
Umbría	229
Argallet	123
Sierra de Crevillente	601
Jumilla-Villena	153
Solana	212
Peñarrubia	99
Argueña	192
Sierra del Cid	113
Tibi	106
Ventós-Castellar	103

currently amount to 16 hm³/year, very much in excess of the aquifer's estimated 0.5–2 hm³/year renewable resources (Bru, 1993; Corchón, 2004).

Pumping is used to irrigate high-value exportable crops, i.e. table grapes, produced with high-efficiency water technology. At an average irrigation dose of 3500 m³/ha/year, these crops used to yield an approximate value of 25,000 €/ha in the past, demanding more than 100 man-day/ha of labor. Over the years, heavy groundwater extraction has caused the water table (Figure 4) to drop at an average rate of 15 m/year in certain areas, at times plummeting 30–40 m within a single year (Pulido-Bosch *et al.*, 1995). Unusual high rainfall led to a certain recovery in the late 1980s and early 1990s, thus alleviating the downward piezometric trend for some years. Currently, water is pumped from depths of 450 to 500 m in some sectors of the aquifer. In addition, intensive exploitation has shown that the aquifer is divided in five apparently independent units, which have experienced different degrees of quality degradation (Andreu *et al.*, 2002). There is also an economic reading: pumping costs for irrigation in the Crevillente aquifer have increased to 0.29 €/m³, well above Spain's 0.12 €/m³ average (DGOH, 2003). Besides, groundwater salinization has had a negative impact upon soil fertility. In consequence, crop value has gradually dropped from about 25,000 €/ha to 12,000 €/ha in the recent past. This means that the cost of irrigation water (about 1000 €/ha/year), has increased proportionally from under 5% to almost 10% of the total crop value.

Despite the spectacular nature of these figures, a world record in their own right, the social and economic system has not collapsed. This experience explains how groundwater remains relatively cheap even in extreme cases of aquifer depletion, and

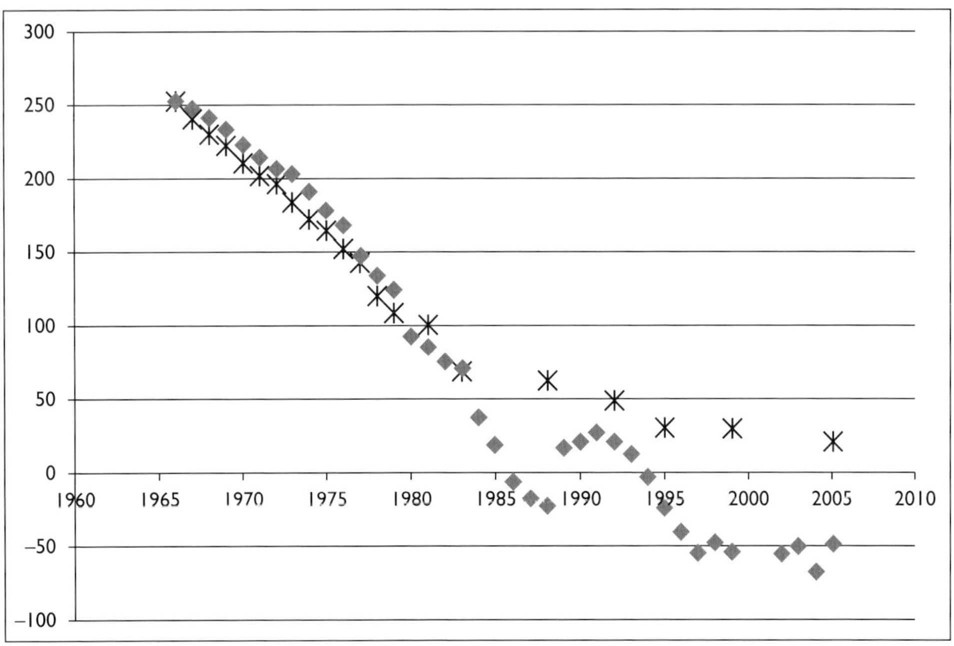

Figure 4 Evolution of piezometric levels in two representative boreholes of Sierra de Crevillente (own elaboration with data from Andreu *et al.*, 2002).

consequently, why dropping water tables do not usually represent an issue of concern for farmers (Llamas & Martínez-Santos, 2005; Garrido *et al.*, 2006). It is true, however, that the loss of crop value associated to water quality issues has become increasingly pressing, and poses a significant mid-term threat to the feasibility of certain crops. This is the reason why Crevillente farmers, together with farmers from small nearby aquifers and urban water supply companies, have managed to obtain the approval of a surface water transfer from the Jucar river. The feasibility of this transfer has been the source of bitter conflicts between scholars, environmental conservation lobbies, politicians and farmer associations, but the infrastructure has been finally built and is currently undergoing a testing phase. Initially the cost of the infrastructure (230 M€ according to the original project) had to be shared among the Spanish Government (32.61%), EU development fund FEDER (34.78%) and the users (32.61%). After the start of the works, however, the users withdraw their co-funding commitment, which meant that eventually the water infrastructure was fully funded with public funds (320 M€, co-founded by the Spanish government and the EU FEDER fund). The reason for the withdrawal of the farmers was that the Government moved the location of the transfer intake downstream in the river to address environmental concerns in the donor area. This change implies that: a. the transferred water will have a lower quality (relative to the original project); b. the urban supply companies in the recipient area will not be interested in buying the transferred water; and c. the cost of elevating water from the origin to the recipient areas will be higher than in the original project (Ortiz & Melgarejo, 2010).

4.2 The Almeria region and the case of Campo de Dalias *plasticulture*

The Almeria province which is located in the south-east of Spain despite its desert like conditions, holds the most productive agricultural province of Spain with a number of agribusinesses that are leaders in their sector. An area of about 28,000 ha of greenhouses, one of the largest in the world for cultivation under plastic or *plasticulture*, highlights the relevance of the sector at national and even at European level. In less than half a century, the region has gone from being one of the poorest regions to become one of the wealthiest in Spain, thanks to the intensive use of scarce groundwater resources and a large deal of human ingenuity. It is a good example of the different policy experiments to deal with the chaos ensued from such intensive groundwater use, in terms of clear environmental externalities like groundwater pollution, saltwater intrusion and the ensuing consequences (Lopez-Gunn *et al.*, 2012a).

The region's groundwater-based economy rotates around the intensive use of three aquifers: Campo de Dalias, Medio-Bajo Andarax and Campo de Nijar. The Campo de Dalias has an area of approximately 800 km^2 with estimated resources of 60 hm^3/yr from the Beninar reservoir (of which only a small proportion is used) and 80 hm^3/yr from the Dalias aquifer. Yet the estimated level of abstractions, around 140 hm^3/yr, almost doubles the estimated renewable reserves. In this area, groundwater has a complex management system, where farmers contribute financial resources or manpower to build a common well which provides water to different plots. Many farmers are organized around Agricultural Transformation Societies, an associative formal body based on shared infrastructure and water use from a well to regularize the situation of the distribution of water and land in private law.

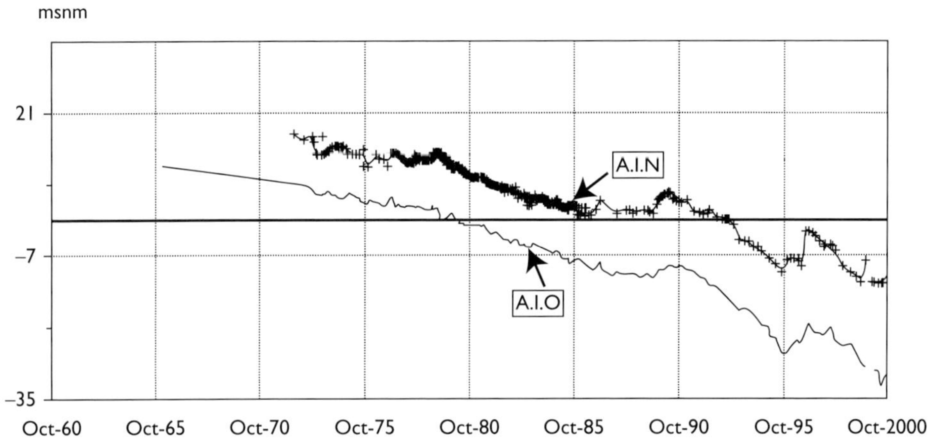

Figure 5 Evolution of piezometric level in representative boreholes in Campo de Dalías aquifer (Modified from Domínguez Prats *et al.*, 2001).

Such intensive groundwater use over a period of forty years has resulted in some noticeable environmental impacts. Water overexploitation is currently located in the lower aquifer which has better water quality and higher storage capacity, while upper layers are no longer being used, causing waterlogging as these layers are recovering their recharge balance. Groundwater wells reach depths of 300 m with pumping costs estimated at 0.13 €/m³ to 0.19 €/m³ (Martínez, 2011). The main strategy adopted to deal with hitting resource boundaries has been to complement existing groundwater resources with additional non-conventional resources. Moreover, there is a plan to diversify the origin of the water, giving particular attention to the nearby reservoir-Benínar, which provides about 3–6 hm³/year, the de-brackishing of the upper aquifer in an emerged wetland for 2 hm³/year, the reuse of wastewater from the main cities for up to about 10 hm³/year and desalination for up to 30 hm³/year. Additionally, water users maintain their claim for surface water from the Ebro river, in the northeast of Spain, which was foreseen in the 2001 Spanish National Hydrological Plan and was revoked in 2004 due to the stiff opposition of the ceding regions. The second main strategy (currently under way) has been based on devising collective agreements on the basis of the existing groundwater groups which exist in all three aquifers. On the back of these collective agreements, farmers are making a more efficient use of groundwater resources combined with other resources like surface water. Furthermore, farmers are now considering other options like desalinated water to mix with groundwater, thus reducing overall abstractions from the aquifer.

4.3 The Upper Guadiana basin: areas of high natural value enter into conflict with irrigation

The Upper Guadiana basin (16,000 km²) is located in central Spain and hosts a series of interconnected aquifers that stand out in three respects: first because of the associated dramatic socio-economic development in the area linked to the conversion from dryland agriculture to irrigated agriculture and second due to the remarkable set of wetlands which constitute the Mancha Humeda natural region, many of which were traditionally groundwater-fed. Third, it is a well-known case globally because it is a remarkable example of so-called 'wicked' policy problems (Rittel & Webber, 1973), i.e. problems that are highly complex, with clusters of interrelated and dynamic interactions that have high levels of uncertainty. It is one of the most obvious examples on the difficulties in taming groundwater chaos in terms of control, a case with many production and consumption externalities (in this case wetland destruction and groundwater level drawdown); and its recent recovery (see below) provides an example as to how nature may plot to puzzle engineers, farmers, managers, environmentalists and academics.

During the 1970s public policies encouraged farmers to tap the 'sea' beneath their feet and a series of soft loans were granted to drill into the Upper Guadiana aquifers. This triggered a remarkable transformation in terms of agricultural models and the deterioration of the existing natural capital (Lopez-Gunn *et al.*, 2011). This meant the economic take off for a region that historically had been fairly poor and disadvantaged. For many years, socio-economic development took place at the expense of the Mancha Húmeda Biosphere reserve, which originally comprised a series of wetlands including Las Tablas de Daimiel National Park. Intensive pumping depleted the aquifer to the point that, by the turn of the century, only 20% of the original area in the eco-region remained funcional (de la Hera, 1998).

There were a number of public policy initiatives to deal with the conflict between intensive groundwater use and the preservation of groundwater-dependent wetlands. In the case of Western Mancha the main initiatives were driven through a series of public policy initiatives, starting with the declaration of aquifer overexploitation. This was followed by a battery of engineering measures, which included drilling a series of 'environmental wells' within the national park in order to pump groundwater to put it into the wetland and the implementation of a water transfer from a neighbouring basin. Structural policy measures were also adopted to address the imbalance between resources and demand. The first attempt (1993–2003), funded by the EU, was based on compensation payments for farmers to cut down on irrigation. This was followed by the Upper Guadiana Water Plan (2008–2012) which aimed at purchasing water rights and stopping illegal water use.

As pointed out by different authors, both initiatives were hampered by poor monitoring, inefficient sanctioning regimes and political constrains (Martinez-Santos *et al.*, 2008). In the more recent years, funding for the Upper Guadiana Water Plan has also been significantly reduced as a result of the economic crisis. However, the aquifer has experienced a spectacular recovery since 2009 (Figure 6). This can be largely attributed to the unusually wet sequence that took place between 2006 and 2010, where rainfall was repeatedly concentrated in extreme events leading to significant recharge episodes. Increased rainfall also limited the need for groundwater abstraction, thus contributing to a dramatic rise of the water table. Some of the groundwater-dependent ecosystems, including springs and wetlands, have been observed to recover over the last two years.

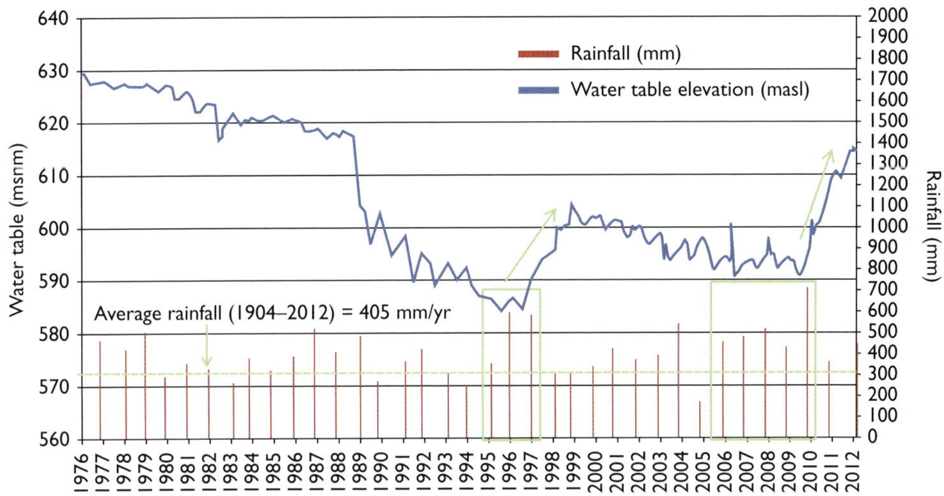

Figure 6 Evolution of piezometric level in a representative borehole in the Western La Mancha aquifer (own elaboration).

4.4 The Loma de Úbeda aquifers

La Loma of Úbeda is an area located in the Andalusian province of Jaén, in the Upper Guadalquivir, where people traditionally practiced extensive agriculture, alternating dry land olive groves with other crops such as vine and cereals. However, over the last 30 years the agricultural area has experienced a dramatic transformation in two ways; first through the shift to a monoculture agricultural model of intensive olive grove cultivation and second, from dryland farming to an intensively irrigated, highly efficient agricultural model through drip and subsurface irrigation of olive groves. This trend in olive grove production can be seen all around the Guadalquivir basin, where olive groves occupy almost 60% of the basin agricultural surface (Rica *et al.*, in press). Between 1997 and 2008, the cultivated area increased by 50%, mainly with irrigated olive groves (Salmoral *et al.*, 2011). In the province of Jaén alone, the olive area in 2007 represented almost 42% of the Andalusian total, of which 30% was irrigated (Sánchez Martínez *et al.*, 2011). Thus, due to the sheer extension of olive grove production, the economy of the area is heavily linked and dependent on olive markets. La Loma itself thus has agriculture as the main source of activity, with approximately 40,000 ha of olive fields.

This spectacular development was made possible by the intensive drilling for private initiative – and in most of the cases without the mandatory license – from two different superimposed aquifers: the tertiary Miocene aquifer (127 km^2) and the Jurassic carbonate aquifer of 626 km^2 (Gollonet *et al.*, 2002). Overall, the Jurassic and Miocene aquifers show an exploitation rate which is unsustainable since CHG (2012) estimates abstractions of around 80 hm^3/yr, versus the 58 hm^3/yr of renewable water resources. While these aquifers are widely exploited, there is little information from the Water Authority on the dynamics and piezometric evolution (CHG, 2012).

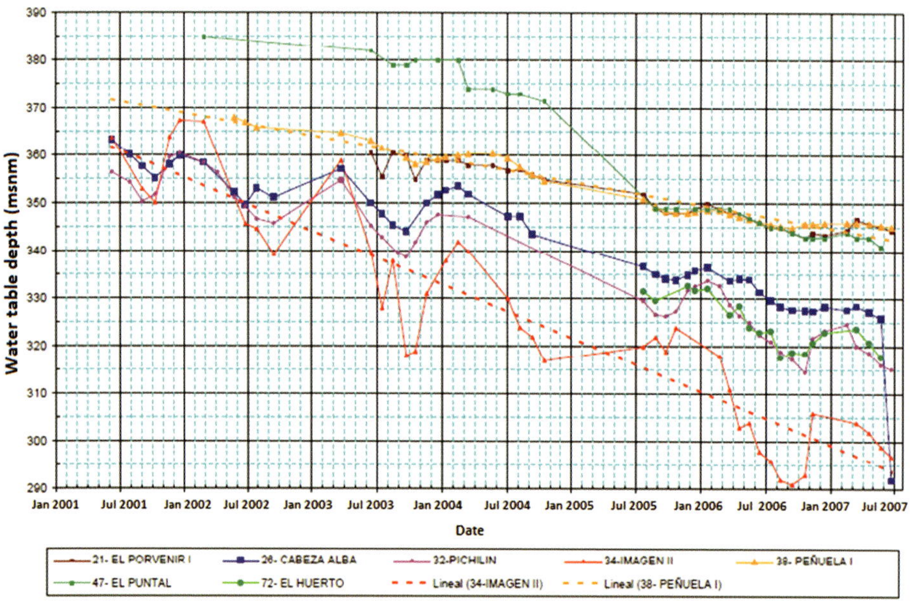

Figure 7 Evolution of piezometric level in representative boreholes of the Jurassic carbonate confined aquifer (Modified from IGME, 2007).

It is known however that in some areas deep wells have reached and are abstracting fossil water that is estimated to be 25,000 years old (Nuñez *et al.*, 2006).

The growth of groundwater irrigation was triggered by a severe drought in 1991–1995, when farmers in La Loma started drilling boreholes to save their crops. Rivers like the Guadalquivir and Guadalimar which flow close to this area, were out of bounds for Jaén irrigators since surface water was already committed to other uses further downstream. The high productivity of irrigated olive trees and the EU Common Agrarian Policies which subsidized olive production made that over the years more and more farmers moved to irrigated olive trees farming using groundwater. Therefore the irrigated surface and water demand have increased notably over the last 15 years: in 1999 there were approximately 12,000 hectares of olive tree groves with an approximate consumption of 23 hm^3/yr. In only three years, by 2002, it is estimated that the area had doubled water consumption with 25,000 ha consuming over 46 hm^3/yr (WWF/Adena, 2006). Currently (2012) the situation is stable with approximately 40,000 ha and 80 hm^3/year of groundwater abstractions (CHG, 2012). The new Guadalquivir RBMP (CHG, 2012) has established the available resource at 46.08 hm^3/yr that is half of the current level of abstractions. Moreover, the dropping of the water levels has implied an increase in the pumping costs: based on fieldwork data, it is estimated that the average pumping cost is 0.20 €/m^3, with an average water use of 1500 m^3/ha/yr; thus, the total energy cost would be around 300 €/ha/yr, representing around 13% of total farm operational costs.

The Guadalquivir River Basin Authority reacted to the uncontrolled water development by opening sanctioning procedures to unlicensed water abstraction and by studying the aquifer functioning and renewable resources. In the meanwhile, farmers are associating to negotiate the legalization of their wells, a cumbersome goal in a basin with an imbalance between demand and supply estimated to exceed 562 hm^3 (CHG, 2012). Another consequence of intensive groundwater use has been a new claim by farmers in La Loma to have access rights to uptake water from the Gualdaquivir and Guadalimar surface water flows in winter. This request, however, is questioned for its impact on the ecological and chemical status of these rivers and on historic downstream water rights.

4.5 Common trends and strategies for managing intensive groundwater use

The comparison of these four case studies allows for the identification of common features (Table 2) and similar trends in the evolution of the strategies to deal with intensive groundwater use, which can be divided into four main phases (Figure 8).

At the beginning of the emergency of groundwater use, the number of wells grew in a paced manner and out of the sight of the competent Water Authorities, who traditionally have been focused mainly on surface water (Phase 1: 'Silent revolution'). Different, sometimes concomitant drivers – the improvement of the drilling technology, EU subsidies supporting irrigated crops, two severe droughts in 1991–95 and 2005–08, the approval of the 1985 Water Act -, led to a frenzy drilling activity, which in turn provoked a sharp increase in the level of groundwater abstraction and the subsequent drop in the water tables.

The Water Authorities reacted to the uncontrolled development using the legal instruments available in the Water Act (Phase 2: 'Authorities' initiative'), but their

Table 2 Comparison of key characteristics of intensive groundwater use in the considered case studies.

	Sierra de Crevillente	*Western La Mancha*	*Campo de Dalías*	*La Loma de Úbeda*
Water quality degradation	✓	✓	✓	✓
Efficient water use	✓	✓	✓	✓
Conjunctive water use	✓	✗	✓	Requested
External resources	✓	✓	Requested	✗
Unconventional resources	?	✗	✓	✗
Users' collective action	✓	✓	✓	✓
Legal pitfalls	?	✓	✓	✓
Recovery of the costs of intensive use	✗	✗	✗	✗
Land use policy favoring irrigation	✓	✓	✓	✓

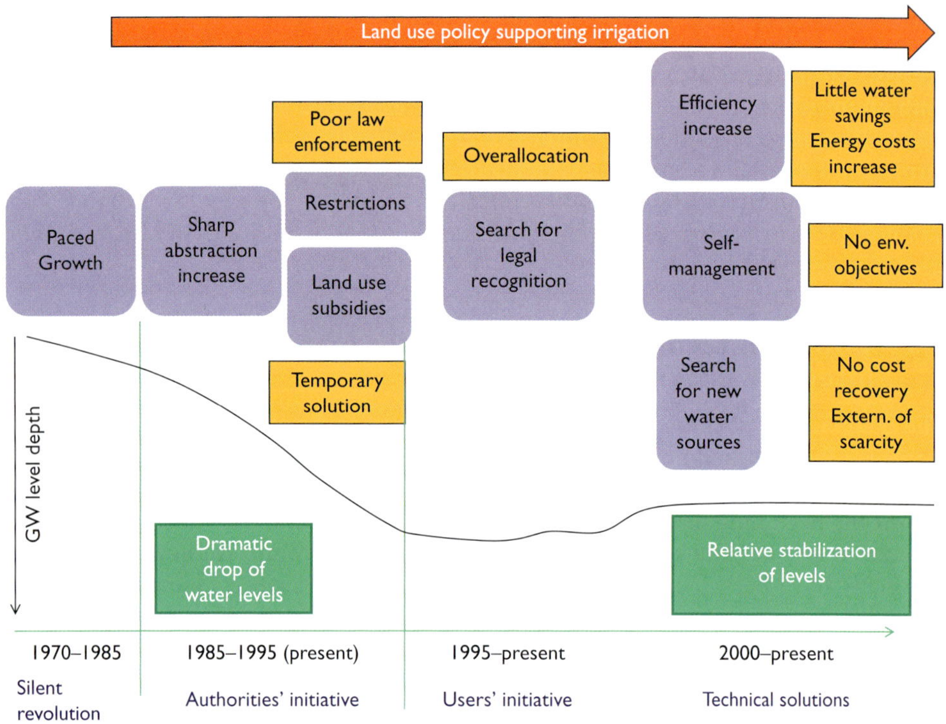

Figure 8 Evolution of intensive groundwater use in agriculture (own elaboration).

initiative was hampered by their limited resources and, due to the economic development brought about by irrigation, by the lack of political backup to take unpopular decisions needed to curb illegal groundwater use. In fact, regional authorities holding competences over agriculture maintained their support to the consolidation and expansion of irrigation. In some cases (e.g. Western La Mancha), the Water Authorities counted also on economic incentives to temporarily reduce the irrigated areas but these measures did not manage to change the crops patterns and therefore had little effectiveness (Martínez-Santos *et al.*, 2008).

Water users soon became aware that their thriving irrigation activity was threatened by the Water Authorities' actions trying to put order in the groundwater abstraction through a command and control approach, and by the degradation of the groundwater resource due to the plummeting of the water table and the decrease in water quality. Thus, they started organizing themselves (Phase 3: 'Users' initiative'). This initiative aimed mainly at: protecting their economic activities from the Water Authorities' legal actions and lobbying for their legal recognition as water users; lobbying for new water resources to complement the shrinking or increasingly more expensive groundwater resources (due to pumping costs); lobbying for economic support to increase water use efficiency; and the creation of a critical mass for optimizing their access to water, e.g. through common wells, the creation of irrigators' advisory services or the peer-control of agreed abstraction rates. This phase has allowed for the progressive regularization of part of the unlicensed groundwater development, which on the one hand means the legitimization of an activity that is providing important incomes for some rural areas, but on the other hand entails granting additional 75-year long water rights in aquifers that are already under severe pressure.

The following phase (Phase 4: 'Technical solutions') includes the implementation of the policies envisaged in the previous phase, as both the Authorities and the users focus their efforts on ensuring the viability of the irrigated hectares that already exist at the lowest economic cost for water users. This includes state-subsidized programs for the modernization of the irrigation systems and the search for additional water resources to complement groundwater resources and ensure the access to cheap (for the users) and good quality water. Depending on the area, this can be surface water transfer from other basins (e.g. Western La Mancha, Almería, Sierra de Crevillente), desalinated water (Almería), treated wastewater (Almería) or conjunctive use with additional surface water (Úbeda, Almería). At a national level, the exemplification of this approach are several large-scale public investment programs: the 2001 Spanish National Hydrological Plan, whose cornerstone was a 900 km long water scheme to transfer water from the Ebro river to several areas along the Mediterranean coast; the 2004 AGUA programme (8 billion €), which foresaw the construction of over 34 desalination plants along the Mediterranean coast; and the 2005 National Irrigation Modernization Plan – 7 billion € – for the modernization of over 1,1 million ha (MAPA, 2001; Lopez-Gunn *et al.*, 2012b). The implemented solutions, however, have some pitfalls: the increase in water efficiency has been at the expense of a sharp increase in the energy costs of irrigation[4] and actually hasn't implied a decrease in water consumption levels (Lecina *et al.*, 2009). Substituting or complementing groundwater abstraction with surface water resources – from other basins or from already fully allocated rivers – in practice means 'exporting' water scarcity problems to other water bodies. Since the beginning of the exploitation of groundwater resources, public land use policies have maintained their firm support to irrigation, first through direct subsidies to 'thirsty' crops and then by subsiding infrastructure to consolidate existing irrigation.

During the last phase, the case studies have experienced a decrease in the pace of groundwater development and, in some cases, even a partial recovery of the water table levels. The reasons for this stabilization possibly are a combination of different concomitant factors: after the end of the 2005–2008 drought, these areas experienced an exceptionally wet period, which caused spectacular recoveries in calcareous aquifers; farmers have hit the ceiling in their capacity to sell their products in a competitive way and, in some cases, perceive the threat of competing non-European areas with lower production costs; and being aware that they have already reached and overpassed the boundaries of the available resources – in terms of quantity and quality –

4 A study comparing water consumption of five modernized irrigation districts in Andalusia showed that between 2001/2002 and 2010/2011 the energy costs had increased by 227% and that the progressive substitution of seasonal crops with more profitable permanent corps could imply a 18% increase of water consumption in the next decade (Montesinos, 2012).

encourages water users to focus on consolidating the already existing activity instead of further increasing the pressure on the resource.

5 CONCLUDING REMARKS

Perhaps one of the main aspects that constrain the implementation of the Water Framework Directive is the fact that many of those decisions that affect water use are made outside the water planning sphere. This, coupled with the practical difficulties that often hamper groundwater management, results in challenges which are not only technical, but also social and economic. Within this context, changing things without changing land use represents a major issue. Water Authorities have little control on the catalysts for intensive groundwater use, including land use. In fact, those Authorities which control land use do not necessarily see water as a priority. Besides, users will not give up their current benefits, even if they are interested in ensuring the quantity and quality of the resource at least to a level that makes it useful for their activity.

ACKNOWLEDGMENTS

The authors would like to thank Dr. Juan Fornés, from the Geological Survey of Spain and Prof. Emilio Custodio for their input to the chapter.

REFERENCES

Aguilera, H. & Murillo, J.M. (2009) The effect of possible climate change on natural groundwater recharge based on a simple model: A study of four karstic aquifers in SE Spain. *Environmental Geology*, 57(5), 963–974.

Andreu, J.M., Estévez, A., García-Sánchez, E. & Pulido-Bosch, A. (2002) Caracterización de la explotación en el sector occidental del acuífero de Crevillente (Alicante) *Geogaceta*, 31, 59–62.

Aragón, J.R. (2003) Hidromitos de la cuenca alta del Guadiana. In: Coleto, C., Martínez Cortina, L. & Llamas, M.R. (eds.) *Conflictos entre el desarrollo de las aguas subterráneas y la conservación de los humedales: La cuenca alta del Guadiana*. Fundación Marcelino Botín. Ediciones Mundi-Prensa. Madrid. pp. 69–101.

Berkes, F. (2009) Evolution of co-management: Role of knowledge generation, bridging organizations and social learning. *Journal of Environmental Management*, 90, 1692–1702.

Blomquist, W. (1992) *Dividing the waters: Governing groundwater in Southern California*. San Francisco, Institute for Contemporary Studies Press.

Bru, C. (1993) *Los recursos del agua: Aprovechamiento y economía en la provincia de Alicante*. Fundación Cultural Caja de Ahorros del Mediterráneo.

CHG (2012) Recursos de la MAS 05.23 Úbeda. *Propuesta del Plan hidrológico de la cuenca del Guadalquivir*. [Online] Documento para la consulta pública. Available from: www.chguadalquivir.es [Accessed 13 December 2012].

Corchón (2004) *Jucar River Basin Authority*. Written Communication March 2004.

Custodio, E. (2002) Aquifer overexploitation: what does it mean? *Hydrogeol. J.*, 10, 254–277.

De Stefano, L. & López-Gunn, E. (2012) Unauthorized groundwater use: Institutional, social and ethical considerations. *Water Policy*, 14, 147–160.

De Stefano, L., Martínez-Cortina, L. & Chico, D. (2012) An overview of groundwater resources in Spain. In: De Stefano, L., Llamas, M.R (eds) *Water, agriculture and the environment in Spain: Can we square the circle?* Leiden, Taylor & Francis. pp. 87–104.

De Stefano, L., Martínez-Santos, P., Villarroya, F., Chico, D. & Martínez-Cortina, L. (2013) Easier Said Than Done? The Establishment of Baseline Groundwater Conditions for the Implementation of the Water Framework Directive in Spain. *Water Resources Management*, 27(7), 2691–2707.

De la Hera, Á. (1998) *Análisis hidrológico de los humedales de la "Mancha Húmeda" y 530 propuesta de restauración de un humedal ribereño: El Vadancho (Toledo).* PhD dissertation. Dept. of Geodynamics, Geology Faculty, Universidad Complutense de Madrid.

Del Saz, S., Fornés, J. & Llamas, M.R. (eds.) (2002) *Régimen jurídico de las aguas subterráneas.* Madrid, Fundación Marcelino Botín and Ediciones Mundi-Prensa.

DGOH (2003) *Valoración del coste de uso de las aguas subterráneas en España.* Technical Report (Draft). Madrid, Dirección General de Obras Hidráulicas.

Domínguez Prats, P., Franqueza Montes, P.A. & González, A. (2001) El conocimiento de los acuíferos del Campo de Dalías y su implicación en la gestión sostenible integral de los mismos. *V Simposio sobre el agua en Andalucía, 25–28 September 2001, Almería.*

Díaz Mora, J. (2002) La clarificación jurídica de los acuíferos sobreexplotados. El caso de La Mancha. In: Del Saz, S., Fornés J.M., & Llamas, M.R. (eds) *Régimen jurídico de las aguas subterráneas.* Madrid, Fundación Marcelino Botín y Ediciones Mundi-Prensa.

Garrido, A., Martínez-Santos, P. & M.R. Llamas (2006) Groundwater irrigation and its implications for water policy in semiarid countries: The Spanish experience. *Hydrogeology Journal,* 14(3), 340–349.

Giordano, M. (2009) *Global Groundwater? Issues and Solutions.* Annual Review of Environment and Resources, 34, 153–178.

Gollonet, J., Rubio, J.C., Martín, C., González-Ramón, A. & Peinado, T. (2002) *Los acuíferos de la Loma de Úbeda, reserva estratégica para el desarrollo del olivar.* Presente y futuro de las aguas subterráneas en la provincia de Jaén. Madrid, IGME.

IGME (2007) *Seguimiento de redes de control piezométrico en el acuífero Carbonatado profundo de La Loma de Úbeda. Jaén.* Informe Final. Julio 2007.

JRBA, Jucar River Basin Authority (2004) *Jucar Pilot River Basin. Report Pursuant to the Water Framework Directive.* Spain's Ministry of the Environment, Spain.

Lecina, S., Isidoro, D., Playán E. & Aragüés R. (2009) *Efecto de la modernización de regadíos sobre la cantidad y la calidad de las aguas: La Cuenca del Ebro como caso de estudio.* Monografías INIA Serie Agrícola. Report number: 26. MAPA (Ministerio de Agricultura, Pesca y Alimentacion) (2001).

Llamas, M.R. & Martínez-Santos, P. (2005) Intensive groundwater use: silent revolution and potential source of social conflict. Guest Editorial. *Journal of Water Resources Planning and Management* 131(5), 337–341.

Lopez-Gunn, E., Zorrilla, P. & Llamas, M.R. (2011) *The Impossible dream The Upper Guadiana system: Aligning Changes in Ecological Systems with Changes in Social System.* On the Water Front. SIWI Special Publication.

López-Gunn, E., Rica, M. & van Cauwenbergh, N. (2012a) Taming the groundwater chaos. In: De Stefano, L. & Llamas, M.R. (eds). *Water, Agriculture and the Environment: can we square the circle?* Taylor and Francis. pp. 227–241.

Lopez-Gunn, E., Zorrilla, P., Prieto & Llamas, M.R. (2012b) Lost in translation? Water efficiency and productivity in Spain. *Journal of Agricultural Water management*, 108, 83–95.

MAPA, Ministerio de Agricultura, Pesca y Alimentación (2001) *Plan Nacional de Regadios Horizonte 2008*. Madrid.

Martínez-Santos, P., De Stefano, L., Martínez-Alfaro, P.E. & Llamas, M.R. (2008) Wetland restoration in the Mancha Occidental aquifer, Spain: a critical perspective on water, agricultural and environmental policies. *Restor. Ecol.*, 16(3), 511–521.

Martínez, F. (2011) *Estudio de la huella energética del abastecimiento urbano de agua de la provincia de Almería (Study of the energy footprint of urban water supply in the province of Almería)*. Master Research on Water and Environment in semiarid area (AQUARID). Department of Hydrogeology and Analytical Chemistry, University of Almería, Spain.

MMA (2000) *Libro Blanco del Agua en España (White Book of Water in Spain)*. Spain, Secretaría de Estado de Aguas y Costas, Dirección General de Obras Hidráulicas y Calidad de las Aguas, Ministerio de Medio Ambiente.

Montesinos, P. (2012) La modernización del regadío en Andalucía disminuye un 20% el consumo de agua pero incrementa un 227% los costes energéticos. [Online] Available from: http://www.iagua.es/noticias/agricultura/12/10/30/la-modernizacion-del-regadio-en-andalucia-disminuye-un-20-el-consumo-de-agua-pero-incrementa-un-22 [Accessed July 2013].

Moreu, J.L. (2002) Los problemas de la legislación sobre aguas subterráneas en España: posibles solucio-nes. In: Del Saz, S., Fornés, J.M. & Llamas, M.R. (eds.) *Régimen jurídico de las aguas subterráneas*. Madrid, Fundación Marcelino Botín y Ediciones Mundi-Prensa.

Nuñez, I., Araguás, L., Moreno, L., Rubio, J.C., González-Ramón, A., Pérez-Zabaleta, E. & Gollonet, J. (2006) Evolución hidrogeoquímica del Acuífero carbonatado profundo de la Loma de Úbeda (Jaén). *Boletín del IGME*, 117, 613–616.

Olson, M. (1965) *The Logic of Collective Action: Public Goods and the Theory of Groups*, Harvard University Press.

Ortiz, I. & Melgarejo, P. (2010) *El trasvase Júcar-Vinalopó*. Una respuesta a la sobreexplotación de acuíferos. Investigaciones Geográficas, number: 51 (enero-abril 2010).

Plummer, R. (2009) The adaptive co-management process: an initial synthesis of representative models and influential variables. *Ecology and Society*, 14(2), Article number: 24. [Online] Available from: http://www.ecologyandsociety.org/vol14/iss2/art24/ [Accessed 14 June 2013].

Pulido-Bosch, A., Morell, I. & Andreu, J.M. (1995) Hydrogeochemical effects of groundwater mining of the Sierra de Crevillente aquifer (Alicante, Spain). *Env. Geol.*, 26, 232–239.

Rica, M., Lopez-Gunn, E. & Llamas, M.R. (2011) The edge of scarcity: Institutional Diversity and incentives in collective action: groundwater user groups and their role in SES resilience. Special Session by Lopez-Gunn, E., Van Cauwenbergh, N. & Mukherji, A. *Collective action in a globalized world XIV IWRA WORLD CONGRESS, September 2011, Porto de Galihnas, Brasil.*

Rica, M., Lopez-Gunn, E. & Llamas, M.R. (2012) An analytical framework on the emergence and evolu-tion of collective action: an empirical case of Spanish groundwater user collective associations ICID. *Irrigation and Drainage journal*, 61, 115–125.

Rica, M., Dumont, A., López-Gunn, E. & Villarroya, F. (in press) Collective action and inaction for water use and water rights reclamation in the olive oil producer region of La Loma de Úbeda, Jaén (Spain) *Water International*.

Rittel, H. & Webber, M. (1973) Dilemmas in a General Theory of Planning. *Policy Sciences*, 4, 155–169.

Salmoral, G., Dumont, A., Aldaya, M.M., Rodríguez-Casado, R., Garrido, A. & Llamas, M.R. (2011) *Análisis de la huella hídrica extendida de la cuenca del Guadalquivir (Analysis of the Water Footprint of the Guadalquivier Basin)*. [Online] Botín Foundation. Available from: http://www.fundacionbotin.org/monografias_observatorio-del-agua_publicaciones.htm [Accessed 13 December 2012].

Sánchez Martínez, J.D., Gallego Simón, V.J. & Araque Jiménez, E. (2011) El olivar andaluz y sus transformaciones recientes. *Estudios Geográficos*, 72(270), 203–229.

Shah, T. (2009) *Taming the Anarchy. Groundwater Governance in South Asia.* Washington D.C., Resources for the Future Press.

Valero de Palma, J. (2011) *Retos Institucionales de la gestión colectiva del agua en España.* Seminario sobre gestión colectiva de las aguas subterráneas en España. Observatorio del Agua de la Fundación Botín.

WWF/Adena (2006) *El Uso Ilegal del agua en España.* Madrid.

Yagüe, J., Villarroya, C. & Xuclá, R.S. (2003) *Proyecto ALBERCA: Modernización de los Registros de Aguas.* Congreso Nacional de la Ingeniería Civil. Madrid. pp. 1853–1861.

Chapter 9

Future Institutions? On the evolution in Spanish institutions from policy takers to policy makers

Elena López-Gunn
Water Observatory, Botín Foundation, Madrid, Spain
Department of Geodynamics, Universidad Complutense de Madrid, Spain

Gema Huelva
Department of Geodynamics, Universidad Complutense de Madrid, Spain
I-Catalist, Spain

Lucia De Stefano & Fermín Villarroya
Water Observatory, Botín Foundation, Madrid, Spain
Department of Geodynamics, Universidad Complutense de Madrid, Spain

ABSTRACT: The chapter analyses the main institutional framework in Spain for the implementation of Integrated Water Resources Management (IWRM) in the context of its 'European equivalent', the European Union (EU) Water Framework Directive. It reflects on the Spanish institutional set up for IWRM, and how far it fits the IWRM paradigm. The second part of the chapter however questions elements of the IWRM paradigm, along with the examination of the underlying assumptions and argues for disruptive institutional innovation in terms of water resources management to deal with the great acceleration of the twenty-first century in terms of resources use and globalization and the proposed River Basin Organizations (RBAs) of the future.

Keywords: Institutions, Integrated Water Resources Management, Water Framework Directive, complexity

1 INTRODUCTION

Spain is a country well known for its water institutions. This chapter however will argue that this valuable historical legacy can become the unwelcome weight of history unless institutions are adjusted to make the most of the old, while not impeding the new, thus taking the leap to the often disruptive (or deep) changes needed for 21st century Integrated Water Resource Management (IWRM). In this context, the chapter will look at three things; it will first review the existing institutional set-up and how far it is well-adapted to meet the IWRM paradigm. In order to do so it will review the main tenets of the IWRM in relation to Spanish institutional arrangements (GWP, 2000; Ait Kadi, 2013, this volume). Secondly, it will outline the main differences (and similarities) between the IWRM ideal and its transposition into the EU under the Water Framework Directive (EU WFD) (EC, 2000). The final section is conceptual, evaluating how far the IWRM concept itself is a product of the twentieth century and whether it has to be ditched or re-assessed in view of the 'exponential times' currently underway. This great acceleration is happening not only globally but is also echoed in Spain which is currently undergoing one of its deepest financial and economic crises, gradually turning into a deep political and social crisis. The chapter reflects on the institutions that might fit these exponential times.

2 CURRENT INSTITUTIONAL FRAMEWORK AND MAIN ISSUES

The implementation of IWRM in Spain in institutional terms consists on the interplay between multiple levels. Spain operates as a full member of the European Union, and is a country which is in the midst of pressures for deeper federalism, pressures which are also mirrored at EU level, as seen in the current Euro crisis which is forcing either more integration or potential disintegration. This, as will be discussed below, raises issues in relation to the clear allocation of responsibility and accountability, and the explicit agreements on the level of government involvement. Where does government responsibility start? When is responsibility devolved or should be devolved to the regions in a nascent federal system? Can this role be shared? What is the range of different collaborative and cooperation mechanisms? Is there a role for emergent community-based organizations? Where do River Basin Organizations fit into the picture? How far and when are stakeholders involved in decision-making? What is the role of citizen science? (Cooper *et al.*, 2007). All these questions remain unanswered at the moment and yet highlight the tensions and opportunities that the current crisis is putting on current institutional structures.

Yet Spain's water organizations and laws have not stayed still, in fact these have been evolving at a faster pace, mostly under the influence of the two parallel processes outlined above at EU and national level, often moving in opposite directions and most importantly, at different scales. On the one hand the homogenization of European water policy arising from the implementation of the European Union (EU) Water Framework Directive which pulls Spain towards high ecological standards and the achievement of Good Status (GS) of all waters, and on the other hand, the decentralization of the Spanish territorial model, which on the ground finds difficulties to implement commitments made at EU level on Good Status for European waters. This is particularly the case when these demanding policy goals impact on the dominant socio-economic model, geared towards intensive water use by irrigation. In addition or maybe because of this clash of visions and priorities between EU goals and existing priorities, the debate gets caught up in politics and territorial identities which increasingly would have to be played out at a water negotiating table which at the moment does not really exist, like for example a 'water senate' that provides a meeting point for different regional interests on water (Lopez-Gunn, 2009; Lopez-Gunn *et al.*, 2012b). This means that the collaborative spaces between regions and the river basin authorities are limited to a few formal forums. At the moment at the basin level there are two formal forums where representatives from the central and the regional governments sit together: the Basin Water Council and the Committee of Competent Authorities. The latter was conceived as a 'cooperation body' and includes representatives from regions, yet considering most decisions that impact on water are sectorial policies which fall under the remit of the regions, this is not sufficient (Hernandez-Mora & Ballester, 2010). This means that in reality most conflicts resort to court cases (Lopez-Gunn & De Stefano, in press).

2.1 The pull of 'eco' water policy institutionalization from the EU Water Framework Directive

The Water Framework Directive (EC, 2000) was approved in 2000 and its regulatory requirements bode challenges but also opportunities for Spain. The main difficulty rested in its ambitious environmental requirements, designed to reverse the gradual deterioration of the state of Europe's rivers and aquifer systems, through a framework law that established high standards to be achieved gradually across the European Union (EEA, 2009). The opportunities come from the fact that the WFD was the partial embodiment of the IWRM paradigm. This partial embodiment was reflected in the adoption of the catchment (or a set of neighboring catchments) as the management unit, as espoused in the IWRM doctrine. It was however partial since the WFD had a fairly 'eco-centric' approach to water management, rather than an integrated approach, i.e. the WFD gave greater weight to the environmental aspects than to other aspects, trying to re-balance the well-established prioritization of sectorial policies in European water management. This is also explained by the European context where there are other directives that cover other aspects where the WFD is conceived in response to EU mandate laid down in the Treaty establishing the European Commission of 'preserving, protecting and improving the quality of the environment, protecting human health, prudent and rational utilization of natural resources, promoting measures at international level to deal with regional or worldwide environmental problems' (Art. 174, Official Journal of the European Communities. C 325/3324.12.2002.). Within this mandate, the WFD would provide the overarching frame for water resources. In addition, both the WFD text and, more recently, the EU Blueprint (EC, 2012), highlight that preserving water is not only about environmental protection, health and well-being, but it is also about economic growth and prosperity. Thus it adopts an ecosystem approach to water management as the basis for sustained economic growth. Yet the past prioritization of sectorial policies had led to the poor state of European waters. The WFD was trying to reverse the balance by requiring all EU member states to achieve and maintain a good status of all their waters (surface and groundwater; transitional and coastal waters) by 2015, as well as to prevent any further deterioration of that status (Figure 1). The ultimate purpose of this 'environmental' focus was to contribute to the 'provision of the sufficient supply of good quality surface water and groundwater as needed for sustainable, balanced and equitable water use, a significant reduction in pollution of groundwater, the protection of territorial and marine waters, and achieving the objectives of relevant international agreements ...' (WFD, Article 1) (EC, 2000). In this sense the WFD represents a shift in the concept of water security in Europe, as it links water security to environmental security, i.e. the good health and functioning of water and water-related ecosystems (Grey & Sadoff, 2007; López-Gunn *et al.*, 2012d).

The adoption however of the catchment scale in the WFD was well aligned with the IWRM paradigm, where the Directive clearly sided with the adoption of a catchment basis model of water management, as well as the inclusion of economic considerations in water management (water pricing, cost-recovery of water services, and the cost-effectiveness of measures) (Table 1).

A critical milestone in the implementation of the WFD was the approval of a River Basin Management Plan (RBMP) for each River Basin District (RBD) by 2009, designed to achieve the WFD objectives within 6-year planning cycles, the first by 2015, the second cycle by 2021 and the third by 2027. The WFD was transposed to the Spanish legal system at the end of 2003, in effect being superimposed on a water management model that had a long tradition. Indeed, Spain (together with the USA) pioneered the catchment management approach in the last century, with the creation of the Ebro river basin authority, in 1926 (Delli Priscoli, 2007; Lopez-Gunn, 2009). Due to this long history of institutions geared for water management on a catchment basis, Spain felt well-positioned to be able to meet the WFD deadline for the adoption of catchment basin plans by 2009. Spain theoretically had a ten year lead time, since in 1998 it had approved all its catchment management plans. It meant theoretically a head start compared to other countries which had not managed water by river basins before the onset of the WFD. Despite the apparent advantage, however, by 2012, three years after all the catchment plans were due, Spain had only presented five out of an expected 10 plans for the mainland plus an additional 8 are for the Canary islands, and Balearic islands. How can this be explained?

Figure 1 WFD as 'eco' WRM: The 'eco' centric balance of the IWRM in the EU WFD.

Table 1 The application of IWRM principles.

IWRM principles	*EU WFD*	*WFD in Spain*
Basin planning	✓	✓
Pollution control	✓	✓
Monitoring	✓	✓
Stakeholder participation	✓	✓
Economic and financial management	✓	X
Flood & drought management	✓ (EC, 2007)	✓
Information management	X	?
Water allocation	X	?

Note: X = not applicable, ✓ = applicable

Figure 2 WFD catchment plans approved for Spain by December, 2012 (own elaboration).

As was explained earlier this push towards 'eco' centred-WRM at a national level coincided with a gradual process of 'regionalisation' of the Spanish Hydraulic paradigm, which historically had water as a central, pivotal issue in economic development, regional identities and political control. Spain is in fact a rich tapestry of regional and historical national identities, as acknowledged in the 1978 Spanish Constitution, which created a quasi-federal territorial model. Thus, contemporary Spain is made up by 17 Autonomous Communities (or regions) plus 2 Autonomous Cities with a democratically elected parliament and president. In relation to water resources, the Constitution established that water had to be managed through River Basin Organizations (RBAs) for the inter-community basins (those shared by two or more regions), and by regional water agencies for intra-regional basins (i.e. those located within a single region). As a result, currently there are 12 RBAs that depend on the central government and 13 regional water agencies.

The decentralization process which started in 1978 is still in flux. Indeed the current economic crisis is providing a stress test on the model. The progress towards a stronger federal model has been mirrored or replicated on regional demands for increased competences by regional governments on water issues. Water resources management is one of the arenas in the tug of war between the central government and the regions over how deep the decentralisation process will go. The parallel political decentralization process – or devolution – has crystallised in directly elected sub-national governments asserting their interests by the development of parallel bureaucratic structures, like regional water agencies and the ability of regional governments in Spain to issue their own laws. This bears consequences for the WFD implementation since it adds layers of complexity for implementation since ultimately the 'competent authority' for guaranteeing compliance with the WFD is the Spanish state, and thus its compliance will be dependent on a robust coordination with regional water agencies, and a good mediation role played by RBAs with regional governments in cases of potential water conflicts, either within regions – e.g. coordination of agricultural and water policy, or between regions.

In this setting, Spain's water administration can be seen as a multi-scale system composed of at least four main levels: The EU itself perceived by some as a nascent federal system, the Spanish central government (and dependent RBAs), the autonomous regions (which in some cases include regional water agencies), the municipalities and finally water users and citizens themselves as the atomistic level. All these layers are caught up between a process of Europeanization (centripetal force) and decentralization (centrifugal force). These different levels play different roles in relation to IWRM, as seen through the 'eco-centric' perspective of the WFD.

At country level, water affairs are the prime responsibility of the Ministry for Agriculture, Food and Environment (MAGRAMA in Spanish), which drafts national regulations, elaborates the National Hydrological Plan (which would harmonize the catchment plans approved in each basin), and ensures international coordination in the case of transboundary basins like the Tagus, the Douro and the Guadiana. The Central Government is responsible through the RBAs for water planning and management – including the elaboration and implementation of catchment plans – in interregional basins. As mentioned earlier, ultimately Spain's government is answerable to the EU for the submission of the catchment plans, their implementation and for the compliance of the WFD objectives. In terms of IWRM, the WFD marked a clear direction for overall national policy on steps to align with EU policy goals on water. It meant however a disruptive influence in terms of previous planning priorities which, according to Spanish water law, were geared and directed towards serving water needs stemming from sector policies (i.e. a supply-led approach). This has meant a change of focus in terms of policy, since the traditional emphasis of water policy used to rotate around supply-based solutions what concerns hydraulic infrastructure until the late 1990s to early 2000 (Lopez-Gunn, 2009). Now the focus has changed (or should have changed) towards a gradual – and difficult – shift in emphasis towards demand management and compliance with European environmental requirements. Spain has more in common with European countries than with emerging and developing economies, where agriculture is an important part of the GDP and employment. However due to the semi-arid climate, agriculture and irrigation represent 68% of the consumptive water use, even if in GDP and employment terms it is an increasingly smaller sector (Martinez Santos *et al.*, this volume; Valero *et al.*, this volume). In this sense, the consumptive water use of Spanish regions remains focused on a paradigm fixed on securing access to water resources, and/or how to increase supply to meet the existing demands from agriculture and growing demands from non-agricultural sectors. The Central Government, although pulled by the eco-centric IWRM approach adopted by the WFD, is still reluctant or finds it politically difficult to re-allocate water resources away from the agricultural sector through command and control regulatory approaches, while dealing simultaneously with centrifugal tensions in water management. As a result, in most basins, water planning has been 'caught up' in political negotiations, quagmires and legal disputes, leading to significant delays in the approval and submission of the new catchment plans.[1]

A lack of coordination and integration in institutional terms partly impedes Spain and its regions from fully achieving the goals posed by the WFD. In Spain, water institutions have a dual and complex personality that in many ways impede integration in the most traditional IWRM sense (across policies and levels) and even less the 'eco-IWRM' approach intended by the WFD. This is still to be achieved by most European countries according to the EU Blueprint where the proposed package would increase coherence between relevant EU policies – in particular the Common Agricultural Policy, Cohesion, Health and Energy (EC, 2012). At national level, agricultural policies (and organisations) have de facto tended to be dominant in relation to decisions made on irrigation.

This 'disjunction' is also reflected at regional level, where the 17 regional governments are in charge of defining, funding and implementing most of the sectorial policies that determine land use in their territory, including agriculture. Thus, the key actors in the achievement and maintenance of good water status required under EU legislation at both national and regional level are poorly integrated in terms of co-thinking and policy integration of *e.g.* crops and their water requirements being matched with water availability. This is crucial since much of the quantitative and qualitative pressure on water resources comes from agricultural use – water withdrawal for irrigation on the one hand, and diffuse pollution from agricultural fertilisers and nitrates on the other hand (Willaarts *et al.*, this volume; De Stefano *et al.*, this volume). Yet, regions, which take most of the decisions on agriculture, are not held responsible before the EU for the achievement of (or failure to achieve) the WFD requirements. Thus it is a clear case of misalignment between duties and responsibilities, which in turn create little or no incentives for re-thinking land use change or the intensification of agriculture. To facilitate the integration of water and land management, a new coordinating body has been established in each River Basin District (RBD) – the Committee of Competent Authorities – which includes representatives from different administrative levels (national, regional and local) and sectors (agriculture, industry, etc.), as well as ports and coastal management. However, these new bodies still have to prove their effectiveness in conciliating often apparently diverging interests and integrating oftentimes opposing interests, like increased irrigation and the re-establishment and prioritisation of ecological flows.

Inter-regional River Basin Organizations (12 in total) are the executive arm of the central administration and are responsible for inter-sectorial allocation, water quantity and quality monitoring and enforcement, the authorization of water and discharge permits and water pricing for *e.g.* agriculture. In intra-regional basins these responsibilities are held by regional water agencies (13 in total). However one of the most problematic 'gaps' in institutional and organizational terms is the existing arenas and procedures for water authorities to interact with regional governments.

1 The deadline for approving new plans expired in December 2009 and, in November 2012 only five RBDs out of 25 have completed the full process leading to a catchment plan. Consequences of this generalized delay include not only legal actions initiated by the European Commission against Spain (in 2010, 2011 and 2012), but also and more importantly, reductions in the time available in most of the river basins to implement the catchment plans.

The nature of both types of organizations is fundamentally different; RBAs are highly corporatist in nature, designed along functional lines to manage water on a catchment scale. The majority of the RBA have a highly bureaucratic nature and rationality, at the expense of political directions which ultimately come from the Central Government via the appointment of its top management positions. This combination of technical and political elements introduces a slightly schizophrenic component in RBAs. What has become an increasing issue however is the strong hierarchical, vertical and formalistic nature of RBAs and, to a lesser extent, of regional water agencies. This is a major weakness on three accounts; first because there are few arenas for the integration of sectorial policies which are not based on 'catchment' lines, like energy which largely operates at national level, or agriculture which is mainly regional in terms of competences but also European and global in terms of agricultural trade. Second, regional water agencies are explicitly more political in nature and in many ways closer to the actions and decisions taken, and third – and particularly pertinent in the context of an 'unraveling federalism' process – as mentioned earlier, apart from the Committee of Competent Authorities, there are almost no arenas or procedures to resolve conflicts between RBAs and regions and also between regions themselves. On top of this, the WFD in itself has entailed a huge administrative and technical burden for River Basin Authorities (RBAs, i.e. RBOs and regional agencies), who had to deal with new concepts, new monitoring networks, new laws, and a new way of working to produce their catchment plans. Despite having had a head start in terms of management by basins, the paradigm shift was much harder to incorporate. The most daunting task for River Basin Authorities, however, has been to harmonize tensions among diverging powers and opposite policy goals especially in intra-regional basins, where catchment plans got to a stalemate and there were no arenas for mediation except via the courts.

Finally municipalities or aggregations of municipalities are the main bodies responsible for urban water supply and wastewater treatment. This makes them instrumental for the compliance with the Urban Wastewater Treatment Directive, which in turn is one of the cornerstones for the improvement of river water quality required by the WFD. For this task municipalities rely on the support of regions and the Central Government, who contribute substantially (often thanks to European regional funds) to the funding of the construction of wastewater treatment plants. Municipalities also hold the key to some of the catchment plan measures (e.g. increasing urban water prices) and define their own urban development model, which has direct impact on water demands. The relationship between RBAs and municipalities is perhaps one of the least explored but probably one of the most interesting ones because of the closeness of municipalities with society, and their potential as agents of change, *e.g.* in relation to water prices for public water supply and sanitation, but also in many rural areas, as a potential mediator between users and water authorities. An interesting institutional innovation experiment is currently being implemented in the Douro basin between the catchment authority and mayors with the development of new Land Custody contracts (Figure 3).

Last but not least, water users constitute an essential level of water management and, together with municipalities, a level (as will be discussed below) that could play a 'revised' role in twenty-first century water management. Water users are *de facto* the main water managers in Spain in day-to-day management. Water users

Figure 3 Land Custody contract: Mayor's school as a public-public partnership in the Douro basin (Huertas, 2012).

are also key because of the importance to understand their interests, and as they are holders of water rights. Their closeness to the resource is pivotal as both a potential obstacle for change but also as a lever for change. Also, the rigidity and perdurance of water rights is an issue, because once granted by the water administration, these water rights are very difficult (and expensive) to be recovered. In Spain, there is a long tradition of water users partaking directly in water management via Irrigation Communities, some dating back a few centuries, others like groundwater users, mirroring the groundwater revolution only half a century old, and other new emerging communities around new water resources like desalination, recharged water or recycled water (Valero *et al.*, This Volume; Lopez-Gunn *et al.*, 2012c). The Spanish regulatory system is well-developed in terms of integrating (consumptive) users into formal procedures for water management. Water users are organized into a very diverse set of associations, which varies both in legal status, size, composition, territorial scope and objectives (López-Gunn & Martínez Cortina, 2006; Rica *et al.*, 2012). These associations serve the objective of facilitating a coordinated use of shared water resources (reservoir; community wells) and participate in water allocation decisions at different levels. At national level, these associations – which have a range of different organisational formats spanning private and public law – interact with the public administration in the National Water Council, which is an advisory body to the MAGRAMA that decides on water projects and plans that are national in scope. Moreover, users and the RBAs sit together in management bodies that decide on technical issues like the volumes of water that can be released for different uses in each water exploitation system of the RBD. Traditional water users (irrigators, hydropower, industry) perceive the WFD as a new 'environmental legislation' that is likely to set new limits to their activities (e.g. for ensuring environmental in-stream flows or allow for the recovery of groundwater levels) and increase their operation costs due to the principle of cost-recovery of water services established by the WFD. One of the main pending tasks is the redefinition of 'water users' to be expanded beyond 'consumptive uses' who are limited to water rights holders (Hernández-Mora & Ballester, 2010). Under the new impetus and legitimisation of public participation provisions under the WFD, other non-consumptive users – e.g. leisure, anglers, etc. – and new, water-oriented civil society platforms are knocking at the door of official participatory bodies or looking for alternative participatory fora from where they can influence decisions, thus slowly changing the people represented at decision-making arenas and the balance in decision-making.

3 THE WATER FRAMEWORK DIRECTIVE AND IWRM IN SPAIN

The Water Framework Directive has been portrayed on occasion as the poster child of IWRM for the European Union and even outside the EU. This section will reflect on this 'poster child' by reviewing the main tenets of both IWRM and its translation into the WFD from the perspective of institutional analysis. IWRM has a series of areas, some of which are directly addressed by the WFD and some of which are left to the member states and their own national regulatory and institutional frames.

The WFD for example shares with IWRM paradigm the emphasis on a number of key functions. In particular it gives clear guidance in relation to pollution control, through a series of water quality European directives with clear water quality goals (e.g. on urban wastewater treatment, nitrate pollution from agricultural activities, groundwater protection, pollution from hazardous substances). It has adopted, as discussed earlier, the basin – or an ensemble of contiguous basins – as the right scale for water planning. It has also incorporated clear monitoring requirements in order to be able to assess compliance with the objectives set on good ecological status for water in the EU, as well as the incorporation of stakeholder participation under Art. 14 of the WFD as a requirement under river basin planning. Finally, it has placed emphasis on economic and financial management through the principle of cost recovery. However, there are some areas where the WFD remains silent, compared to the IWRM paradigm for the case of Spain. An example of this is the issue of water resource allocation, which goes to the heart of the deeply embedded political processes hidden under the IWRM talks and which often is at the core of the difficulties in its implementation (Allan, 2003). Equally, as will be discussed below, the WFD remained silent or did not get overtly involved or concerned with information management, for example in terms of the structural and/or functional relationships between data, information, knowledge, and wisdom (to be discussed below). As will be discussed in the next two sections these two areas not incorporated into the WFD are probably two of the most crucial – even critical – for its actual implementation. The recent EU Blueprint (2012) in many ways is a reflection on these aspects and the analysis of which aspects should be included.

According to the Global Water Partnership (GWP, 2000) 'IWRM is a challenge to conventional practices, attitudes and professional certainties. It confronts entrenched sectorial interests and requires that the water resource is managed holistically for the benefits of all.' Thus IWRM is a 'political process' which requires robust institutions. According to Cap net, evidence across the world has shown that the reality of IWRM implementation falls short of its aspirations. Other institutions like GWP are more positive since it is described as a process (Ait Kadi, this volume). Evidence points to a common malaise in lack of IWRM implementation.

In the case of Spain some of the issues in relation to implementation or malaise are applicable, and some are not (Table 2). For example, the RBAs in Spain (and regional water agencies) due to their long history some dating back to the beginning of the twentieth century, are well recognized amongst stakeholders and have clear roles. However most of the other issues would be applicable to Spain since many refer

Table 2 Evidence from review of IWRM, including the case of Spain (own elaboration).

Issues	*Evidence from Review IWRM*[2]	*Spain*	*Comment*
Roles	Lack of clear role for the RBAs	Role of RBAs is questioned by some sectors/regions	Need to 'review' role?
Independence	Lack of autonomy for the RBAs	Heavily conditioned by political appointments at senior levels	
Model of engagement	Lack of recognition of the role of stakeholders Lack of recognition of the RBA among stakeholders	Stakeholders recognized in formal processes but limited to water right holders	Need to review definition of 'stake holders' and 'timing' of participation (rather soon than late, monitor incorporation of views)
Financial capacity and autonomy	Financial management is not done at basin level and therefore the opportunities to use financial tools are limited. Lack of human and financial resources of the RBAs	Financial management at basin level is very limited and also often self-financing tools are not fully applied. Lack of stable human and financial resources.	Review financing scheme. Diversity of human resources (less corporatist?)
Functions	Lack of resources and responsibility limit the RBA engagement with the full range of water resource management functions	Functions clearly outlined but no existing procedures to align water policy with other sectorial policies (lack of policy coherence)	Devise specific procedures for policy coherence
Coordination	Lack of cross-sectorial coordination	Very limited cross-sectorial coordination, water as weaker sectors *vis-à-vis* e.g. energy or particularly agriculture	
Flexibility	Lack of adaptive management in the RBAs	Rigid institutions in rigid institutional framework; long lead time for response (except for e.g. droughts)	
Monitoring and control	Monitoring and enforcement hardly practiced	Good monitoring for compliance with e.g. EU Directives but need for improved overall monitoring and compliance framework	

2 Please see Ait Kadi (2013, this volume) and UN (2012) *Status Report on the Application of Integrated Approaches to Water Resources Management*. UN Water Report.

to strong governance systems. Thus in terms of incremental institutional strengthening, RBAs provide a good starting point for institutional reform and capacity building. However, as will be discussed in the last section, maybe incremental is no longer enough. Indeed maybe disruptive reform (big bang reform) is what is needed to really catapult Spanish water management into the twenty-first century, and the 'transformative' changes now being called for (Meuleman, 2013; Berkhaut, 2011). In the case of Spain, the main issue with the implementation of the WFD has been its 'reductionist' approach to water resource management functions, somehow focusing on the ultimate indicator set by the Directive (good status) instead on addressing a wider, and more nuanced frame of ecosystem services or more recent discussion around the nexus water/food/energy. Although the Directive leaves freedom to the member in terms of implementation, this freedom cannot jeopardize the main objectives.

Thus the traditional functions of RBAs have been redrawn in Spain trying to shift the balance back to nature, but along the process creating a reaction from the system against this narrow definition of ecosystem services, which are much more nuanced since it includes *all* water services. The WFD has also set a clear target but does not help draw a path for transformational governance which is the *sine qua non* for societies to accept or internalise changes in lifestyle or expectations. The best case is reflected by the Segura basin. The graph below shows what compliance with the WFD with no change in the current economic development model would require for the Segura catchment in terms of drop in water abstractions, which are well above the natural physical boundaries of the system, and which can only continue due to the mining of groundwater resources and the use of water imported from outside the basin.

Thus Spain so far has lost the opportunity provided by the WFD for addressing water management as a complex system, which would necessarily incorporate for example, other sectorial uses beyond the dominance of existing uses, like agriculture and hydropower to seek a diverse economic model, which considers all sectorial uses including agriculture, domestic water supply and sanitation, mining, industry, fisheries, tourism, energy and transport as highlighted in the EU Blueprint (EC, 2012) or even opportunities under the new digital economy, Thus the functions (and institutional structures and priorities for water rights have remained 'frozen' in time and highly rigid to the priorities of the twentieth century, when agriculture was the economic motor of a largely rural society and hydropower was key to ensure energy security in an autarchic, isolated economy.

A good test on the implementation of IWRM via the WFD (excluding allocation and information aspects) is the current Program of Measures in the plans presented and in draft form in compliance with the WFD. The incorporation of the WFD in Spain was articulated in the so-called Instrucción de Planificación Hidrológica (BOE, 2008). This exercise highlights the priorities identified in catchment plans prepared in order to comply with the WFD to meet its objectives for GS for 2015 or 2027 under some circumstances.

An analysis was undertaken on the current draft program of measures of the approved and pending catchment plans. The analysis gives a coarse view of the measures since measure categories vary from basin to basin; and because the classification of a specific measure into a specific category e.g. modernization of irrigation systems as a measure 'to meet environmental objectives' sometimes is not exempt from controversy. It is interesting that in both shared and internal basins, the distribution of fund-

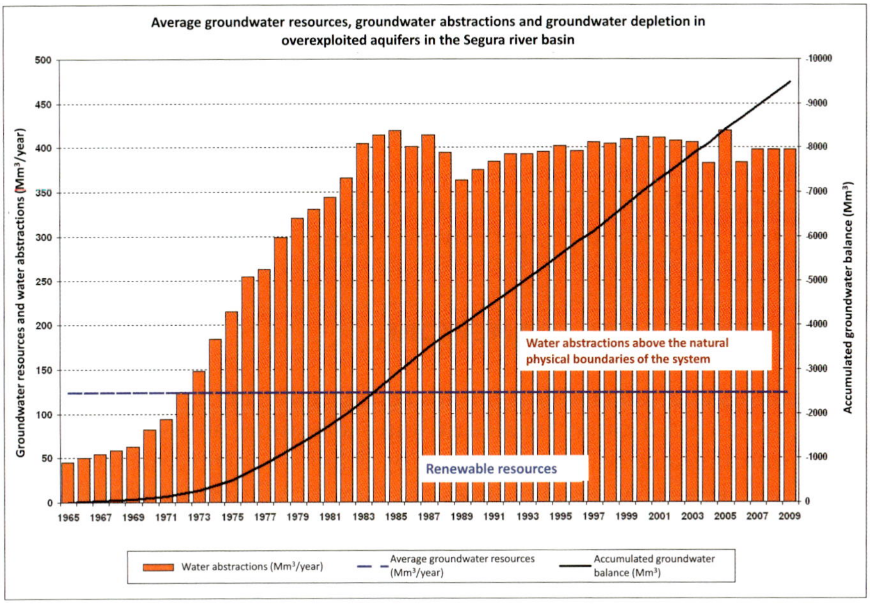

Figure 4 The Segura basin: Reduction in water abstraction needed to comply with WFD (Cabezas, 2011).

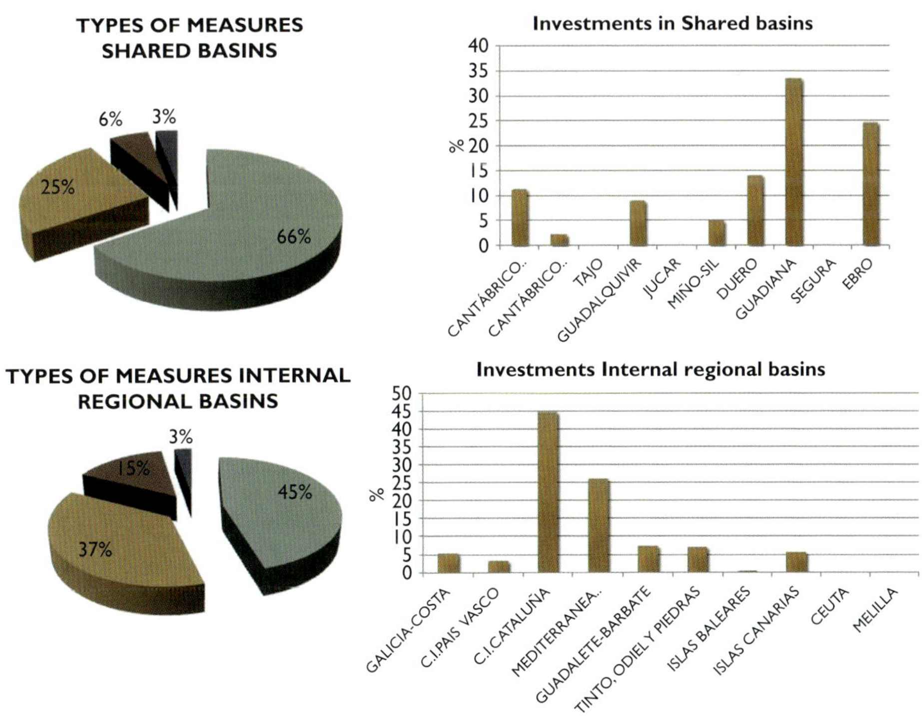

Figure 5 Type of Measures in Spanish catchment plans elaborated to comply with EU Water Framework Directive.

ing in relation to water governance is equal to 3%. Considering that infrastructural measures are more expensive, this amount falls short considering the increasing evidence and awareness that many of the issues related to water problems in Spain are related to water governance as argued by Rogers & Hall (2003) and Llamas *et al.* (2012).

Note that in the case of shared basins, 66% of measures are to meet environmental objectives, 25% to satisfy demands, and 6% to prevent extreme events. In the case of internal basins 45% of measures are to meet environmental objectives, 37% to satisfy demands, and 15% to prevent extreme events.

4 POLICY AND LEGAL FRAMEWORK

An important issue in water management and therefore also in the implementation of IWRM, as discussed earlier, relates to water allocation, an issue that the WFD left to individual member states to manage (by definition a Directive sets 'what' has to be achieved, leaving to member states the 'how'). Adjusting water allocation to the IWRM paradigm can require hard decisions to be made *e.g.* between conflictive demands. These are often at the heart of a new water policy and the reform of water law and institutions which is an area that sovereign member states in the EU would probably not let the EU touch, as it is in fact out of the scope of the EU treaty. In this sense water legislation is key, since it crystallizes policy by clarifying the entitlement and responsibilities of users; the roles of the state in relation to other stakeholders, the procedures to follow in order to formalize the water allocation system; and the legal basis and legal status for the different water management institutions and water user groups.

A distinctive feature of the Spanish legislative framework is its diversity. This diversity refers on the one hand to a multi-level legal framework, from the supranational level (European Union) through to national laws, regional laws and local bylaws, and on the other hand to water rights, covering the span from fully private to state concessions, and all types of water encompassing both conventional and non-conventional water sources. In terms of IWRM and also the WFD Spain has a number of international or transboundary basins. In this context Spain has signed the bilateral Albufeira Convention with Portugal for shared rivers. In terms of IWRM what is relevant is the proposal under the next round of WFD planning to eventually have a common plan produced jointly by Spain and Portugal based on a catchment basis. At the country level Spain has a national Water Act, dating back to 1985 (with main reforms in 1999 and 2003), and also some regional water laws which are a result of the decentralization in progress. Some of these laws, e.g. the Andalusian one, are opening the path for politically sensitive issues such as water pricing, water trading and water banking. The full adaptation of the Water Act to the WFD is still in progress and it has consisted more in making partial adjustments to the existing legislation than revising it in depth to fully align it with the principles and guidelines of the Directive.

In relation to water access and typology of water rights, Spain has a great diversity in water rights, with the whole spectrum from public to private to collective water rights co-existing within the current legislative framework. Under the

1985 Act all the waters are included in the public domain and water is bundled to land. In the case of groundwater, the acknowledgement of water rights pre-dating the Water Act, however, has led to a coexistence of private and public right that complicates groundwater management. Water generated through new technology, like desalinated and reclaimed water is also part of public water resources and requires a public concession to be exploited. In terms of water trading, the 1999 reform opened the possibility to trade water rights associated to public concessions (including temporary water rights). The enforcement of the existing regulatory framework faces several challenges, which, to be addressed, require a combination of political will, political skill and a sufficient level of resources. The most impelling challenges include the monitoring of water uses, especially in groundwater; updating water books and registers (revision of concessions; what to do with illegal uses); and the creation of a sanctioning system that rewards compliance and makes it more beneficial to comply with the law than to break it (López-Gunn *et al.*, 2012b).

In relation to water rights, as stated above, on IWRM and water allocation issues the WFD does not position itself on their nature but it does state clearly the equal duty – independent of the juridical nature – to protect water resources (Huertas, 2011; Poveda, 2011). However, the flexibilization of a rigid rights system either through administrative revision of water rights or through fully regulated water markets seem to be potential venues for reducing pressure on closed and often over-allocated water systems to reserve water to improve the status of water bodies. Ultimately, the different regulatory levels and the diverse typology of water rights come together and have to be harmonized in catchment plans. These Plans are normative documents themselves, and thus mean that many parties use delay tactics to slow the plan rather than freeze water rights for the duration of the water planning period. Furthermore, once approved, these individual catchment plans will have to be followed by the elaboration of a National Hydrological Plan, where centripetal and centrifugal tensions will play out in full.

However in the big overall picture, regarding the main elements that would dramatically improve the strength or robustness of the legal and policy framework, there are two main aspects: firstly, the excessive rigidity and administrative hurdles of the legal system, rather than nimble, adaptive legislation and secondly, the central role of adequate monitoring and sanctioning regimes and the essential element of involving users in these regimes in terms of ensuring buy-in, legitimacy and co-management for it to be implemented. This means a change in thinking away from paternalistic administration to a citizen-led administration.

5 FUTURE INSTITUTIONAL SET UP AND MAIN ISSUES

This section will introduce the idea that all the concepts and analyses undertaken above are based on an incremental mode of change in the 'nirvana' process of IWRM (Molle, 2008). However, the section focuses on how in the midst of exponential times, it is possible or more likely for disruptive changes to occur or emerge. In Spain, an equivalent to the current crisis has not been witnessed since the last century. In the last century across Europe this led to dramatic political convulsion and change. The European Union is undergoing a deep crisis of identity and Spain together with a number of member states is at the brunt of this crisis of identity. Yet, what does this mean for water resources management? Indeed what does it imply for the IWRM paradigm? According to Kuhn (1962) (Table 3) scientific progress is the result of 'development by accumulation', i.e. when normal science is interrupted by periods of revolutionary science. In this line, are there enough anomalies in the IWRM paradigm to merit changes? This section will argue that to 'speed up' the implementation of IWRM it is fundamental to ask new questions on the main tenets of IWRM, to move beyond the normal problem-solving of 'normal science'.

Thus this section will look at the discrepancies revealed under the IWRM paradigm and its implementation, in line with the work presented in this book by Giordano & Shah (this volume). In order to do so it will reassess not just the central tenets of

Table 3 Evolution of IWRM Paradigm (based on Kuhn 'normal science' cycles).

Pre-paradigm	*Consolidation*	*Normal science*	*Crisis*	*Revolutionary science*
1980s–1990s		2000–2010		Post IWRM? (Water security, ecosystem services, planetary boundaries, flat world,...)
No consensus Several incomplete and incompatible theories	Agreed methods, terminology, assumptions	Puzzles solved within current paradigm	Anomalies revealed	Underlying assumptions re-examined, a new paradigm is established

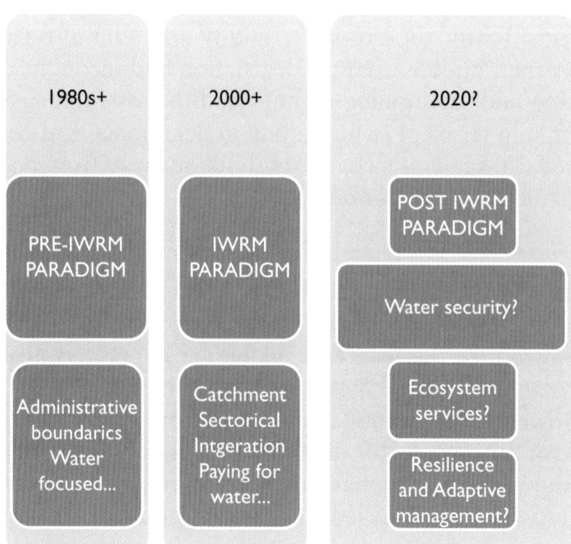

Figure 6 Paradigm shifts in IWRM (own elaboration).

IWRM but also wider, deeper trends from the ongoing global and national crisis and what it means for institutions.

The first issue refers to scale and the unit or subject of intervention. In the case of IWRM the main scale is centered on the catchment as the natural basis. However this will inherently bring problems of integration in relation e.g. to policy coherence since by definition and a priori it fixes the scale which makes it in many ways incompatible or more difficult to be integrated with other policy sectors like energy or agriculture. Energy for example is structured according to networks (Figure 7), whereas agriculture operates increasingly at a global level, though e.g. global food trade. Thus an issue to consider would be for catchment plans to somehow incorporate a network approach to planning (i.e. a network of institutions and policies) that somehow can communicate and engage with these different horizontal and vertical scales without losing its catchment physical focus and boundaries. Thus the catchment of the twenty-first century has both a 'physical' space and a 'virtual' space that somehow have to 'be integrated'. Yet there is little guidance or know-how in this area.

The second issue refers to public participation. It is a key cornerstone of IWRM yet with the coming of 'information society' and the concept of social innovation, there is a huge untapped potential that goes beyond traditional public participation. It is more about visioning, and engaging with citizens as sources of innovation, as co-producers of services. It refers to the so-called 'hidden wealth' of citizens/clients. This means looking at citizens not just as a source of demand but also as a valuable asset. For example, the opportunities for undertaking so-called 'deliberative polling' to decide when there are competing public policy demands under tight budgets, as a way to prioritize spending while generating ownership and public support for choices made (Box 1). This goes more to the 'active' participation aspects of the WFD which

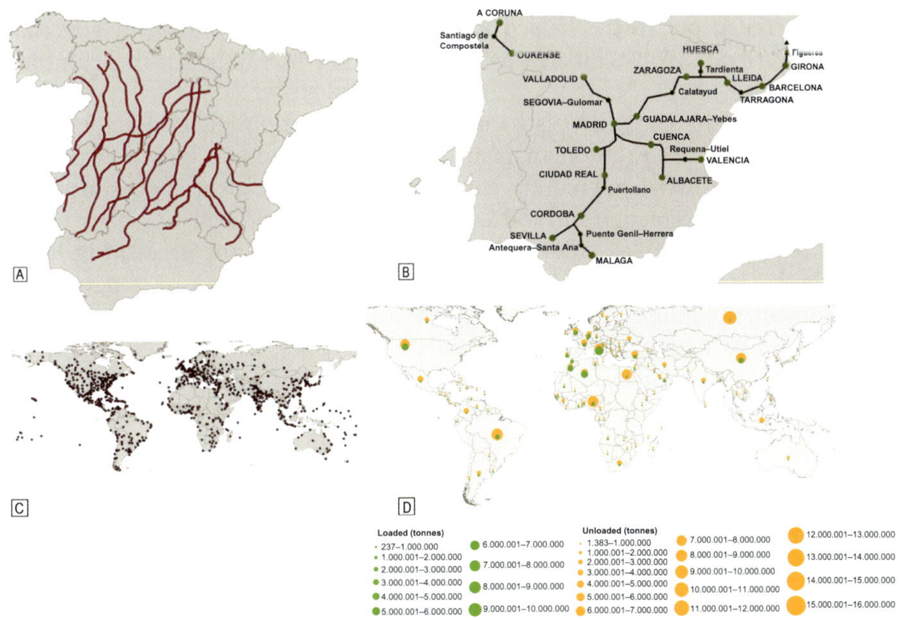

Figure 7 A. Spanish network of public cattle transhumance routes. B. Spanish network of high-speed trains. C. World airports map. D. Loaded goods by boat from Spain and unloaded goods by boat in Spain (Own elaboration, data from Gobierno de España, 2011).

Box 1: Example: Spending reviews: Beyond water as an economic asset towards water as a multiple asset

Part of 21st century IWRM would be based on expanded, transparent and open spending reviews where annually there is a determination on the types of spending; looking at a comprehensive analysis of sources of revenue and programs and initiatives would come under scrutiny in the competition for budget resources.

have so far been barely tapped except for a limited number of cases in the Ebro and Catalonia.

In this sense it is a fundamental change because it is only through the support of society that transformative change can happen to shift current water resources management – which at the moment is often beyond the physical boundaries – back to 'sustainable resource use'. In this context, the chapter has discussed the existing arrangement for water users' participation. However what is needed is a quantum leap in public policy, towards new innovative partnerships, like public-public, or private-private or private and community partnership. These would be able to influence the policy parameters, the administrative guidelines and contractual conditions. These would have to be built as new types of 'social contracts' into the structure of government, building a relationship, as much as managing a contract (Warner, 2004). As mentioned before, examples of these are the public-public Territory Custody contracts being trialed out in the Douro basin (Huertas, 2011). Intelligent governance devolves power and involves citizens as meaningful partners in policy through the use of technology.

Third, the nature of IWRM is focused on Organisations (RBAs) achieving efficiency and effectiveness in the management of water resources. However the frame provided by the IWRM is incredibly cumbersome and complex. It goes against the grain of ease of implementation and a more 'nimble' approach to policy making. Although reality is complex, the concepts and policies do not necessarily have to amplify these complexity issues. Thus it is fundamental to look at a new concept of 'government by design', combined with a strategic evidence-based approach, to find new ways to measure effectiveness of public policy. In this context there are useful stress tests like for example, extreme droughts, energy shortages or increases in food prices at the global level or in the case of Spain, for key strategic crops like wine, olives, horticultural crops, or the environment. In this context one should look at regulation as outcome-based regulation, rather than focusing on pure procedural matters. It is thus important to find ways to sidestep silo mentality, and focus on process compliance such as on assessing performance outcomes. Thus, as stated above, the right regulatory framework in this sense is 'integrated' via technology to develop a collaborative approach through sensors and apps to engage citizens in implementation. Smart regulation goes beyond outcomes to look at incentives in participation and incentives for behavioral change and existing social norms. Translate examples like

Apps for Democracy to Apps for water democratization, transforming big data into open data. Data-driven policy like *e.g.* good water accounting allows for more rational public policy debates. For example, RBAs could engage in the development, with the participation of society, of measurable meaningful and understandable indicators that evaluate performance. With the onset of Information and Communication Technology (ICT) there is greater scope for more real time data, that yet can be made easily accessible and digestible. Technology may act as an enabler, to monitor the behavior of governments and bureaucrats, and for streamlining information on the line of calls for e-government and open government. Thus monitoring and accountability become a two way flow to develop inclusive institutions rather than the perceived 'extractive' institutions that benefit the politically powerful elite by taking resources from the majority of society for the benefit of a few (Acemoglu & Robinson, 2012).

Here a major role will be played by concepts like 'open government', where technology and society will be key. Technology is a critical enabler, where it can facilitate a holistic, service-oriented approach to engaging communities of citizens and users. For example advances in cloud computing, open data and big data, where citizens engage and become (voluntarily) part of the RBAs central functions *e.g.* of monitoring and control. This approach was recently announced as a way forward by the Director of the EEA (McGlade, 2012). In terms of finance the current model favors capital-intensive infrastructure, large and expensive, even when it might not represent the most optimal or most cost-effective option from the perspective of uses when there is commitment to ensure users payment. Oftentimes the second-best solution might be more realistic and practical since these are more likely to be implemented as compared to best solutions, the latter might never materialize and/or get caught up in political wrangling (Molden *et al.*, 2010).

Fourth, the paradigm of IWRM starts with the word 'integrated' which by definition implies a level of systemic thinking. However, it is a half-baked paradigm as long as it does not accept or internalize non-linearity of the system. That is, the complexity of the system (and how water for example is linked to other systems like food systems, energy networks, etc.) means that there are interdependencies that are emergent and difficult to predict. Governments (and RBAs) have to acknowledge and internalize this complexity rather than assume and operate as if all problems are amenable to simple policy prescriptions. In this context the example of the major investment on irrigation modernization is a classic example of a simple policy prescription that had unintended consequences i.e. where investment to generate water savings ultimately resulted in more intensive water use (Gleick *et al.*, 2011, Lopez-Gunn *et al.*, 2012a). This is because the IWRM of the twenty-first century has to take a quantum leap in thinking to help address inherently 'wicked problems'. These are large, intractable problems like the case of the Segura basin, in the southeast of Spain, where its water economy is running on empty, since it is consuming more water than what is naturally available in the basin and where there are multiple dimensions, intangible values and multiple stakeholders beyond those formally represented that do not necessarily share convergent views or goals. Wicked problems require the integration not just of different sectorial policies but much more fundamental, the integration of different insights, expertise and knowledge from different organizations in and outside the Government and RBAs. A much more 'permeable' type of RBA organization which can pool this knowledge and discover and trial potential solutions is now needed.

The current institutional structure in Spain is poorly designed for this mode of governance. Yet the potential is already there with a great number of user organizations, an active civil society (e.g. 15-M which is the Spanish equivalent of the youth movement that spread throughout Northern Africa and eventually developed countries like Israel and the USA) and a latent private sector in different areas of water services (from water treatment, supply, desalination, monitoring and IT, etc.).

IWRM as a process has to be open 'recognizing that insights and good ideas are not the monopoly of single agencies or of governments acting alone' It is a 'search and discover' approach which is currently the default mode in RBAs i.e. the adoption of adaptive governance based on experimentation (Gunderson & Light, 2006; Holling & Meffe, 1996). Instead of embracing this, it is seen as a reactive knee-jerk reaction, as contingency planning for rigid structures. Instead 'search and discover' has to be adopted and legitimized not as alien to IWRM but as a normal culture to address uncertainty built into the DNA and the *modus operandi* of RBAs in dealing with complex problems on a day-to-day basis, and avoid 'optimizing' at departmental level.

6 CONCLUSIONS AND RECOMMENDATIONS

The current default mode of Spain under IWRM and the onset of the WFD can be marked as reactive. Spain had a strong first-mover advantage since the first RBAs were created almost a century ago. However, this element of complacency also stopped a healthy element of institutional innovation: Creating the RBAs of the twenty-first century, a blank slate that had no inertias, or acquired habits and vices. Yet what should the RBA of the future look like? IWRM and WFD are useful in this respect under the great acceleration of the twenty-first century, but may not suffice because the scale of the issues and the solutions has also increased exponentially.

The definition of institutions is in itself useful to ground the debate on the origin of institutions as a socially embedded system of rules. In this context, it is evident that organizations like RBAs are a special kind of institution, with additional features, that involve (a) criteria to establish their boundaries and to distinguish their members from non-members, (b) principles of sovereignty concerning who is in charge, and (c) chains of command delineating responsibilities within the organization RBAs and their institutional frame have to adapt to changing social norms and exponential times.

Thus, there are two main paths to future institutions: one incremental and one disruptive (where one does not exclude the other). Under a business-as-usual IWRM, the goal is how to strengthen the existing implementation. Under a re-thinking of future institutions and disruptive institutional innovation, it requires an approach of structure-following-strategy, and thus it finds the current institutional structure wanting in terms of facing the great acceleration of our times. For example, what if RBAs in Spain were thought anew on the basis of strategy forms structure? The model proposed by Chandler (1969) offers some interesting ideas for this exercise. Under this mode of thinking it is wondered how strategy and structure in institutions can best be aligned, how it impacts issues related to institutional capacity, institutional incentives and institutional structures? The challenge however to sum up the arguments is to think up institutions that evolve from 'policy takers' to 'policy makers' capable of addressing exponential times in future water management.

Recommendations

In terms of recommendations for future institutions, the main conclusions would be:

- Catchment scale vs. virtual, networked catchment scale (invest in awareness, trained human capacity, but also infrastructure and institutions).
- The role of society and enabling technology.
- RBAs 2.0 (save lives, shift into an emergency mode of operation with possibly different ethical principles and standards than in regular mode).
- Financing and Open Government (provide help for self-help and to strengthen preparedness. Honor local habits of solidarity).

Strategy forms structure (Based on Chandler, 1969)

Aligns the organization to best-followed strategic direction.
Allows for clearly defined roles and responsibilities.
Clarifies who makes which decisions.
Enables clear accountability.
Minimizes handoffs that affect the users and citizens experiences. Minimizes the users and citizens 'run-around.'
Minimizes handoffs that create confusion over who is responsible for which outcomes.
Pulls together the people who most need to work closely with each other.
Allows information to flow unrestricted to those who need it.
Creates manageable spans of control.
Is augmented by informal channels of cross-boundary communication.

REFERENCES

Allan, T. (2003) *IWRM/IWRAM: A new sanctioned discourse?* SOAS/KCL Occasional Paper number: 50. Water Issues Study Group; School of Oriental and African Studies/King's College London University of London.

Acemoglu, D. & Robinson, J.A. (2012) *Why Nations Fail: The Origins of Power, Prosperity and Poverty.* Profile Books Limited.

Ait-Kadi, M. (2013) *Integrated Water Resources Management (IWRM): the international experience.* This volume.

Berkhout, F., Wieczorek, A.J. & Raven, R. (2011) Avoiding environmental convergence: A possible role for sustainability experiments in latecomer countries? *International Journal of Institutions and Economics,* 3(2), 367–385.

BOE (2008) *Royal Decree on Hydrological Planning and Ministerial Order approving the Hydrological Planning Regulation* (Instrucción Técnica de Planificación Hidrológica). Agencia Estatal Boletín Oficial Del Estado.

Cabezas, F. (2011) Las aguas subterráneas en la cuenca del Segura Nacional "El papel de las aguas subterráneas en la política del agua de España" Observatorio del Agua de la Fundación Botín; 1 de febrero de 2011; Madrid. [Online] Available at www.fundacionbotin.org/agua.html) [Accessed 15 June 2013].

Cooper, C.B., Dickinson, J., Phillips, T. & Bonney, R. (2007). Citizen science as a tool for conservation in residential ecosystems. *Ecology and Society*, 12(2), 11.

Chandler, A. (1969) Strategy and Structure: Chapters in the History of the American Industrial Enterprise. MIT Press (MA) (first published January 1st 1962).

Delli Priscoli, J. (2007) *River Basin Organizations: North America* [Online] Available from: http://www.transboundarywaters.orst.edu/research/case_studies/Documents/RiverBasinOrganizations.pdf [Accessed 15 June 2013].

EC (2000) *Water Framework Directive- Directive 2000/60/EC of the European Parliament and of the Council of 23 October 2000 establishing a framework for Community action in the field of water policy.* [Online] Report number: OJL327, 22.12.2000. Available from: http://eur-lex.europa.eu/LexUriServ/LexUriServ.do?uri=CONSLEG:2000 L0060:20011216:EN:PDF [Accessed 14 July 2013].

EC (2007) *Addressing the challenge of water scarcity and droughts, from the European Commission adopted.* Report number: COM(2007)414. Review of the EU policy on water scarcity and droughts – Communication. European Commission.

EC (2012) *The blueprint to Safeguard Europe's Water resources.* Report number: COM(2012)673. [Online] Communication from the Commission. Available from: http://ec.europa.eu/environment/water/blueprint/index_en.htm [Accessed 15 December 2012].

EEA (2009) *Water resources across Europe- confronting water scarcity and drought.* EEA Report. Report number: 2/2009.

Gobierno de España (2011) Tráfico de mercancías en 2011 (toneladas). *Anuario estadístico de Puertos del Estado.* Ministerio de Fomento.

Grey, D. & Sadoff, C. (2007) Sink or Swim? Water security for growth and development. *Water Policy*, 9(6), 545–571.

Gleick, P.H., Christian-Smith, J. & Cooley, H. (2011) Water-use efficiency and productivity: rethinking the basin approach. *Water International*, 36(7), 784–798.

GWP (2000) *Integrated Water resources management.* TAC Paper. Report number: 4. Stockholm.

Gunderson, L. & Light, S.S. (2006) Adaptive management and adaptive governance in the Everglades ecosystem. *Policy Sciences*, 39, 323–334.

Hernández-Mora, N. & Ballester, A. (2010) *Public participation and the role of social networks in the implementation of the Water Framework Directive in Spain.* [Online] Ambientalia, Special Issue: Ten years of the Water Framework Directive: An Overview from Multiple Disciplines. Available from: http://www.ambientalia.org [Accessed 8 July 2013].

Holling, C.S. & Meffe, G. (1996) Command and control and the pathology of natural resource management. *Conservation Biology*, 10, 328–337.

Hernández-Mora, N. & Ballester, A. (2010). Public participation and the role of social networks in the implementation of the Water Framework Directive in Spain. *Ambientalia*, Special Issue: Ten years of the Water Framework Directive: An Overview from Multiple Disciplines. Available from: http://www.ambientalia.org [Accessed 8 July 2013].

Hogson, G. (2006) What are Institutions? *Journal of Economic Issues*, 40(1), 1–25.

Huertas, R. (2011) *Retos y oportunidades en la gestión colectiva de las aguas subterráneas. [Challenges and opportunities in the collective management of groundwater].* Personal communication. Seminar on collective action for groundwater management in Spain, Botín Foundation Water Observatory, Madrid.

Huertas, R. (2012) *Contratos de Custodia del Territorio.* Presentation at CONEMA (Oct 2012). Personal communication. November 2012.

Kuhn, T.S. (1962) *The Structure of Scientific Revolutions* (1st ed). Chicago, University of Chicago Press. pp. 172.

Lopez-Gunn, E. (2009) Agua para todos: the new 'regionalist' twist in the Spanish hydraulic paradigm. [Online] *Water Alternatives,* 2(3), 370–394. Available from: (http://www.water-alternatives.org/index.php?option=com_content&task=view&id=45&Itemid=44) [Accessed 13 July 2013].

Lopez-Gunn, E. & Martinez-Cortina, L. (2006) Is self-regulation a myth? Case study on Spanish groundwater associations and the role of higher level authorities. *Hydrogeology Journal*, 14(3), 361–375.

Lopez-Gunn, E., Zorrilla, Prieto, P. & Llamas, M.R. (2012a) Lost in translation? Water efficiency and productivity in Spain. *Journal of Agricultural Water management*, 108, 83–95.

López-Gunn, E, Cabrera, Custodio, E., Huertas, R. & Villarroya, F. (2012b) Institutional reform in Spain to address water challenges. In: De Stefano, L. & Llamas, M.R. (eds) *Water, Agriculture and the Environment: can we square the circle?* Taylor & Francis. [Online] Available from: http://www.fundacionbotin.org/fb-water-agriculture-and-the-environment-in-spain-can-we-square-the-circle-seminarios-internaci-787798051267687.htm [Accessed 8 July 2013].

López-Gunn, E., Rica, M. & van Cauwenbergh, N. (2012c) Taming the groundwater chaos. In: De Stefano, L. & Llamas, M.R. (eds) *Water, Agriculture and the Environment: Can we square the circle?* Taylor & Francis. [Online] Available from: http://www.fundacionbotin.org/fb-water-agriculture-and-the-environment-in-spain-can-we-square-the-circle-seminarios-internaci-787798051267687.htm [Accessed 8 July 2013].

López-Gunn, E., De Stefano, L. & Llamas, M.R. (2012d) The role of ethics in water and food security: Balancing utilitarian and intangible values. *Water Policy*, 14, 89–105. [Online] Available from: http://www.fundacionbotin.org/libros-de-seminarios-internacionales_publicaciones_observatorio-del-agua.htm under Numeros especiales [Accessed 8 July 2013].

Lopez-Gunn, E. & De Stefano, L. (in press) Between a rock and a hard place: redefining water security under decentralization in Spain. In: Garrick, D., Anderson, G., Connell, D., Pittock, J. (eds.) (2014) *Federal Rivers: Water Management in Multi-Layered Political Systems.* Cheltenham, Edward Elgar Publishing.

Llamas, M.R., De Stefano, L., Aldaya, M., Custodio, E., Garrido, A., López-Gunn, E. & Willaarts, B. (2012) Capítulo 1- Introducción. In: De Stefano, L. & Llamas, M.R. (eds.) *Water, Agriculture and the Environment: Can we square the circle?* Taylor & Francis.

Martinez Santos (2013) Integrated Water Resources Management: State of the art and the way forward. This volume.

McGlade (2012) Oral presentation at IAASA (October 2012) Worlds within reach: From Science to Policy 10th Anniversary Conference [Online] Available from: Laxenburg 24–26 Austria http://conference2012.iiasa.ac.at/person.html?code=mcglade&scode=s1s1 [Accessed 8 July 2013].

Meuleman, L. (2013) *Transgovernance Advancing Sustainability Governance; XIX.* Springer.

Molden, D., Lautze, J., Shah, T., Bin, D., Giordano, M. & Sanford, L. (2010) Governing to Grow Enough Food without Enough Water - Second Best Solutions Show the Way. *Water Resources Development*, 26, 249–263.

Molle, F. (2008) Nirvana concepts, narratives and policy models: Insight from the water sector. *Water Alternatives*, 1(1), 131–156.

Poveda, J.A. (2011) *Uso conjunto a través de la gestión colectiva e incentivos para un uso eficiente (Agua-Energía) [Conjunctive use through incentives and collective management for efficient use (Water-Energy)].* Seminar on collective action for groundwater management in Spain. Botín Foundation Water Observatory, Madrid.

Rica, M., Lopez-Gunn, E., & Llamas, R. (2012) Analysis of the emergence and evolution of collective action: An empirical case of Spanish groundwater user associations. *Journal of Irrigation & Drainage*, 61, 115–125.

Rogers, P. & Hall, A. (2003) *Effective water governance.* TEC Report. Report number: 7. Global Water Partnership, Stockholm.

Valero, J., Thuy, L. & Lopez-Gunn, E. (2013) *The institutional organization of irrigation in Spain and other Mediterranean countries.* This volume.

Warner, J. (2004) Water, Wine, Vinegar, Blood: on Politics, Participation, Violence and Conflict over the Hydrosocial Contract. *Proceedings of the Workshop on Water and Politics: Understanding the Role of Politics in Water Management, 26-27February 2004, Marseille, France.* World Water Council.

Chapter 10

Urban water, an essential part of Integrated Water Resources Management

Enrique Cabrera
ITA, Universitat Politècnica de València, Spain

Emilio Custodio
Groundwater Hydrology Group, Technical University of Catalonia, Barcelona, Spain

ABSTRACT: The necessity of managing water in an integral way is currently not argued. It is considered unavoidable. This comes forth from the scarcity and intense concurrence for good quality fresh water in many areas of the world and the important political and social stresses this generates inside a country and between different territories. Integrated management is linked to the concept of considering the river basin as a unit for management, or exceptionally the aquifer system. Human beings cannot change this physical reality except by performing major works to link distinctive basins, but even in these cases physical boundaries do exist. Due to the relatively small fraction of water resources humans use directly to attend their needs, a preliminary analysis of water resources quantity problems – water quality is a different point of view – tend to ignore urban water as an essential part of a water resources planning when compared to the much larger water volumes required to produce food, fibers, and energy, and to sustain important ecological services, especially in semi-arid countries. However, the situation is very different when an analysis is made taking into account the real value of water for human supply and the different urban cycles the water passes through in a river basin or linked river basins. This is due to the strategic importance of adequate water supply for the human being to attain desirable standards of health, sanitation, life, and economic activity, as well as due to the important impacts and conditioning that urban cycles impose on the river basin. From a quantity point of view, water scarcity is mostly linked to agricultural demand since in general water supply can easily meet the water demand for direct human in a wide territorial context, although quality problems may become a limiting factor in areas that have limited financial resources. Important urban problems may be solved through making adjustments in the agricultural water use domain. Only an integrated analysis of all the involved factors, giving the due weight to urban water, may minimize problems and provide reasonable solutions to current stresses. This integrated analysis has to consider both quantity and quality aspects, and also the involved energy, ecological and territorial implications. Although the discussion has a general scope, comments are largely based on the experience in Spain.

Keywords: Urban water, Integrated Water Resources Management, urban metabolism, concurrence for resources, Spain

1 INTRODUCTION

In the global town, 'what happens in Madrid affects Milwaukee', said US President Barack Obama in May 2012 in the Camp David (Maryland) summit. This has been an appealing slogan drawing attention to the world's economic globalization in a particular moment when European economic turbulence was centered in Madrid. It is clear that half a century ago what happened in Madrid was irrelevant in Milwaukee. The many and profound changes in international behavior of the last decades can be amalgamated together under the term globalization.

Water policy was a pioneer in integration and wide-scoped point of view, although the river basin – or artificially interconnected river basins, or in some cases the aquifer system – set a natural limit to 'globalization'. In this case shorter distances (just the watershed scale and not, as in economic business, an intercontinental scale) favored that the concept of globalization – beyond local boundaries – could be put ahead some decades in advance when referring to water resources. The year 1936 may be considered the start of the change in scale of water policies from local to wide-scoped. This year, the impressive Hoover Dam started to operate; being 221 m high and 380 m wide, and able to fully change the face of Las Vegas. In 1931, when construction works started, Las Vegas was a 5000 inhabitant town located 50 km downstream the dam site. During the four years of construction an average of 4500 workers were employed, most of them uprooted men. Harsh work and severe climatic conditions, which explain more than a hundred of casualties (Viollet, 2005), favored the concomitant spread of leisure activities, bringing into life especially casinos and night clubs. The Hoover Dam changed the future of the area, an uncertain one at that time (Morello, 2009). Las Vegas is currently home to two million people.

Important water management activities started in Spain early in the 20th century at a river basin level – a pioneering achievement at that time –, when 'River Water Confederations' of water users (in theory, not as much in real practice) were created in the late 1920s. These mostly intended to foster irrigated agriculture for improving social development of the country. However, true water globalization did not arrive to Spain until a few decades afterwards, in the late 1940s, after the Spanish Civil and the Second World War, inside governmental economic development plans. This was the start of a dramatic economic growth requiring lots of water, and also marked the start of water-intensive use and the associated social stresses between neighbors due to growing water scarcity, deteriorating water quality and expanding ecological damage. In the 1950s the construction of large dams flourished for several decades, until fading out by the end of the 20th century (Figure 1). Most of these dams were for hydro-energy and irrigation water supply. Only a few were for urban supply, the exceptions being the dams for the supply of Barcelona and Madrid, the two largest cities of Spain.

In the same period, intensive use of groundwater started (Llamas & Martínez Santos, 2005; Custodio, 2012), mostly for irrigation, but also to supply towns, tourist areas, and some industrial settlements. A lot of water was made available, in parallel with a noticeable economic growth, albeit with little incentive for an efficient water use due to the low prices that had been prevalent after generous public subventions. At that time the common water policy was to continually increase water offer instead of controlling water demand, especially for agriculture and some industrial settlements. This was not so much the case in towns or cities, due to the involved added

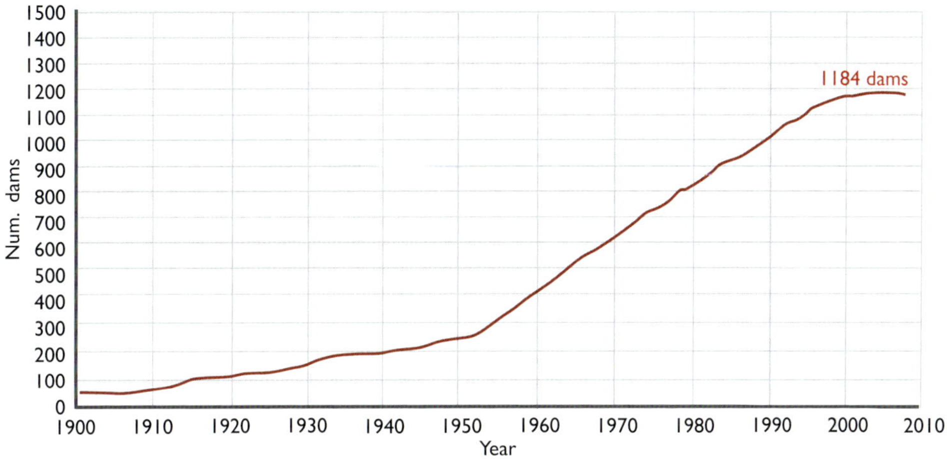

Figure 1 Number of large dams constructed in Spain (Yagüe & de Cea, 2008).

costs to be covered by the companies and users, and in some cases the need to harshly compete for available water resources with other seemly insatiable water users. In earlier times many old towns were suffering water shortages and poor quality due to lack of adequate infrastructures. This had promoted in-house water savings as a culture, but not in the new areas and quarters, or with the municipal water suppliers. They didn't care enough about losses and leakages since they were easily compensated by applying for cheap water due to the important public subventions. Both in towns and in factories, water quality degradation was on the rise, thus giving way to more heavily polluted urban and industrial effluents. The increasing concern on this problem led to the 1995 Sanitation Plan, when water quality recovery started to be seriously considered. However, currently the problem is still not fully solved, despite of the added pressure imposed by the European Water Framework Directive of the year 2000, transposed into the Spanish Water Act in 2003.

Agricultural use of water followed a similar pattern, especially since 1960, when chemicals were intensively and extensively applied. Poor control and low prices fostered an excessive use. This lasted three decades during which agriculture was economically very rewarding and markets absorbed the production. During this time inner consumption was in the rise but especially was the export of food products. The need to increase land productivity and to obtain high quality and good-looking products, and also well-sized fruits, were key parameters that induced the use of agrochemicals and the progressive abandoning of the agricultural practices that were common in the first half of the 20th century. Nitrogen fertilizers were used at application rates above what was really needed, which is currently the main cause of diffuse (non-point) nitrate groundwater pollution in large areas, and consequently of many wells, springs, and streams (Custodio *et al.*, 2012).

As a reaction to this, during this time the term ecological agriculture was born. This term was unknown until about 1960, since most agricultural production till that time was mostly under close-to-ecological conditions. The dramatic agricultural development went in parallel with a spectacular development of feedstock, mostly pigs and poultry, to feed the increasingly demanding urban areas. But the net result of this development was a harsh concurrence for the increasingly scarce water resources, especially around the main urban areas. Local conflicts rose up between farmers and urban water suppliers, including tourist areas, new urban development area for secondary residence, and the industrial belts around. This was not only due to water quantity – a large part of Spain is under semi-arid conditions, although not necessarily deprived of water resources (Garrido & Llamas, 2010) – but also due to pollution; in other words, conflicts arisen from both quantity and quality problems. Public policies have often favored agriculture; since many Spaniards have more or less rural roots and still preserve memory of this background, doing so brings in a high number of votes. Besides this there are also the powerful agricultural lobbies, which are much more socially active than urban supply entities. Some facts reinforce this claim, as low salinity demanding agriculture was the preferential destination of good quality water during the 1960s and 1970s in exchange of more saline and poor quality groundwater for urban supply, as was the case of Las Palmas de Gran Canaria in the 1960s.

In some way water globalization in Spain started in 1950, when the different water uses were considered together, taking into account not only quantity but quality as well, however water quality conscience was retarded and low-placed. Water quality did not start gaining its place until the 1990s, disregarding some exceptions. The importance of water quality is currently widely recognized, albeit not always effectively translated into practice.

An equilibrated consideration of quantity and quality issues is the only pathway to solve combined problems in global analysis. This recognition was common to many countries and leaded toward Integrated Water Resources Management (IWRM) as the only solution to the newly posed problems. IWRM, as it is currently understood, was born in the 1960s (Biswas, 2004).

IWRM is especially a complex way of thinking in countries like Spain, where a traditional water culture exists, which is deeply rooted in the population and the institutions. This originates resistance and reluctance to the introduction of new approaches. In fact, the integration of water problems with administrative tasks and responsibilities led to undesirable results, especially after the autonomous administration of the Regions (Autonomies) was introduced in the Spanish Constitution of 1978. This is the result of inadequate norms and the insufficient corrective means that allow to fully develop the administrative and managerial advantages of the subsidiarity principle, a highly promoted – albeit poorly respected – basic principle of the European Union. Clear abuses, political deviations, poor behavior of authorities, and vested interests have led to current water atomization of the Spanish water administration (Molinero *et al.*, 2011), where short-sighted objectives and political opportunity dominate the scene (Aldaya & Llamas, 2012). A reaction to correct these deviations is slowly starting to appear in recently promoted legal changes. A big loser in this evolution is urban supply, which is currently underfinanced and threatened by future supply failures.

In a globalized analysis of water problems, which is the basis of IWRM, not giving the due role and weight to urban water is a serious error. When water problems began to be combined to find solutions, in the 1950s, world population was close to

2400 million, while it is currently over 7000 million. The challenge is to guarantee sufficient good quality water to a population that consumes and pollutes increasingly. The unstoppable flow of population from rural areas to urban areas currently worsens the anthropic impact on the natural water cycle since, among others, this increases surface runoff and reduces rainfall recharge to aquifers in many areas, but also increases in-town recharge due to leakages, and besides increases environmental degradation and water resources pollution. This situation may probably deteriorate in some areas due to climatic change, although global change – modifications of land use, population and living standards – may dominate. This makes it necessary to take into account urban water in any integral analysis. This global problem, which is currently more acute in developing and emergent countries, has been the case in Spain in the last decades, compounded recently by a high rate of population immigration from abroad.

All this is not new. A few years ago, the Global Water Partnership, which is especially active in promoting integrated water policies, devoted a full monograph to this issue (Rees, 2006). This chapter intends to further elaborate on this issue from a Spanish point of view. Spain is a heterogeneous country with large semi-arid areas, where most of urban and industrial areas are near the sea coast – with the exception of Madrid –, and has important agricultural and feedstock developments to be taken into account, even if they only employ about 3% of the labor force (Garrido & Llamas, 2010).

2 URBAN METABOLISM

In the preceding introduction it has been shown that urban water cannot be considered as the small fraction of total water quantity demand that is directly consumed by human beings. Taking urban water into account is necessary in IWRM, but it is also difficult in many cases. The scenario has to recall what is really a town immersed in a natural environment, and this can be accomplished through the old concept of urban metabolism (Wolman, 1965). It has been recently put forward again to a first plane by Novotny *et al.* (2010).

In what follows, urban water also comprises water for industries in the urban and peri–urban environment. However, water consumed in auto feeding production factories, energy production, and mining are not included.

A town may be compared to a big black box with inflows and outflows (Figure 2), whose sustainability depends on its inner processes – metabolism – since they condition both the inputs and the outputs. Flows are well known and commonly are grouped into five inflow blocks and three outflow blocks.

The five inflow blocks are:

- Water: Surface water, groundwater, reclaimed water, desalinated water.
- Energy: Electricity, motor fuels, domestic gas-currently dominated by natural gas –, and coal, among others. Electric energy origin depends on the energy mix applied by each country and moment.
- Chemicals: From household detergents to pharmaceuticals applied by urban, industrial, and agricultural activities.

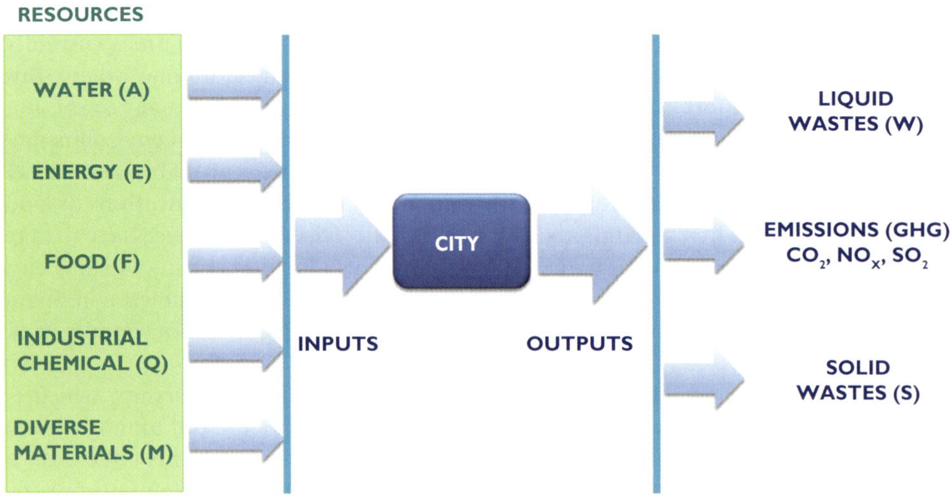

Figure 2 Lineal flow chart of urban metabolism (Novotny *et al.*, 2010).

- Diverse materials: It is the most heterogeneous group and includes construction materials and raw materials, goods, and the equipment needed by urban activity.
- Food: From the point of view of the town, this food is mostly imported.

Virtual water embedded in food could be included among the water inflows, but the concept of urban metabolism has the objective of making an evaluation of urban sustainability and from this point of view including it in the final budget has no clear meaning.

The three outflows blocks are:

- Wastewater: From urban and industrial effluents to highly polluted storm waters.
- Emission of greenhouse gases: They are mostly generated by urban traffic, but also by local factories and domestic burners. Towns are actually responsible for the global emission of 80% of these gases (Hoornweg *et al.*, 2011).
- Solid waste: This includes the different kinds of domestic trash, waste construction materials from building demolition, and furniture and rejected goods such as domestic appliances, computers, TV sets, etc.

The analysis of urban metabolism of a town takes into account the inflows and outflows, as well as the circumstances specific to each case. From the sustainability point of view the analysis of the environmental impacts the town generates, and the economic risks involved, is especially important, which means putting the accent on the guarantee of supply. Dividing the global 'town' in different sub-systems is convenient to analyze the flows, as well as following the different flows together with their interactions, but finally they all have to be combined.

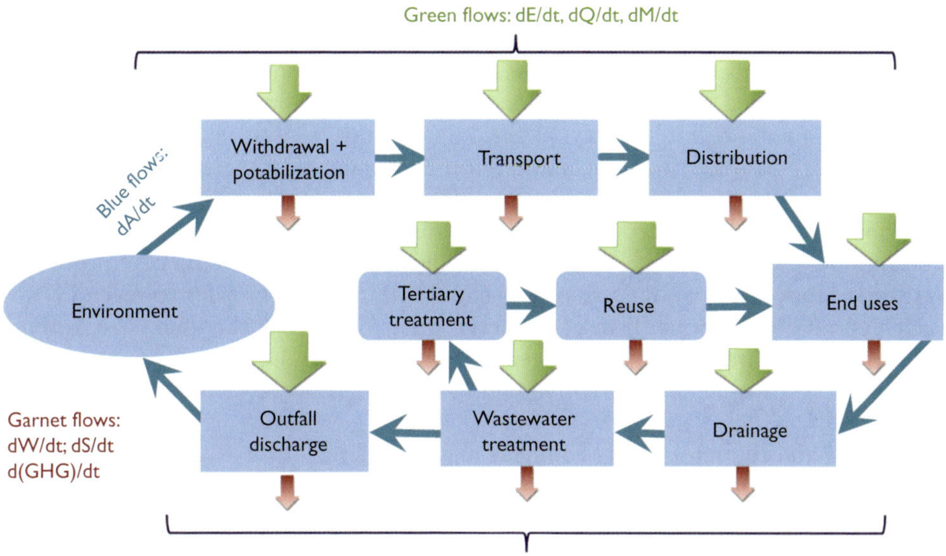

Figure 3 Stages and characterization of urban water metabolism (own elaboration).

In what follows the focus will be the integrated analysis of water resources (IWRM) and thus the main interest refers to characterizing 'water' inflow and 'wastewater' outflow, but without forgetting their interrelations with the other inflows and outflows. Figure 3 shows this synthetically. Each of the urban cycle stages have inflows and outflows derived from the other flows. The green arrows pointing to the stages represent the inflows, the thinner and garnet-colored ones correspond to outflows, and the blue arrows represent water flows between the stages.

Once all these arrows are characterized, the water metabolism of the town under consideration is known. With these characteristics, indicators can be defined to better synthesize the degree of town sustainability from the water point of view. Thus, the water use presented in Figure 3 can be analyzed to lessen or optimize the per capita water use in the town, making it more sustainable. This is the reason by which the traditional line model represented by Figure 2 should be substituted by other cyclic models (Jønch-Clausen, 2004) that favor reuse, as in Figure 3. Water reuse is especially interesting and effective to reach increases in sustainability (Asano, 2006).

There are many strategies that allow improving urban water metabolism. Urbanization, especially when the town is short of green space, greatly favors urban flooding. All the strategies can be grouped in the following three broad categories:

- Minimal alteration of the natural environment existing before urbanization.
- Saving of natural resources by minimizing inflows.
- Reusing the resources as much as possible by transforming progressively the traditional lineal metabolism in other more circular (recycling) cycles as shown in Figure 3; in other words, marching towards the auto-sufficiency paradigm.

Current understanding shows the need to advance along with these strategies. Thus, the need and importance of green areas and constructions, of natural space restoration, of savings and reuse, and using storage space in local aquifers are currently not contested. But to enhance these actions an economic framework is needed, as well as improved and innovating technology. The economic framework can only be attained after integrated analysis, since only in this way the externalities derived from business as usual can be accounted for. For example, if material and especially human damage from a flood are accounted for, action contributing to lessen this hazard can be analyzed and accounted for; then applying these measures will often result in economic profits. To achieve this, technology is a major asset and need, to which sufficient labor and economic resources should be devoted in order to reach IWRM under optimal conditions. From the energy point of view, if the goal is to attain a neutral urban water cycle, energy generation from all outputs containing organic matter has to be upheld. Figure 4 summarizes the path to be followed to attain the paradigm.

Towns and the natural environment may interact in diverse forms, and in both directions. The analysis of urban water metabolism allows to characterize the relationships in the left to right direction. From one part, and in a much more general framework, the analysis quantifies water needs and the impacts that water withdrawals for urban uses transfer to the environment and, what is more important, the possibilities of improvement by disaggregating the urban cycle in its different stages, as well as the identification of the strategies more suited to each case. This is an analysis to be carried out inside a general framework. For example, reusing urban wastewater will not be supported when considered separately from the other components due to its high cost, which, even if supportable, is not negligible. This happens when the value and the unit cost of water in the natural environment is erroneously taken as nil (EEA, 2012). Similarly, valuing the impact of contaminated urban and industrial waste disposals in the receiving natural environment will point out the importance of treating all urban waters properly and justify the accompanied costs. This is crucial for IWRM.

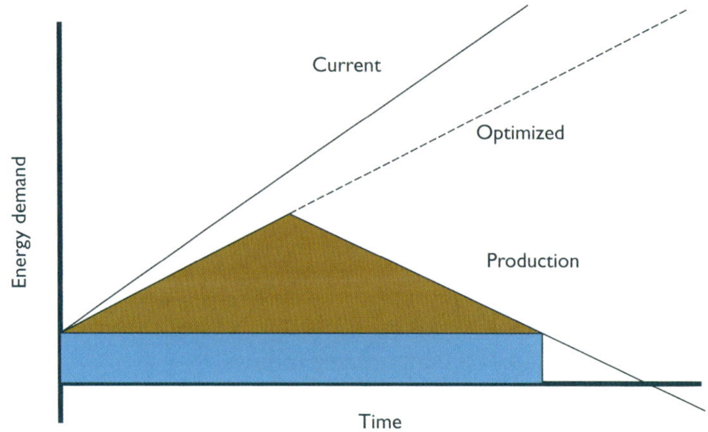

Figure 4 Energy consumption in a town after the strategy to be followed (GWRC, 2008).

The interrelation between the urban environment and the natural environment is actually much more complex. Thus, developing metrics to quantify the interdependence of these flows is a convenient pathway. This is a broad and complex field that requires much more attention than that it has received until present. Some recently developed metrics point to the relation of water network leakages with the embedded energy that is lost with them (Cabrera *et al.*, 2010), some also quantify the combined effect of land use and climatic changes on water consumption, terrain cooling in summer (House-Peters & Chang, 2011), or quantify the capacity for collecting water through urban gardening (Zhang & Guo, 2012). The main conclusion is that all these flows are deeply interconnected, and thus the need to carry out global analyses is highlighted. Figure 5 shows the well-known relationship between water, energy and climatic change.

In summary, the characterization of the different urban water metabolisms of the towns existing in a river basin easies global analyses and the need to manage water resources in an integral way becomes clearer, since urban water management conditions that of the whole river basin and *vice versa*. Considering that externalities – the positive or negative consequence of urban water management on the whole river basin – may be more or less important depending on the management actions that are carried out, it is not possible to analyze only a part but the whole water system.

Quantitative characterization of urban metabolism is a complex issue and also a current need. In fact it is one of the most novel work areas of the TRUST Project (Transitions to the Urban Water Services of Tomorrow) founded by the 7th European Union Framework Program, in which the first author of this chapter participates. A first approach to the analysis is described in one of the deliverables (Morley *et al.*, 2012). This is a concept that cannot be easily applied to agricultural water use since the impacts of urban use are much greater than those derived from agricultural use. What follows is an example.

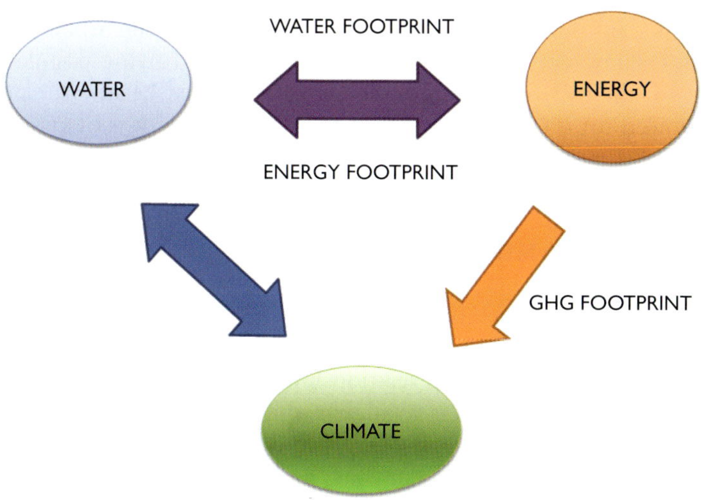

Figure 5 Water – energy – climate interconnections (DHI, 2009). GHG = Greenhouse Gases.

In the urban and agricultural areas of Valencia, in eastern Spain, with an urban density of 6000 inhabitants/km^2, total water requirements are about 250 L/person/day, equivalent to 0.547 hm^3/km^2/year, while water demand for an orange-tree field is 0.4 hm^3/km^2/year. This is less than the former value but of the same order of magnitude. When considering the other unit inputs the figures for other demands are several orders of magnitude lower. For electric energy, considering 5700 kWh/person/year and 0.25 kWh/m^3 of irrigation water, the respective figures are 34.2 and 0.1 GWh/km^2/year, or a ratio of agriculture to urban of 0.0034. The difference is still more conspicuous when considering the use of motor fuels given the enormous difference in the number of vehicles. The differences are still greater for other inputs. Considering chemicals, urban consumption of soap and detergents, at a rate of 10 kg/person/year, is 60 ton/year/km^2, and adding to this cosmetics, perfumes, pharmaceuticals, drugs, paints, plastics, synthetic materials, and others the total input may be estimated to be over 500,000 kg/year/km^2 (Harnick, 2010) of materials that carry with them a high contamination load. They appear today as concerning emergent contaminants in rivers and aquifers (Barceló & Petrovic, 2008). All this supports the need to integrate urban metabolism in IWRM.

The same conclusions can be reached when taking into account outflows. The comparison has only a meaning for water, albeit urban and agricultural activities pollute in a very different form. Cities and towns produce highly polluting discharges, although many of them are point sources and consequently more prone to be controlled, albeit very numerous. Conversely agriculture produces mostly diffuse pollution, which is much more difficult to be controlled, but that generally carry a much lower contaminant charge, but on a very large territory. The second of the outputs is solid wastes. This forms one of the major urban problems and amounts to approximately 0.5 ton/person/year, or 3000 ton/km^2/year. In agriculture the amount is almost negligible or at most a small quantity in the case of intensive agriculture under plastic cover. The greatest difference is for the third output, the emission of greenhouse effect gasses. While cities produce the 80% of the planet's emissions, agriculture provides a sink for these emissions, mainly in the vegetative growth period.

The result is that the metabolic analysis of the agricultural use of water is much more easy and its unit impact is clearly much lower than for the urban case. This fact and the need to guarantee the health of a continuously growing population are the main reasons for the increasing importance of urban water, besides the political one, and stress the need for it to be included in the top-down analysis of any IWRM.

3 RELATIONSHIP BETWEEN NATURAL ENVIRONMENT AND TOWN

The characterization of the metabolism of a town helps in rationalizing global analyses, since it allows quantifying the transfers of the impacts generated by urban activities to the natural environment. This relationship is very important in both directions. Thus, when better quality water is available, this helps a lot in urban metabolism since the energy consumption to convert it into drinking water is much less. This is because energy-intensive treatments, such as those using membranes, can be avoided or lessened. Since what is dealt with is urban water, the water–energy interaction is only considered in elemental form.

The natural environment and town relationship transcends the immediate water quality problems that river basin management may transfer to populations, or those already mentioned when dealing with water quality in the urban metabolism. Some interesting relationships appear. One of them is differential land subsidence due to piezometric level drawdown due to intensive groundwater abstraction, which may break down pipelines containing sewage or potential contaminant fluids – crude oil, motor oils, petroleum products –, as it is the case in some areas of Guanajuato, Mexico. Intensive groundwater development has also been put forward to explain the triggering – not the energy accumulation – of the May 2011 earthquake of Lorca, Murcia, Spain (González *et al.*, 2012). However, this is marginal and the real causes are controversial and much more complex. An earthquake is accompanied by large debris flows of construction materials and solid wastes from demolished buildings, but also by the disruption or urban water supply systems.

Thus, river basin management may have important consequences in urban metabolism, the most conspicuous impacting water quality. A clear example is nitrate excess in agriculture. Mining activities in the watershed may also have serious consequences and condition IWRM. This is the case for Barcelona, the second largest city in Spain, where potash salt mining since the 1920s in the mid Llobregat River Basin badly affected river and aquifer water salinity, obliging to adopt expensive special measures although having limited results. This is the consequence of a lack of integral vision of the river basin in the mid of the 20th century, when mining production was a protected sectorial affair and urban water supply had not enough power to protect its water resources (Marcé *et al.*, 2012; Custodio, 2012). There are no sound economic studies about the consequences. Potash exploitation produced and continues to produce benefits, but is accompanied by important costs to population, agriculture and industry due to the high river water and related groundwater salinity, to which costs of investments to control the effects and to improve water quality should be added. For long-term benefits to compensate for these costs, a too high and unacceptable discount rate should probably be applied.

4 URBAN WATER AND INTEGRATED WATER RESOURCES MANAGEMENT

From what has been said before it is clear that urban water is an essential part of IWRM when considering the two-way impacts of the town on the basin where it is located, and *vice versa*. This is highlighted by the following synthesis:

- Watershed management impacts on urban water
  - Quantity impacts
    - Multi-use water management, particularly important during drought periods
    - Aquifers' reserves depletion
  - Quality impacts (drinking water)
    - Organic water pollutants, such as problems with THM (three-halo-methanes) for superficial waters
    - Non-point nitrate groundwater pollution
    - Aquifer water salinization

    - Other impacts
        - Floods and dam failures
        - Subsidence due to groundwater head drawdown
- Urban water impacts on the watershed
    - Quantity impacts (urbanization modifies the natural hydrologic cycle)
    - Quality impacts (main water pollution comes from urban water)
    - Other watershed impacts (thermal contamination, biodiversity changes) Biodiversity changes may be of paramount importance (Vorosmarty *et al.*, 2010).

The interrelations that justify including urban water in the integrated analysis of the river basin from an economic point of view can be framed inside the following scenarios (Rees, 2006):

- The negative externalities that arise from the uncoordinated use of water and land resources.
- The opportunity costs which arise when factors of production – including water, land, and capital – are employed for low-value/benefit purposes.
- The negative externalities and opportunity costs which arise from the non-coordinated provision of interdependent basic services.
- The cost savings which can occur by widening the range of management options.

An example is given below for each of these four groups. Their analysis will allow the evaluation not only of the additional economic costs of not-solving the problem with a broad point of view, but also to quantify the corresponding social, political and environmental impacts.

1 Negative externalities derived from the non-coordinated use of water: Aquifer contamination due to intensive use of fertilizers and other agrochemicals produce important externalities. The need to maintain or attain a high drinking water quality for urban use, generally by means of denitrification, has a cost that often is much higher than the lost income for the farmer due to the crop reduction as a consequence of not using or reducing the use of fertilizers. Thus it seems that a deal should be arranged on a win-win basis. There are some experiences, such as those involving the Danish farmers and the companies responsible for urban water supply to Copenhagen (Henriksen *et al.*, 2007). Andrews & Zabel (2003) carried out a survey of 435 cases of what they called co-operative agreements to reduce nitrate load to aquifers in order to improve groundwater quality for drinking purposes, mostly in Germany (with emphasis in Bavaria and Hesse), but also in other Central and Northern European countries, by transferring economic resources from suppliers to farmers. Interesting good results are mentioned, although not always since time required to see the results is often long and natural situations are complex. More experience is needed but this is an interesting activity to consider in IWRM linking urban and agricultural interests. This is what is behind the good agricultural practices considered in the EU Directive on Nitrates and the Water Framework Directive (Custodio *et al.*, 2012), although economic instruments are not defined.
2 Opportunity costs derived from a non-efficient use of water, and for purposes with lower priority/value: The reform of the Spanish Water Act (Ley 46/1999)

considers the possibility – albeit used only in some cases – to re-assign water uses through deals and water rights exchange centers in dry periods and droughts (Embid, 2008). In this example of IWRM, the priority user (urban water) always plays a relevant role.

3 Externalities and opportunity costs derived from an uncoordinated water use: Inadequate management of storm urban runoff. Its disposal into the natural environment under good conditions has an important economic cost, but this cost is less than the damage of a direct disposal, as externalities and lost opportunities, taking into account the contaminant load that it carries and generates. Since clean water for drinking, fishing, and swimming becomes scarce, infusing rainwater with contaminants and toxins before dumping it into waterways is an increasingly bad idea. This explains why cities have traditionally applied public funds to finance storm water infrastructures. However this is becoming increasingly untenable from the economic point of view and it seems inevitable that private investment will be needed as cities contemplate future storm water management planning. Saving the rain is a comprehensive storm water management program linked to 'green' infrastructure projects and significant storm water storage projects. Focusing attention on the employment and the economic development impact of green infrastructure projects can help to build broader constituencies of support for smart investments.
4 Savings derived from analyzing all possible solutions: A good deal of the European Structural Funds of the last decades has been spent in the construction of head reservoirs in the towns when the correct use would be the repair of the distribution network losses. However those in charge of fund management had no responsibilities inside the municipal administration, did not know the real problem, and were pressed to look for an easy and fast project to get the funds, especially if this meant politically visible and rewarding constructions. Cases of this kind have been numerous, especially in Southern Europe. So, the European Union passed to the Court of Auditors to perform a deep revision on the projects carried out (ECA, 2010).

All of this shows the absolute need of wide-spectrum water policies. IWRM offers a broad panorama of water resources and allows finding a solution with the optimal cost/benefit ratio. Once identified, the solution has often a clear local character. Its analysis follows an inverse path, bottom-up, which honors the well-known Dubos strategy 'think globally, act locally' (Hirsch & Moberg, 1989), as the starting point of IWRM.

5 BASIC RULES TO GET URBAN WATER TO BE PART OF IWRM

The need to integrated water management is clear, but how to carry it out in practice may be quite difficult. While the feeling of this need is relatively recent, water management is as old as mankind. This means that management implementation is conditioned by history, which may introduce serious handicaps, such as existing water rights, atomized and uncoordinated water administration, local and political interests, lack of environmental education, poor transparency, laws developed under a context very different from the current one, and vested interests. All this is a heavy load that prevents giving flexible answers to current dynamic problems. Besides, since the political and hydrological circumstances of each country are so different, there are neither common nor magic formulas to ease the path from compartmentalized to global procedures. Thus it is no surprise that lots of papers deal with this complex issue (Pochat, 2008; Jønch-Clausen, 2004; Rees, 2006; Grigg, 2008; UNW & GWP, 2007; Binney *et al.*, 2010), which besides admits an infinity of shades that depend on the importance of the problem to be considered. Looking for improvement of water supply guarantee to the metropolitan area of a large city is not the same as improving water quality in its surroundings, even if both objectives are linked in some way (Rees, 2006).

However, it is possible to synthesize the way to address a water problem from an integrated point of view in the following four stages:

- Right identification of objectives; this is needed to find the most convenient pathway towards a solution.
- Correctly define the matters for which the partners are responsible and competent in order to achieve the wanted objectives.
- Coordinated action of the involved institutions. This is a key issue, especially in the urban context, since problems are every time more interlinked. As an example, land planning also involves water policies, but often the two responsible administrations for land planning and water ignore each other, even in cases when they are under the same ministerial department. The same can be said of the management of water and energy, two of the most important natural resources.
- All possible strategies have to be explored and evaluated, after this one should opt for the most appropriate.

The third stage is without doubt the most complex. Institutions are led by politicians and policy makers who have different points of view and interests, generally considering a too short time scale – most often the lapse between elections – when it comes to environmental issues. It is no surprise that this aspect is always present in any work that deals with issues concerning urban water. The most important organization dealing with urban water, the International Water Association (IWA) has created the Spatial Planning and Institutional Reform Working Group (Binney *et al.*, 2010).

All that's been said above cannot be carried out without the involvement of all stakeholders, which is also a topic always present in all works related to IWRM (Wortzel, 2009; Jønch-Clausen, 2004; Rees, 2006; Grigg, 2008; UNW and GWP, 2007; Binney *et al.*, 2010) and that focus on water resources governance. This topic will not further be dealt upon here since the current chapter focuses on the need to manage urban water in a more general framework. In this context, the Water Urban Utilities are in a pivotal position to promote watershed management programs, cooperate in improving stream flows and bring together various local stakeholders to safeguard aquatic ecosystems for the mutual benefit of all (Grigg, 2012).

6 CONCLUSION

Globalization is the term that most probably synthesizes best the changes produced in the last years. Figuratively speaking, the world is currently much smaller than it was in the past, and consequently what happens in our antipodes is known practically in real-time.

More specifically, we now share problems, from the economic markets to climate change. We share the environment and thus its related problems as well to some extent.

While problems have been globalized, and consequently their correct solution also needs integral analyses, the institutions and the solutions have not been adapted to this new context. This explains the continuous environmental degradation that has been experienced until now in most areas, with few exceptions. This is not another form of the 'tragedy of the commons' (Hardin, 1968) since there is the possibility to control and to compensate the damage to some extent. However, this needs time, commitment and IWRM. This is what underlies the European Union Water Framework Directive of the year 2000. Even if current results are not as good as expected, this has been an important step forward and it allows planning further steps (EU, 2012).

The water world, especially in the countries where water is scarce, has a millennial culture that becomes a handicap to the adequate institutions and solutions when placed in a new context. In this situation, Integrated Water Resources Management appears as the strategy that will enable to successfully face the new challenges, and to move from competition for the resources to coordinated and cooperative use. This is largely the case for agriculture and cities in many occasions. However, water policy is still far away from integrated considerations and the path to be followed to attain this can be assumed to be long. This chapter, centered on urban water, has made clear the need of carrying out global analysis as the only strategy to find adequate solutions and it has coarsely outlined the guidelines to be followed to reach this. While the diagnosis is unchallenging and easy to set, the changes needed to implement action, and especially the implementation order, are subjects that admit many diverse pathways. With independence of priories and their temporal ordering, action and governance are needed urgently. Problems grow with time, and the future of coming generations is at stake.

The metropolitan area of Eastern Pima County, Arizona, is developing and implementing a Regional Optimization Master Plan as a US$ 720 million program to modernize and upgrade the two major RWRD metropolitan treatment facilities, in order to improve the water quality for recharge and reuse, develop a state-of-the-art water quality laboratory, incorporating solar and rainwater harvesting structures, while helping to develop habitat and bird nesting features. In conclusion, good and poor practices of IWRM (Box 1 and 2) are frequent all around the world albeit sustainability and future generations will require increasingly the first and eradicate the second ones.

Box 1: Good practice: Recent undertakings to manage urban water

As part of Kansas City's federally-mandated Overflow Control Program, a 744-acre green infrastructure project is underway in one watershed to reduce combined sewer overflows. Green infrastructure is used to intercept storm water, keeping it out of the combined sewer system, reducing the overflow and the amount of excess water that gets pumped and treated. To further enhance the green infrastructure efforts, the City is working with residents and neighbors to make improvements on their own properties by reducing water consumption and reducing the amount of storm water that leaves a property through runoff or direct connections to the sewer system.

Box 2: Poor practice: Lack of agreement to reuse treated wastewater

The plant of Pinedo treats the wastewater of the city of Valencia, which are subsequently disposed into the sea through an outfall. About 120 hm^3/year of water are being treated here, which is enough to irrigate 25,000 ha of orange trees that currently are served with water from the Turia River, which also supplies urban water to Valencia with a high supply guarantee. During the 2004 and 2005 drought, irrigation flows were not guaranteed and then the possibility of constructing facilities to be able to irrigate the area with treated wastewater (reclaimed water) was put forward. The amortization period of these works would be rather short since they involved only a pipe about 12 km long, elevated 20 m. Financing was available through European and Spanish funds; operation and maintenance, about 0.04 €/m^3, was to be paid by the farmers. While the drought period lasted, irrigators showed good interest in the project, which was launched. However, as soon as rain resumed they shifted their interest to continuing with the concession of river water, the water used by them traditionally and without energy cost since it is served by gravity. The town of Sagunto, was also dependent on water from the Turia River and already linked by a pipe. With farmers using river water for irrigation, an additional supply to Sagunto to cover new needs (5 hm^3/year) was not guaranteed without taking water from farmer's concession. To avoid this conflicting situation, probably a not very conflicting one, a 8.4 hm^3/year sea water desalination plant was commissioned and the reclaimed water project was cancelled.

This is a lost opportunity for increasing available regulated water quantity at a lower cost. However things may be different when water quality considerations are taken into account, since reclaimed water is slightly more saline, their quality is not so well guaranteed, and there is the fear of rejection of agricultural products on the markets once the origin of applied water is known. On the spot solutions, under the pressure of special circumstances, in this case an exceptional drought, often lack a sound analysis and do not consider the time to adapt to new (unexpected) circumstances. In the 1970s, subsidized reclaimed water from the Las Palmas de Gran Canaria wastewater treatment plant was pumped to the Almatriche agricultural area in exchange of groundwater supplied to the farmers for town supply. In this case the reclaimed water salinity and chemical quality was poor for soil irrigation as the treatment plant suffered intermittent problems leading to excess salinity and pollution due to poor operation of the sewage network. The result was that the soils were spoiled due to salinization and alkalization. As a consequence, new developments in the field of reclaimed water use have been slow and serious urban sewage network surveillance efforts were introduced. Later experience in salinity reduction through inverse osmosis allowed the distribution of reclaimed water to other areas of the island.

REFERENCES

Aldaya, M.M. & Llamas, M.R. (2012) *El agua en España: Bases para un pacto de future.* Water in Spain: basis of a pact for the future. Fundación Botín, Madrid.

Andrews, K. & Zabel, T. (2003) Achievements of co-operative agreements. In: Brower, F.M., Heinz, I., Zabel, T. (eds) *Governance of Water-Related Conflicts in Agriculture.* Boston, Kluwer Acad. Publ. pp. 45–65.

Asano, T. (2006) Water recycling – a relevant solution? In: Rogers, P., Llamas, M.R. & Martínez-Cortina, L. *Water Crisis: Myth or Rality* London, Taylor & Francis. pp. 261–284.

Barceló, D. & Petrovic, M. (eds.) (2008) *Emerging contaminants from industrial and municipal waste: occurrence, analysis and effects.* Berlin, Springer Verlag.

Binney, P., Donald, A., Elmer, V., Ewert, J., Phillis, O., Skinner, R. & Young R. (2010) *Cities of the future: Program, spatial planning and institutional reform.* Discussion Paper for the World Water Congress, September 2010. London., International Water Association.

Biswas, A.K. (2004) Integrated water resources management: A reassessment. *Water International*, 29(2), 248–256.

Cabrera, E., Pardo, M.A., Cobacho, R. & Cabrera, E. (2010) Energy audit of water networks. *Journal of Water Resources Planning and Management*, 136, 669–677.

Custodio, E. (2012) Low Llobregat aquifers: Intensive development, salinization, contamination, and management. In: S., Sabater, A., Ginebreda & D., Barceló (eds.) *The Llobregat: The Story of a Polluted Mediterranean River.* The Handbook of Environmental Chemistry, 21. Berlin, Springer. pp. 27–50.

Custodio, E., Garrido, A., Coleto, C. & Salmoral G. (2012) The challenges of agricultural diffuse pollution. In: De Stafano L. & Llamas M.R. (eds.) *Water, Agriculture and the Environment in Spain, can we square the circle?* Fundación Marcelino Botín. Leiden, CRC Press. pp. 153–163.

DHI (Danish Water Institute) (2009) *Linking Water, Energy & Climate Change.* A proposed water and energy policy initiative for the UN Climate Change Conference, COP15, in Copenhagen 2009.

ECA (2010) *Is EU structural measures spending on the supply of water for domestic consumption used to best effect?* European Court of Auditors. Luxembourg, Publications Office of the European Union.

EEA (2012) *European waters – current status and future challenges: Synthesis.* European Environmental Agency. Report number: 9.

EU (2012) *A blueprint to safeguard Europe's water resources.* Communication from the Commission to the European Parliament, the Council, the European Economic and Social Committee and the Committee of The Regions. European Commission, Report number: COM (2012) 673.

Embid, A. (2008) *Asignación del agua y gestión de la escasez en España: Los mercados de derechos de aguas.* Semana Temática, number: 7. Tribuna del Agua. Expo 2008, Zaragoza.

Garrido, A. & Llamas, M.R. (eds.) (2010) *Water policy in Spain.* Boca Raton, CRC Press.

González, P.J., Tiampo, K.F., Palano, M., Cannavó, F. & Fernandez, J. (2012) The 2011 Lorca earthquake slip distribution controlled by groundwater crustal unloading. *Nature Geoscience*, 5, 821–825.

Grigg, N.S. (2008) Integrated water resources management: balancing views and improving practice. *Water International*, 33(3), 279–292.

Grigg, N.S. (2012) Water utility leadership and its key role in protecting aquatic ecosystems. *J. Am. Water Works Assoc.*, 104(10), 50–59.

GWRC (2008) *Water and energy: Report of the GWRC Research Strategy Workshop.* Global Water Research Coalition, London.

Hardin, G. (1968) The tragedy of the Commons. *Science*, 162, 1243–1248.

Harnick, P. (2010) *The future of the european chemical industry.* KPGM International Report. UK.

Henriksen, H.J., Rasmussen, P., Brandt, G., von Bulow, D. & Jensen, F.V. (2007) Public participation modelling using Bayesian networks in management of groundwater contamination. *Environmental Modelling & Software*, 22, 1101–1113.

Hirsch, J.G. & Moberg, C.I. (1989) *René Jules Dubos: A biographical memoir.* National Academy of Science. Washington D.C.

Hoornweg, D., Sugar, L. & Gomez, C.L.T. (2011) Cities and greenhouse gas emissions: Moving forward. *Environment and Urbanisation*, 23(1).

House-Peters, L.A. & Chang, H. (2011) Modeling the impact of land use and climate change on neighbourhood-scale evaporation and nighttime cooling: A surface energy balance approach. *Landscape and Urban Planning*, 103, 139–155.

Jønch-Clausen, T. (2004) *Integrated water resources management (IWRM) and water efficiency plans: Why, what and how?* The Global Water Partnership, Technical Committee. TEC Background Papers, number: 10.

Llamas, M.R. & Martínez-Santos, P. (2005) Intensive groundwater use: Silent revolution and potential source of social conflicts. American Society of Civil Engineers (ASCE). *Journal Water Resources Management*, 13(5), 337–341.

Marcé, R., Honey-Rosés, J., Manzano, A., Moragas, L., Catllar, B. & Sabater, S. (2012) The Llobregat river basin: A paradigm of impaired rivers under climate change threads. In: Sabater, S., Ginebreda, A. & Barceló, D. (eds.) *The Llobregat: The Story of a Polluted Mediterranean River.* The Handbook of Environmental Chemistry, 21. Springer. pp.1–26.

Molinero, J., Custodio, E., Sahuquillo, A. & Llamas, M.R. (2011) Groundwater in Spain: Legal framework and management issues. In: Findakis, A.N. & Sato, K. (eds.) *Groundwater Management Practices.* Leiden, CRC Press, 123–137.

Morello, L. (2009) *Las Vegas gambles with an uncertain water future.* Energy & Environment, The New York Times, November 10, 2009.

Morley, M.S., Kapelan, Z. & Savić, D.A. (2012) *Deliverable D54.1: Integrated Decision Support Framework.* [Online] TRUST project. Available from: www.trust-i.net [Accessed October 2012].

Novotny, V., Elmer, V., Furumai, H., Kenway, S. & Phills, O. (2010) *Water and energy framework and footprints for sustainable communities.* IWA World Water Congress. Montreal, September 19–24.

Pochat, V. (2008) Principios de gestión integrada de los recursos hídricos: Bases para el desarrollo de planes nacionales. [Integrated water resources management principles: Bases for developing national plans]. Global Water Partnership.

Rees, J.A. (2006) *Urban water and sanitation services: An IWRM approach.* Global Water Partnership, Technical Committee. TEC Background Papers, number: 11.

UNW & GWP (2007) *Roadmapping for advancing Integrated Water Resources Management (IWRM) processes.* United Nations- Water & Global Water Partnership. Danish Government, Copenhagen.

Vorosmarty, C.J., McIntyre, P.B., Gessner, M.O., Dudgeon, D., Prusevich, A., Green, P., Glidden, S., Bunn, S.-E., Sullivan, C.A., Reidy, Liermann, C. & Davies, P.M. (2010). Global threats to human water security and river biodiversity. *Nature*, 467, 555–561.

Viollet, P.-L. (2005) *Histoire de l'energie hydraulique.* Paris, Presses de Ponts.

Wolman, A. (1965) The metabolism of cities. *Scientific American*, 213(3), 179–190.

Wortzel, A.W. (2009) *Water supply: is it still a local decision?* Journal AWWA, 101(6), 48–51.

Yagüe, J. & de Cea, J.C. (2008) *Evolución de las presas en España.* Revista de Obras Públicas, number: 3493.

Zhang, S. & Guo, Y. (2012) An explicit equation for estimating the stormwater capture efficiency of rain gardens. Journal of Hydrologic Engineering, 10.

Section 3

Selected case studies on Integrated Water Resources Management

Chapter 11

Integrated water resources in Peru – The long road ahead

A. Julio M. Kuroiwa
Director of the National Hydraulics Laboratory and Associate Professor at the School of Civil Engineering, Universidad Nacional de Ingeniería, Lima, Peru

Luis F. Castro
Junior Instructor and Graduate Student, Universidad Nacional de Ingenieria, Lima, Peru

Marielena N. Lucen
Water Resources and Environmental Auditor, Ministry of Energy and Mines, Lima, Peru

Julio I. Montenegro
Junior Instructor and Graduate Student, Universidad Nacional de Ingenieria, Lima, Peru

ABSTRACT: This chapter describes the current water resources situation in Peru, recent efforts to promote governance in water resources management and challenges the country faces in this area. The physical territory and the variability of water availability are described and water uses are discussed. Special emphasis is given to the extremely low water availability in some areas of the Pacific Basin, where population is the highest and surface water availability the lowest. Examples of the impact of economic growth on consumption habits are given. Use of water in different economic activities is described. Progress in the water footprint estimation of selected agricultural products is also presented. Agriculture is being studied in detail as it consumes 80% of the water resources of the nation and a few examples of 'cash per drop' are given. Groundwater resources are especially important in the Pacific Basin from the economic viewpoint and two representative cases are included. The change of the energy matrix is also affecting water uses, as ethanol (made of sugar cane) is incorporated into gasoline by law. Environmental and social issues are pointed out, and the degradation of large areas of land and the pollution of water resources at a large scale is discussed. It is concluded that a number of factors such as population growth, economic growth, change in the energy matrix, the lack of regulation enforcement, and a lack of planning have led to an increasing pressure on land and water resources. In some cases, damage to the environment and water resources might be irreversible. The background history and concepts under which the National Water Authority was created and the roles assigned to this institution are briefly described and discussed. This entity has an enormous challenge in securing one of the nation's most valuable resources: Water.

Keywords: IWRM, Peru, water overuse

1 INTRODUCTION

Peru with a surface area of 1.285 million km² and an estimated population of 29.46 million as of 2010 is located in Western Central South America approximately between parallels 0° and 18°21'S and meridians 68°40'W and 81°30'W. Seen its location, Peru's weather should be tropical. However, the presence of the cold Peruvian Current, the Andes cordillera, the South Pacific High (a subtropical anticyclone) and the Amazon Jungle have created a wide variety of climates and a very uneven water distribution throughout the country. In general, precipitation is zero near the Peruvian coastline and grows in the East direction. Moreover, Peru's population and economic activities are concentrated along the coastline, on the Pacific Basin, where water resources are most scarce. Even economic activities that are large water consumers, such as agriculture are more intensive along the Coast.

2 WATER DISTRIBUTION – SPATIAL AND TEMPORAL VARIABILITY

The Andes Cordillera rose because of the interaction of the South American Plate and the Nazca Plate. The Continental Divide is not far from the Pacific Ocean, which has caused the formation of 53 very steep river basins with narrow mountain valleys west of the divide. Rivers have deposited sediments along the coastline forming land suitable for agriculture, although in some cases salinity is too high. Agricultural valleys are formed in the vicinity of the watercourses. The hydrographic region composed by these river basins will be called the Pacific Basin. On the other hand, east of the divide, surface waters flowing to the Atlantic Ocean, mostly in narrow mountain valleys and then in the Amazon Jungle, are finally collected in the Amazon River before they leave Peru's territory. This watershed is called the Amazon Basin. Finally, an endorheic basin is shared by Peru and Bolivia. Waters are collected at Lake Titicaca before they flow into Bolivian territory. This will be called the Titicaca Basin. Figure 1 shows Peru's main hydrographic regions.

Ordóñez & Vera (2012) estimated the average annual precipitation in Peru's main hydrographic regions. Although precipitation variability is high in each region these averages can give an idea of the distribution of precipitation in the Peruvian territory. It can be seen that in the Amazon Basin precipitation is almost eight times greater than in the Pacific Basin. In reality, the range of precipitation within each basin is very broad. A study by INRENA (1995) estimated the water availability per hydrographic region. Population was updated in a report by ANA (2009). Approximate numbers of water availability and population are also given in Table 1. In essence, the Pacific Basin is home to 65% of the population but only produces 1.8% of the water resources of the nation. Water availability is 2040 m³/person/year. On the other hand, the Amazon Basin produces 97,7% of the Peru's surface water, has only 30% of the population and offers a water availability of 232,979 m³/person/year, more than 100 times more than the water availability of the Pacific Basin.

Figure 2 shows a map of mean annual isohyets based on mean annual precipitations with data collected between 1970 and 2000. It is evident that precipitation grows significantly in the East direction. Precipitation is higher in the lower reaches of the Amazon Basin, where it can be greater than 3000 mm/yr.

Figure 1 Hydrographic division of Peru. The three main basins are shown (own elaboration from Instituto Geográfico Nacional data).

Table 1 Characteristics, precipitation, water availability and water availability per capita in Peru's main hydrographic regions.

Basin	*Area (10^3 km²)*	*Average annual precipitation (*) (mm)*	*Water availability (MCM/yr)*	*% Water availability*	*Population*	*% Population*	*Water availability (m³/person/yr)*
Pacific	279.7	274.3	37,363	1.8	18,315,276	65	2040
Amazon	958.5	2060.8	1,998,752	97.7	8,579,112	30	232,979
Titicaca	47.2	813.5	10,172	0.5	1,326,376	5	7669
Total	1285.2		2,046,268	100.0	28,220,764	100	72,510

Note: (*) These values were estimated in Ordoñez & Vera (2012). Other values were extracted from ANA (2009), first cited in INRENA (2009). Population data was updated in ANA (2009).

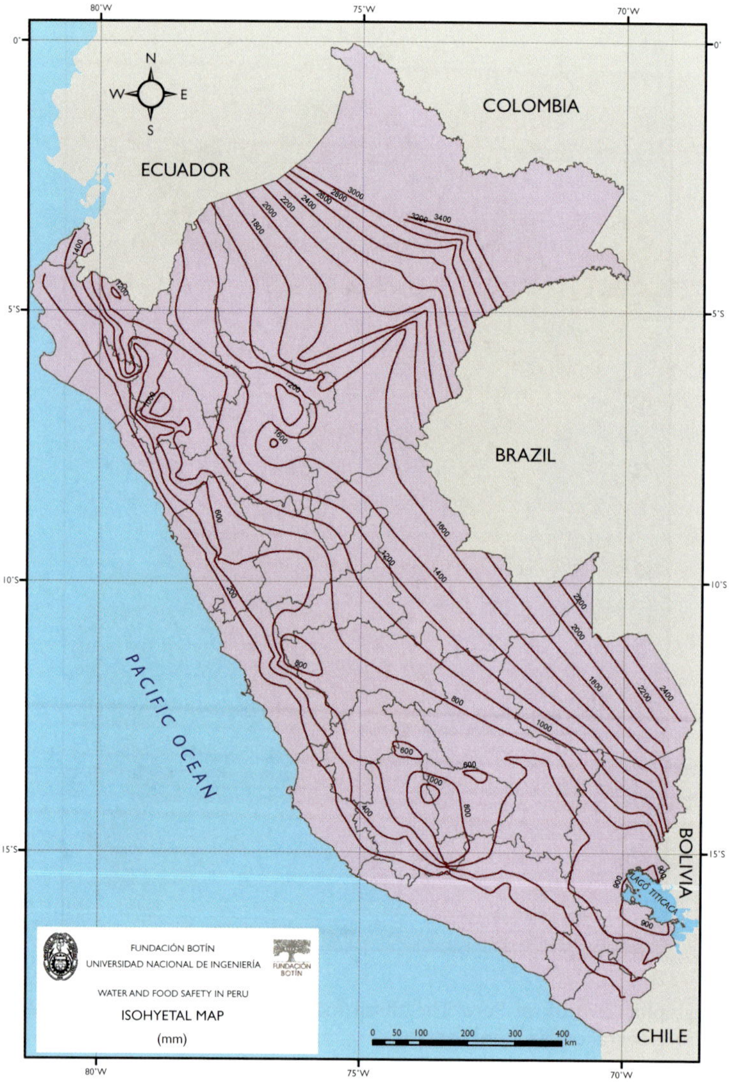

Figure 2 Isohyetal map of Peru. Annual precipitations are shown in mm (own elaboration from Instituto Geográfico Nacional (IGN, 2012) and el Servicio Nacional de Meteorología e Hidrología (SENAMHI) data available from ANA (2010).

Large water bodies have been identified and their surface areas and storage capacities have been estimated. A national inventory indicated that lakes in the Pacific Basin have a total storage capacity of 1996 Millions of Cubic Meters (MCM). Closed systems can hold up to 226 MCM. In the Amazon Basin, lakes can hold up to 4610 MCM and in the Lake Titicaca Basin lakes have a capacity of 149 MCM (INRENA, cited in ANA (2009)).

For illustrative purposes, Peru can also be divided in three distinct regions that extend north to south: The Coast, a very narrow strip, mostly desert-like with flat to

Figure 3 Peru's main natural regions (own elaboration from Instituto Geográfico Nacional and el Servicio Nacional de Meteorología e Hidrología (SENAMHI) data available from ANA (2010).

undulating relief; the Sierra, a very mountainous territory where semiarid conditions prevail, and the Jungle, that can be divided into Upper Jungle, a humid mountainous area between 400 m above sea level (m.a.s.l) and 1000 m.a.s.l., and the Lower Jungle, whose altitudes fall below 400 m.a.s.l in which the weather is much more humid and warmer. The Lower Jungle is considered unsuitable for intensive agriculture because the layer of organic soil is very thin and rapidly washes out after the original vegetation has been cleared. The majority of Peru's statistics refer to this so-called 'natural regions' as geographic, economic and cultural differences evident throughout the country. The Coast is composed, in essence, by the lower reaches of the Pacific hydrographic region, below elevation 500 m.a.s.l.

Most of surface runoff occurs in the rainy season. Approximately 80% of water flows between December and April in the Pacific Basin. It has been estimated that the volume of excess flow during the rainy season is approximately 17,452 MCM. This amounts for 50% of the total annual volume and cannot be used before it enters the ocean (ANA, 2009).

Severe droughts have caused major social and economic impacts. In 1983, during an extremely intense El Niño Southern Oscillation (ENSO) event, one of the most severe droughts in decades affected Peru's SE region while in the Peruvian NW region, near the Peru-Ecuador border, the largest floods in decades destroyed crops, roads (including the Pan-American Highway) and properties.

To satisfy the increasing demand of the Pacific Basin, a number of large projects to divert water from the Amazon Basin have been conducted. For instance, the project Marcapomacocha consists of diverting water from the upper Mantaro Basin into the Rimac River Basin where Lima, Peru's capital is located. Water is collected in lakes that are near the divide and the collected water is diverted to the Rimac River Basin through a tunnel. This is a joint project between SEDAPAL, Lima's water supply and sanitation company, and EDEGEL, a private power generation operator. In 2005, EDEGEL used more water than it was allowed to, and SEDAPAL urged them to decrease the release of water from the reservoirs because the 2005–2006 rainy season might have been lower than average. Eventually, there was not a significant water shortage, but it helped establish better rules of operation between both companies.

The Olmos Project is currently underway. It consists of diverting water from the Huancabamba River, the Tabaconas River and the Manchara River, in the Atlantic Basin, into the Quebrada Lajas and the Olmos River to irrigate 43,500 Ha (in a First

Figure 4 Limon Dam and reservoir. It was built on the Huancabamba River, in the Amazon Basin, to store water and divert it to the Olmos River, in the Pacific Basin (Kuroiwa, August 12, 2012).

Table 2 Population of Peru's seven largest populated metropolitan areas.

City	*Altitude (m)*	*Basin*	*Geographical area*	*Population (2007)*
Lima-Callao	154 (Downtown Lima)	Pacific	Coast	8472,935
Arequipa	2335	Pacific	Mountains	749,291
Trujillo	34	Pacific	Coast	682,834
Chiclayo	29	Pacific	Coast	524,442
Piura	29	Pacific	Coast	377,496
Iquitos	106	Amazon	Lower Jungle	370,962
Cuzco	3399	Amazon	Mountains	348,935

Phase) and to produce energy that will serve the cities of the Western Slope (Gobierno Regional Lambayeque-Peru, 2012). Diverting water from the Amazon Basin has proven to be troublesome in other areas, where project developers have faced opposition from local farmers and citizens who thought that they were going to be left with less available water for their present and future uses.

In general, central and southern coastal basins are very steep and narrow. Therefore, it appears that it is not very practical to build in stream reservoirs, as the storage capacity might be too low and incoming sediments from the young Andes Cordillera may fill them rapidly.

3 POPULATION DISTRIBUTION

Most of Peru's population is concentrated along the coastline where population growth rates are higher. Although the birth rate has diminished in the Pacific Basin, migration from the Sierra to the Coast accounts for a relatively high population growth rate. For instance, the Metropolitan Lima and Callao areas have the lowest fertility rates in the nation at 2.1 births per woman. However is has one of the highest population growth rates at 2.1% (INEI, 2012).

People from the Highlands migrate to the Coastal cities and the Jungle mainly searching for job and educational opportunities. In general, people are migrating from the Sierra (Mountains) to the Coast and in a minor proportion, to the lower Jungle and from rural areas to urban areas.

Almost one third of Peru's population resides in the Lima-Callao metropolitan area (pop. 8.5 million), followed by Arequipa with a population of approximately 750 thousand. Lima's population is 11 times larger than the second largest city. Notice that Peru's 5 largest cities are located in the Pacific Basin (Table 2).

4 A FEW FACTS ABOUT PERU'S ECONOMY AND ITS INFLUENCE ON FOOD CONSUMPTION

The Gross Domestic Product (GDP) is steadily growing since the 1990's. New laws protecting private investment attracted foreign investors and encourage local entrepreneurs to invest in a number of economic activities. While foreign investors invested mainly in mining, telecommunications and banking, local investors invested in agriculture, farming, industries and services. Peru's GDP is estimated at US$ 176.7 billion as of 2011 and is considered a middle income country (World Bank, 2012). The contribution of all sectors to Peru's GDP is given in Table 3 (Scotiabank, 2008). Estimated private investment is given below in Table 4.

Figure 5 shows the evolution of the Peru's per capita income (World Bank, 2012) in the period 1960–2010. Notice that it has sharply increased in the last 10 years. The per capita income has grown three times in the decade of 2000–2009 from approximately US$ 2000 to US$ 6000. The economic growth also causes an increase in the consumption of food, energy, goods and services. For instance, Gestión (2011) reported that consumption of chicken had been growing at an annual rate of 10% in the past few years and had reached the maximum historic per capita consumption of 35 kg. MINAG (2012) reported steady growth of per capita consumption of chicken, an important source of proteins for Peruvians. Figure 6 shows that there might be a strong correlation between per capita income and per capita consumption of chicken. This increase in the demand also occurs in other food products and eventually increases the use of water.

Table 3 Percentage of GDP by sectors (Scotiabank, 2008).

Sector	*% GDP*
Agriculture	8
Fishing	1
Mining	6
Manufacturing	15
Construction	6
Commerce	15
Electricity	2
Taxes	10
Services	37

Table 4 Private investment by sectors estimated in 2010 (Banco Central de Reserva del Peru 'BCRP', 2010).

Economic activity	*Investment (Millions of US$)*	*Percentage of private investment (%)*
Agriculture	203	1.6
Fishing	92	0.7
Mining and Oil industries	5333	41.8
Manufacturing	1523	11.9
Electricity, gas and water	1513	11.9
Construction and Infrastructure	2305	18.1
Commerce	935	7.3
Services	861	6.7
Total (Million US$)	12,765	100.0

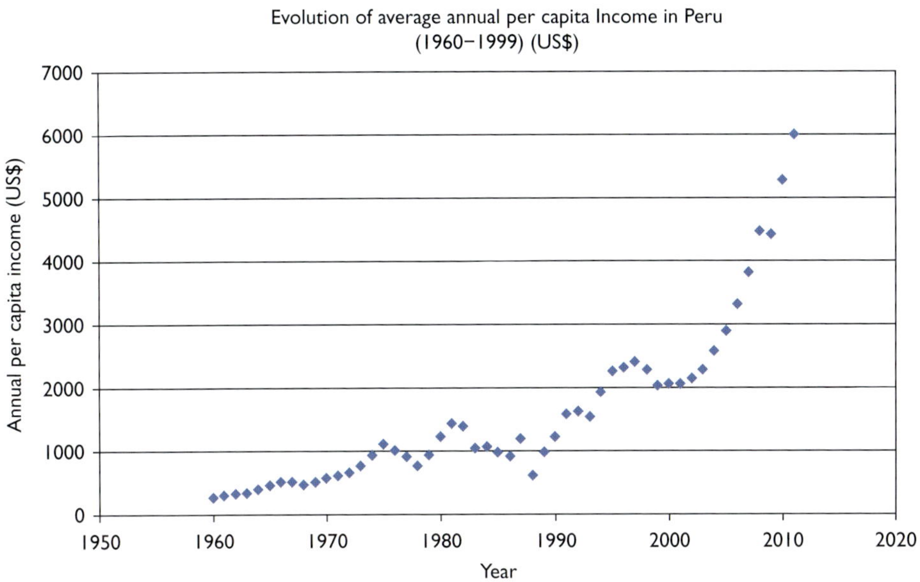

Figure 5 Evolution of Peru's per capita income between 1960 and 2010 (own elaboration from data by World Bank, 2012 and MINAG and DGIA-MINAG, 2012).

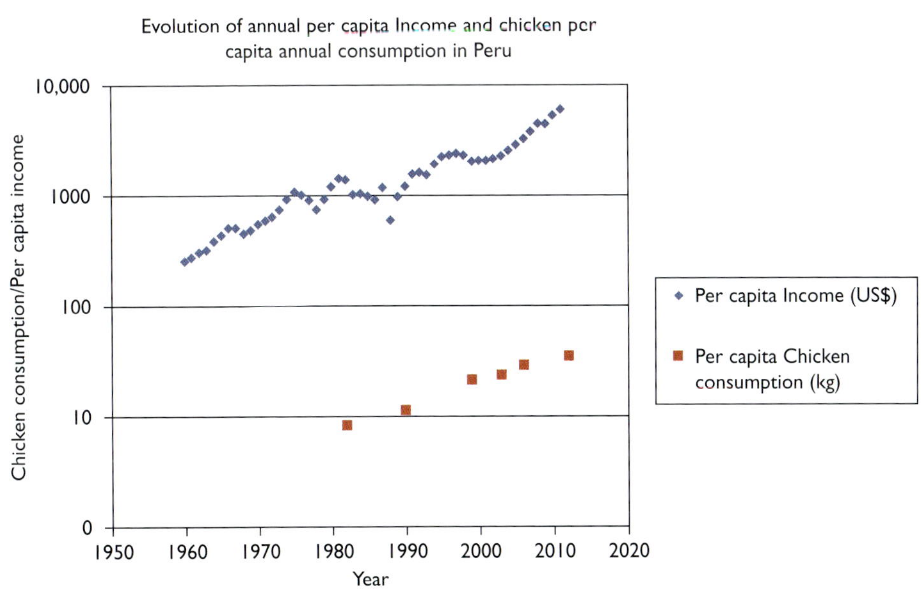

Figure 6 Evolution of per capita income and per capita chicken consumption (own elaboration from data by World Bank, 2012 and MINAG and DGIA-MINAG, 2012).

5 WATER USES IN PERU

5.1 Agriculture

Agriculture is, by large, the greatest consumer of water in Peru. A study by INRENA (in ANA, 2009) indicated that approximately 80% of Peru's water resources are used in this economic activity, although recent figures indicate that the percentage may be as high as 88%. Growth in the demand of export agricultural products and internal consumption has led to an increase in the production of certain crops. Although more efficient irrigations methods are being used, particularly for the export products, an increase in water use is creating severe stress in some valleys, leading to conflicts among users. In general, agriculture is particularly inefficient, with average efficiencies of 35–40%.

Because agriculture is by large the greatest consumer of water in Peru, special attention is being given to this economic activity. UNI is conducting a study of the Water Footprint with the auspices of the Botín Foundation. The 11 main agricultural products have been identified and their blue, green and grey footprints have been estimated (Kuroiwa *et al.*, 2013).

To estimate the Water Footprint (WF) of agricultural products, the method proposed by the Water Footprint Network was used. Data from the Ministry of Agriculture (MINAG, 2010) was used to compile production data, cultivated area, yield, and farm prices of the crops that were studied. Evapotranspiration was estimated using data made available at its web site by Peru's National Meteorology and Hydrology Service (SENAMHI, in Spanish). Data used were: Daily precipitation, maximum and minimum temperature, solar radiation, wind velocity and relative humidity. Evapotranspiration was calculated using the FAO Penman Monteith method (FAO, 2006) and effective precipitation was estimated using the FAO/AGLW formula. Crop coefficients and root depths were obtained from MINAG. Irrigation efficiencies were variable and depended on the crop and location. In this ongoing study, the WF was estimated in the period 2007–2010, for which data was available. The Blue, Green and Grey WF's for the 11 main agricultural products were estimated at the regional and national level. Results are presented below in Table 5.

Table 5 indicates that the WF may vary with time, depending on the agricultural production. In absolute terms, water use in 2010 fell more than 58% with respect to the previous year.

Rice crops have the highest WF in the agrarian sector with 5091 hm^3, of which 62.5% is Blue WF, 23.3% is Green WF and 14.2% is Grey WF. This crop uses 3438 hm^3 of Blue WF and all this water is consumed in the Coast region. Coffee WF is the second

Table 5 Blue, green and grey WF for Peru's 11 main agricultural products.

Year	*Blue WF (hm^3/yr)*	*Green WF (hm^3/yr)*	*Grey WF (hm^3/yr)*	*Total WF (hm^3/yr)*
2007	6920	8679	2195	17,794
2008	5353	4434	8071	17,858
2009	7278	9061	2352	18,691
2010	5589	4212	2011	11,812

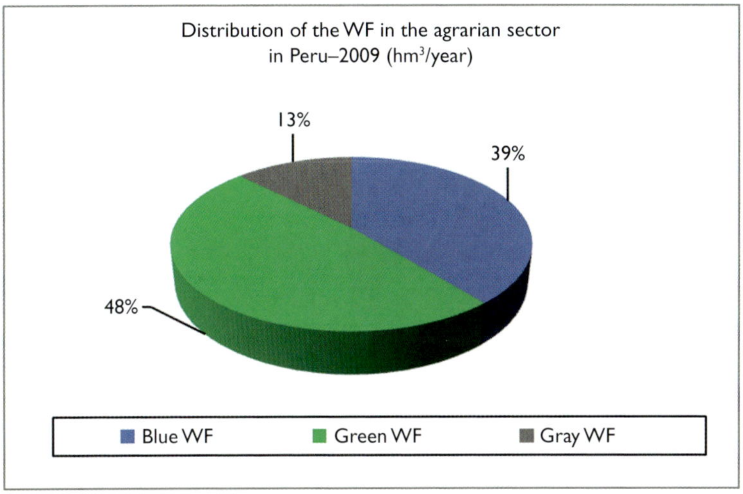

Figure 7 Distribution of the total WF for Peru's 11 main agricultural products (own elaboration).

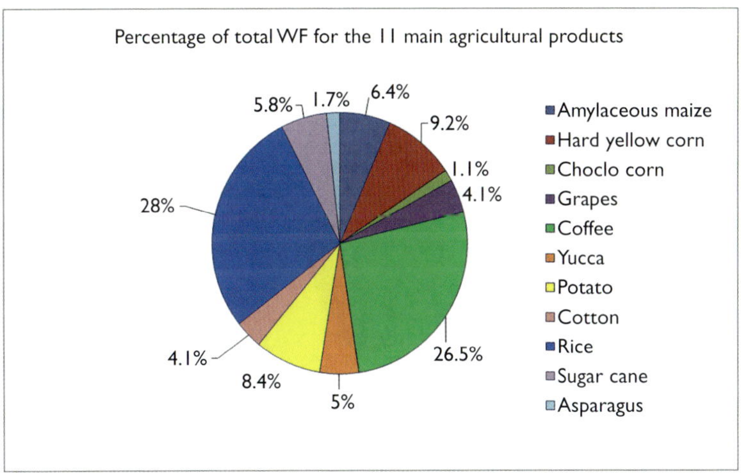

Figure 8 Percentage of the total WF of Peru's 11 main agricultural products (own elaboration).

largest consumer of water with 4813 hm³. However, all the consumption is Green WF as this crop grows in the Upper Jungle, where average annual precipitations are in the order of 2000–3000 mm/yr.

Blue water represents 39% of the water consumed by the crops. The largest amount of Blue water consumption occurs in the Coast (Pacific Basin). In this arid region, irrigation occurs by traditional methods such as flooding, rills and also water efficient methods such as sprinkler irritation and drip irrigation are used. Groundwater is extensively used, particularly during the dry season. This has caused persistent lowering of the water tables in some valleys such as Ica and La Yarada.

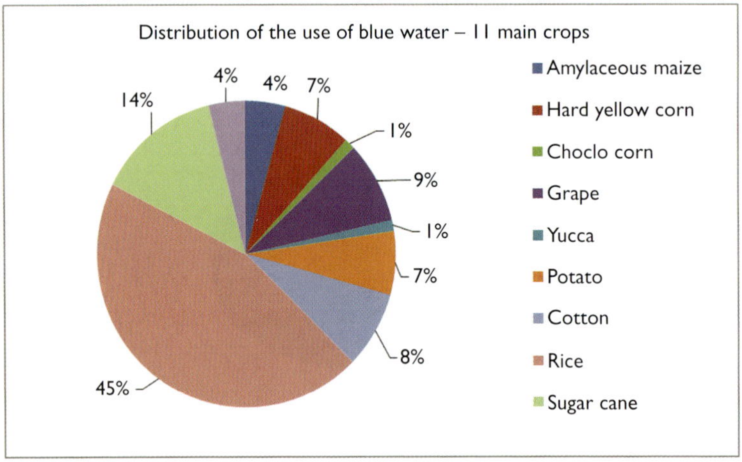

Figure 9 Distribution of blue water use in the 11 selected crops. Notice that coffee does not use blue water (own elaboration).

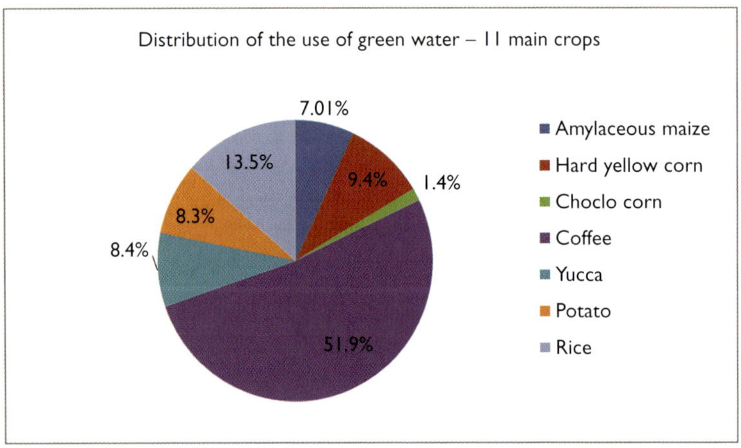

Figure 10 Green water use distribution of the 11 selected crops. Notice that asparagus, grapes and cotton do not use green water. These crops are grown in the Coast (Pacific Basin) (own elaboration).

In the latter, even saltwater intrusion has been detected. See Section 6 – Use of Groundwater in Peru.

Green water represents 48% of the total water use in the main crops. Most of the crops that use green water are grown in the Sierra and the Upper Jungle. In most cases, crops are both rainfed and irrigated. In the highlands of Peru, crops are grown during the rainy season (December–March). For instance, potatoes are rainfed between January and March, and when rains diminish, crops are irrigated. Rainfed-only crops, such as coffee, are grown in the Jungle. This crop in particular consumes 51.9% of the green water of the products that have been analyzed.

Table 6 Blue, green and grey water use in Peru's three natural regions.

Natural region	*Blue water (hm³)*	*Green water (hm³)*	*Grey water (hm³)*	*Total (hm³)*
Coast	4096.37	393.92	1031.17	5521.46
Sierra	1319.22	2959.81	1358.03	5637.07
Jungle	869.39	3242.86	1267.87	5380.12
Total	6284.99	6596.58	3657.08	16,538.65

Table 7 WF and EWF of three agricultural products: Vines, rice and coffee.

Crop	*WF (liters/kg)*	*EWF (US$/m³)*	*Water*	*Natural region*
Vines	559.68	2.891	Blue	Coast
Rice	1501.00	0.302	Blue	Coast and Jungle
Coffee	21,394.30	0.090	Green	Upper Jungle

In Table 6, blue, green and grey water estimates are given on a natural region basis. 65 Percent of blue water is consumed in the Coast, 21% of the total volume is consumed in the Sierra and the Jungle only uses 14% of the total blue water. Green water is mostly consumed in the Sierra (45%) and the Jungle (49%). Grey water consumption is higher in the Sierra (37%), followed by the Jungle (35%) and the Coast (28%).

The ongoing study has also identified the regions that consume the most water. The San Martin region consumes the most water with 2059 hm³. In this region, 53% of water consumption is green, 23% is blue and 24% is grey (Kuroiwa *et al.*, 2013).

Finally, the WF and the Extended Water Footprint (EWF) of a few selected products is presented. The EWF indicates the average gross benefit obtained per cubic meter. The vine WF is approximately 560 L/m³ and the EWF is US$ 2891/m³. Rice WF is 1501 L/m³ and the EWF is 0.302 US$/m³. Both crops use Blue water and are mainly grown on the low reaches of the Pacific Basin (Table 7). It can be clearly seen that it is more rational to grow profitable crops that use less water, such as vines, than growing rice, that uses much more water and produces low profits. Coffee uses large amounts of water and the EWF is very low, close to 0.1 US$/m³. However, in this case, all of the water use for this crop is green and this crop is grown in the Amazon Basin, where rain is abundant.

5.2 Municipal water supply

The second use is municipal water supply. Somehow industrial use of water is included because a number of industries are located in urban areas and, although many industries have their own water supply, usually wells, located within their facilities.

Table 8 Number of connections, percentages of water, sewage and wastewater treatment coverage, percentage of micro-measurement, and percentage of non-billed water.

Metropolitan area	*Number of connections*	*Water coverage (%)*	*Sewage coverage (%)*	*Wastewater treatment coverage (%)*	*Micro measure-ment (%)*	*Unaccounted water (Non-billed)*	*Company*
Lima-Callao	1,194,879	88.1	83.7	13.3	70.1	37.5	SEDAPAL S.A.
Arequipa	201,144	82.2	71.6	16.1	64.2	35.9	SEDAPAR S.A.
Trujillo	135,883	84.1	71.2	80.1	37.7	45.7	SEDALIB
Chiclayo	133,767	84.0	75.8	89.2	9.3	41.6	EPSEL
Piura	163,824	82.7	64.9	50.6	19.9	55.9	EPS Grau
Iquitos	56,684	68.2	47.5	0.0	23.7	57.9	SEDALORETO S.A.
Cuzco	57,497	96.7	85.8	75.4	78.2	46.0	SEDACUSCO S.A.

Municipal use is also inefficient. Efficiencies of 35 to 50% have been reported in large cities. This is due to the existence of very old pipelines, unmetered water consumption and clandestine connections. Most effluents are left untreated, polluting freshwater bodies, watercourses and the ocean. Table 8 shows water and sewage coverage, wastewater treatment coverage, micro-measurement coverage and non-billed water in the largest Peruvian cities.

It should be pointed out that although Lima-Callao is by far the largest of all of Peru's metropolitan areas, only 13.3% of the wastewater is treated. The rest is released into the Pacific Ocean, the Rimac River or the Chillon River, thus polluting the environment.

Total investment necessary to achieve the projected goals by 2015 is US$ 4042 million. Projected annual investment in rural and urban areas is US$ 550 million to accomplish the Millennium Development Goals by 2015.

5.3 Mining

Mining is a very important economic activity in Peru, as it accounts for an estimated 41.8% of the private investment in 2010 (BCRP, 2010) and produces 6% of the GDP (Scotiabank, 2008). However, it only uses 2% of the water resources of the nation as was estimated by INRENA (1995). These figures might have changed slightly as the mining production has increased significantly in the last few years. This activity is very water efficient. Efficiencies of up to 90% have been achieved (De Piérola, 2008).

Illegal mining, however, has caused extensive environmental damage and even negative social impacts in large areas, particularly in the Amazon Jungle, although other areas are being negatively affected. In Section 7 of this document, Environmental Issues, the case of Huaypetue is shortly described. Other uses have not been extensively studied.

5.4 Energy

Energy is mainly used for a number of activities such as transportation of people and goods, lighting and other domestic uses, commerce and industries. Energy consumption is mainly oriented to the following sectors in Peru: Residential-commercial, transportation and industry. Approximately 90% of energy is used in these activities. Each sector takes approximately 30% of the total. The rest of the energy is used in agriculture, public services and non-energetic uses. In the last years, the residential-commercial use proportion has diminished and the transportation and industrial use has grown (Alejos, 2011). Peru has an urgent need to increase its sources of energy because of population growth and the fast economic growth of the last two decades. Table 9 and Figure 11 show the evolution of total energy consumption in Peru in the decade 2000–2009.

Table 9 Estimated annual energy consumption in Peru in Terajoules.

Year	*Energy (TJ)*
2000	462,885
2001	446,811
2002	456,279
2003	459,664
2004	495,537
2005	477,175
2006	498,121
2007	518,982
2008	535,392

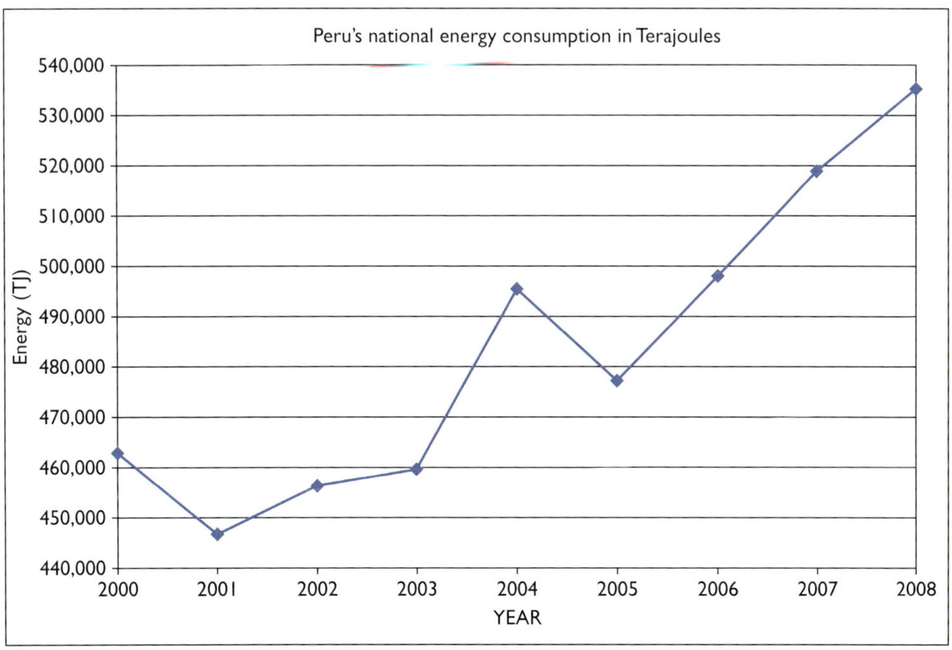

Figure 11 Evolution of total energy consumption in Peru in the decade 2000–2009 (own elaboration).

Liquid hydrocarbons have been Peru's main source of energy. However, the Camisea Project has allowed the use of natural gas as energy source for the transport sector and for domestic use since 2005.

Water is mostly used in the production of electric energy, although is also used to cool off thermal and nuclear power plants. During the decade 2000–2009, the demand for electric power steadily rose at an annual rate of 4% and during the period 2005–2009 the growth rate was above 5%. Electric energy consumption rose from 17,620 GWh in 2000 to 29,807 GWh in 2009. Power had also increased from 2630 MW in the year 2000 to 4320 MW in the year 2010, meaning that both the demand for energy and power has almost doubled. The production of energy by source in Gigawatt-hours is given in Table 10.

Notice that sugarcane bagasse is beginning to show in the electric energy matrix in 2009. A number of new projects in which sugarcane is used to produce energy are underway. In Piura, located in Peru's NW, the Romero group through Agricola del Chira S.A. company has invested US$ 160 million in the Project 'Caña Brava', which produces 350,000 liters of ethanol per day. This production is entirely exported to the European Union and when production increases, a part of it will be used for Peru's market. In addition, 12 MW of electric power are produced with the sugarcane bagasse, a by-product of ethanol production (Agronoticias, 2009). Gamio and Garcia (2011) indicate that this project may not be sustainable in the long term because it may not be possible to irrigate the 30,000 Ha used in the production of sugarcane.

The theoretical estimated hydropower potential for Peru was 206,107 MW and the technically feasible potential was estimated in 61,832 MW (MEM, 1968 and 1979). In reality, it may be much less as new environmental and social considerations may render some projects unfeasible. No evaluations of hydropower potential were made in the lower Jungle, as its development may require the flooding of large extensions of forests, altering the natural environment. In addition, geology is not very favorable in the lower jungle (Kuroiwa & Romero, 2008).

The Peruvian Government has given a number of incentives for the construction and operation of Renewable Energy Small Power Plants (RER, in Spanish). A Small Power Plant is defined as one whose maximum output is 20 MW. Concessions are given to investors that propose the use of renewable energy power plants such as hydropower plants, eolic parks, biogas, biodiesel, etc. New investments are being made by local and foreign companies in this economic activity. In the Pacific Basin,

Table 10 Production of electric energy by source in Gigawatt-hours.

Electrical energy source	*2000*	*2001*	*2002*	*2003*	*2004*	*2005*	*2006*	*2007*	*2008*	*2009*
Sugarcane bagasse	0	0	0	0	0	0	0	0	0	2
Diesel	511	107	109	251	858	59	120	65	342	184
Residual	650	466	472	616	1188	950	827	448	685	579
Coal	394	339	846	859	2170	831	881	840	909	929
Natural gas	669	744	1006	1230	994	4062	4260	7314	9313	9261
Hydraulic	15,410	16,807	17,224	17,732	16,693	17,101	18,671	18,589	18,010	18,752
Total	17,634	18,463	19,657	20,688	21,903	23,003	24,759	27,256	29,259	29,707

most projects are run-of-river and are not expected to interfere with other economic activities such as agriculture.

6 USE OF GROUNDWATER IN PERU

Groundwater is mostly exploited along the Coast, as surface water is insufficient to cover the current demand, especially during the dry season (May–October). Water extraction from aquifers is an ancient practice. The Nazca's already took advantage of underground flow. The Cantayoc groundwater filtration galleries are an example of adaptation in a very dry environment (Rodríguez Zubiate, 2005). Surface water resources are abundant in the Jungle and apparently, there are no active wells in this area. Use of groundwater has only been documented for the Pacific Basin. Information about the use of groundwater in the Amazon Basin and the Lake Titicaca Basin is scarce and very inaccurate.

ANA, the National Water Authority has identified three areas in which the exploitation of groundwater might have reached critical levels, as the water table level decreases with time, meaning that the water use is higher than the recharge. Approximately 31,599 wells are in use in the Pacific Basin, mainly in the lower areas of the 53 river basins. The annual potential exploitation of groundwater has been estimated at 2700 MCM and its use has been estimated at 1500 MCM (ANA, 2009). However, in some areas, use rates exceed the rate of recharge. Overexploitation and salinity intrusion are among the main problems that farmers face who use groundwater resources. The cases of groundwater use in two valleys in which problems have occurred due to overuse are described below.

6.1 Ica Valley Villacuri and Pampas de Lancha aquifer

A number of agricultural lands have been developed near the Ica River, which has an intermittent regime. Approximately 80% of the annual volume flows during the rainy season. During dry years the flow is zero during the dry season. The aquifer is fed by the infiltration of surface flows of the Ica River, a watercourse that collects water from a basin whose area is 7187.5 km^2. The aquifer consists of layers of unconsolidated porous soil mostly composed by high permeable fluvial and eolic deposits that favor infiltration of water to the aquifers (Peña *et al.*, 2006).

The Villacuri – Pampas de Lancha aquifer is located in the Ica region, just south of the Lima-Provinces Region. The rising prices of export products and the investment of entrepreneurs have led to an increased production of asparagus, grapes and paprika hot pepper. Groundwater production has been estimated at 543 MCM per year. The need for labor in the agricultural sector has generated employment among the Ica population and has attracted migrants from the highlands, particularly from neighboring regions. However, the water table has been descending continuously due to flows lower than average. ANA has been demanding a diminution in the rates of exploitation.

6.2 La Yarada – Tacna case

La Yarada is an agricultural area located near the coastline just north of the Peru-Chile border. The water table has descended at a rate of 0.45 m/year in this area and

ANA has declared a maximum limit of groundwater extraction. A study conducted by ANA concludes that the annual recharge is 45.04 MCM whereas annual groundwater use has been estimated at 62.77 MCM. Saline intrusion represents a major problem as the wells are located near the ocean. An increase in the salinity has been registered in a few wells that were monitored by ANA. Simulations conducted by Rojas Rubio (2005) showed that if the current situation is maintained, saline intrusion might affect the wells in the short term. Construction of new wells is prohibited and the exploitation rate has been decreased up to 3 km from the coastline.

The cases presented in this section illustrate that there is an increasing trend of use of groundwater and it may be reaching its exploitation limits in some regions with intensive agriculture. In the case of La Yarada, overexploitation may even reduce water quality, possibly damaging the crops in this agricultural area.

7 ENVIRONMENTAL ISSUES

A large numbers of watercourses and water bodies are heavily contaminated due to uncontrolled human activities. The main sources of contamination are: Untreated municipal sewage, uncontrolled acid mine effluents, agriculture by-products, etc. It has become evident that illegal mining is a large source of pollution in recent years. This is not necessarily a minor source of pollution, as heavy machinery is being used to extract and transport minerals. Two relevant cases are described below.

7.1 Huaypetue area – Madre de Dios river basin

The Madre de Dios River collects water from tributaries located on the Eastern Slope of the Andes, mainly from the upper and lower jungle. Annual precipitation ranges between 2000 mm and 3000 mm. Agriculture is a marginal activity. The organic soil layer is very thin in this area and nutrients are rapidly washed away in a few seasons. Underneath the layer of organic soil, layers of clay make the soil unsuitable for agriculture. However, this area attracts people from the Sierra because illegal exploitation of gold appears to be a very profitable activity. Extensive areas of the area are being rapidly deforested, even using heavy machinery used in legal mining activities. Dredges and large pumps have been brought to extract sediments from the bottom of watercourses. Massive amounts of mercury are used to separate gold from the rock matrix. There is no control of the effluents that are left untreated. It has been estimated that 32,000 Ha have been deforested. However, the illegal miners face no opposition because of the economic benefits of gold exploitation. Fish is perhaps the most important source of proteins of this region. When animals and human beings are exposed to mercury in its most basic form approximately 2–7% might be absorbed by digestion. However, when methyl mercury is formed, up to 100% of it might be absorbed by the small intestine. Methyl mercury is formed when mercury is dragged along sediments and water. The Ministry of Health has been testing fish for mercury and its derivatives. It has been found that three species of fish contain excessive amounts of mercury and whoever consumes more than 2 kg of the Mota fish might ingest 24 times the maximum permissible limit. Social impacts have also been reported. Exploitation of children that are exposed to mercury gas and liquid

Figure 12 Deforestation and water pollution due to illegal mining activities in the Madre de Dios region located in SE Peru (Ministry of Energy and Mines, 2011).

mercury and sexual exploitation of minors has also been documented. (Álvarez *et al.*, 2011) In the long term, certification of origin for gold products may help reduce the deforestation and pollution of the Amazonian rainforest.

7.2 Tambogrande Valley

The Tambogrande Valley is located in the Piura Region, in Northern Peru, near the Peru-Ecuador border. This has traditionally been an agricultural area that produces limes and mangoes. The Manhattan Minerals Corporation (MMC), a Canadian-based mining company, through its Peruvian subsidiary, Empresa Minera Manhattan Sechura (EMMS), was given a concession for extracting gold in the Upper Basin of the Tambogrande River. The project was approved by the Peruvian Government and licenses and permits were granted to start exploitation of gold. The estimated initial figure was 24.18 million kg of gold and 0.28 million kg of silver. Local farmers showed strong opposition to the project, leading to protests and unrest. NGOs supported the farmers in their efforts to make MMC give up the original project. Five people died during the protests and property of MMC was vandalized. Finally, the company's concession rights were terminated in 2003. (De Echave *et al.*, 2009). A few months later, farmers and people who migrated from areas in which mining is the prevalent activity, started digging for gold, as the ore layers were located near the surface. Illegal miners used sodium cyanide and other chemicals for separating gold from the rest of the minerals. Effluents were left untreated and collapse of cyanide pads caused pollution of streams, contaminating the Quiroz and Chipilico rivers. Currently, the agriculture is at risk in the Tambogrande valley as 10,000 Ha have been compromised. Urgent measures are needed to solve the environmental problems that threat the main economic activities in the valley (El Peruano, 2012).

The last case shows that environmental problems may arise when an unexploited ore deposit is left behind. Farmers first strongly opposed a legal mining project to protect their farms. However, after the mining project was cancelled, a few farmers together with miners from other regions started uncontrolled extraction of gold, polluting the waters that irrigate their crops. At the end, because minerals were near the surface, ore exploitation is a relatively easy task and the final result is extended contamination of land and water resources. There is a need to act immediately to stop illegal mining in the Tambogrande Valley. Some damage may be irreversible.

8 CONFLICTS RELATED TO WATER USES

Scarcity of fresh water in the Coast and the Mountains and the increasing use of this resource have led to conflicts among users. Peru has created a system for defending its citizens' rights against the abuse of the Government or large companies or institutions called People's Defensorship (Defensoría del Pueblo in Spanish) (www.defensoria.gob.pe). This institution constantly reports conflicts that occur in Peru every month. Environmental issues are usually the main cause of conflicts (51% as of May 2009). The majority of these conflicts are mostly related to water use (Kuroiwa, 2012).

Recent conflicts have paralyzed the execution of mining projects, such as the Conga Project, whose main investor is Newmont Mining Corporation (NMC), a US-based company. Compañía de Minas Buenaventura and the International Finance Group (IFC) are NMC's partners. An investment of US$ 4800 million was initially estimated. Protestors have strongly opposed this project for the extraction of copper and gold. According to the protestors, less water will be available if the project is carried out because the watershed will be significantly altered and will be left without ponds that collect water. Farming and ranching is traditionally the most important activity in terms of job creation. However, the project included the construction of several reservoirs which may guarantee water allocations during the dry season. As of October 2012, the Cajamarca Region president, Gregorio Santos, kept his position of not allowing the continuation of the Conga Project. Other mining projects have been cancelled or postponed because of social conflicts, such as Tia Maria in Southern Peru. The cancellation of mining projects will affect Peru's economic growth in the medium and long term.

9 WATER GOVERNANCE

During the past thirty years water resources in Peru have been managed under the General Water Law (GWL) of 1969. At that time the Peruvian Government was a military and nationalist regime that encouraged the land reform in order to eliminate Large Farm Estates ('Latifundios') and national elites in Peru. Due to this fact, the GWL was centered on hydraulic development, irrigation rights, agriculture development, and did not incorporate multi-level (local, regional and national) strategies necessary to manage the nation's water resources. After the International Water and Environment Conference in Dublin in 1992 many efforts have been made to introduce a new approach for the water resources management in Peru. One of the most remarkable attempts was held in 2004 when the government formed a multi-level (local, regional and national) committee in order to establish a water resources plan. After five years, the Government of Peru promulgated in March 2009 the new Water Resources Law (WRL) and one year later the Water Law Regulation (WLR), accompanied with a water management national plan.

Large water policy reform as was the case of Peru was mainly guided by the following drivers:

In the case of Peru, the general water policy reform was driven by the need to update the GWL. The GWL was based on the management of the quantity regardless the quality and the environmental degradation. It did not recognize the economic value of the resource and did not recognize the basin as an administrate unit. The law promoted progress in the Pacific basin ignoring the other two main basins (Atlantic and Titicaca). An example of that period was that during the 1950s and 1960s the two largest dams of Peru where constructed in the Andean Region in association with irrigation infrastructure in the Pacific region using the water of the Andean region to develop only one portion of the county. Having at that time, in the coast large plantations and in the rest of Peru small irrigated scheme, this disparity is clearly reflected in the high rate of migration from the Andes to the coast that triggered the social conflict to compete for the water resources (MINAG, 2009a).

Another driver was the necessity to decentralize water management (participation of users; national, regional and local government in the decision process). The first step was the establishment of the National Water Authority (ANA) in charge of managing the water resources per basin. Previously, the institutional arrangement of the GWL was chaotic, because the Ministry of Agriculture was in charge of all water permits. Due to the position of the Ministry to grant all the permits, the focus was centered on agricultural uses. The consequence was the marginalization and no recognition of the private and non-agricultural users, therefore the other national authorities as ministries started to manage the water in their own way and with their own offices for example (Figure 13). The Ministry of Mining and Energy started to collect fees for this use; the water quality was under the Ministry of Health which is characterized by its weak operational capacity due to budget shortages; the medical uses were managed by the Ministry of Tourism; and the monitoring inspections were under the Board of Ministries, etc. (Donoso *et al.*, 2004).

Furthermore, international agreements had driven the water policy reform in Peru. Some examples of such international agreements were conferences where Peru had an active participation such as the UN Conference on the Environment (Stockholm), UN Water Conference in Mar del Plata, Dublin Principles of 1992, Rio Declaration and Agenda 21, and World Summit on Sustainable Development. A particular international commitment that helped to reform all government bodies was the United States – Peru Trade Promotion Agreement (PTPA) of April 12, 2006. Peru's Congress ratified the Agreement in June 2006 and a protocol of amendment in June 2007. On December 14, 2007, President Bush signed into law the PTPA Implementation Act which approved the PTPA. The PTPA entered into force on February 1, 2009.

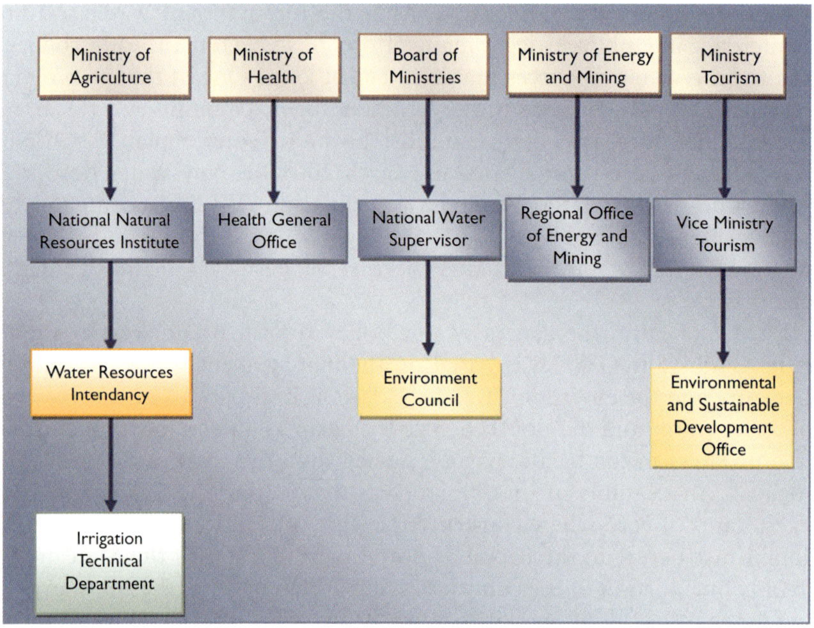

Figure 13 Public institutions related to water management at the national level according to the Water Law of 1969 (own elaboration).

The PTPA is a comprehensive free trade agreement based on US Environmental Protection Agency (EPA) requirements. Due to the necessity to show commitment to the preservation of the environment and demonstrate the quality of Peruvian human resources, the Ministry of the Environment was created. This entity was created on December 2007 and since then all the ministries and offices in charge of the protection of the natural resources were restructured including water resources, having as a result the creation of the National Water Authority and the subsequent new water legislation. Finally, the civil society pressure played an important role in the water policy reform. In the case of Peru, there is a large disparity present in access to water between the poorest and wealthy users (United Nations Development Program, 2010), the lack of water infrastructure and the community concern about the reduction and degradation of water resources (e.g. Pastoruri glacier).

Attempts to change the attitude regarding water management were conducted in Peru. The most remarkable was the extinct PRONAMACH project that was a joint effort with international funds and the Ministry of Agriculture. The aim of this project was to manage water at the catchment level. PRONAMACH was implemented in 5 pilot basins: Chira-Piura, Chancay – Lambayeque, Jequetepeque, Chillón – Rímac – Lurín, Río Santa and Tambo – Moquegua. At the end, the project only focused in small irrigation infrastructure, and did not achieve the catchment management. The only pilot basin that achieved important results was the Chira-Piura basin, but only because additional funds of a German cooperation agency helped to the continuity of the project. According to Hendriks (2005) it is unusual that attempts such as the previously described couldn't help establish enough synergies to change the pattern in the water management.

9.1 The National Water Authority 'ANA'

The National Water Authority is the highest technical and regulatory authority of the National Water Resources Management System. The National Water Authority is responsible for operating the system under the provisions of the Water Resources Law (WRL) Art. 14 (MINAG, 2009b). Although ANA is still attached to the Ministry of Agriculture it is defined by WRL as an autonomous technical specialized body. The National Water Authority relevant tasks involve:

1 Development of the policy and national strategy of water resources and the national water resources management plan, leading, supervising and evaluating its implementation, which must be approved by Supreme Decree by the President of the Council of Ministers;
2 Formulate rules and procedures to ensure an integrated, bottom-up and sustainable water resources management, and efficient water use;
3 Coordinate and organize the actions to implement the National Water Resources System;
4 Provide technical opinion about water availability for the implementation of projects; promote programs and activities to protect the water resources, and their quality through technology and research;

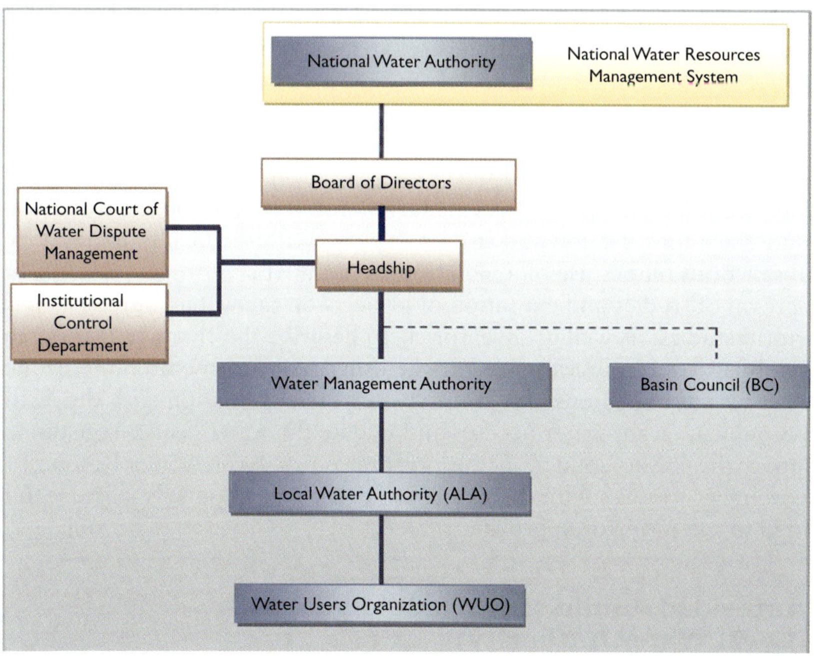

Figure 14 National Water Authority according to the Water Resources Law of 2009 (own elaboration).

5 Promote knowledge exchange in water resources management, sustainable use and conservation with national and international organizations related to the subject;
6 Exercise jurisdiction in matters of water management, developing management actions, monitoring, control and surveillance to ensure the conservation of water sources, goods and services related to them and the public water infrastructure;
7 Exercising the power to impose sanctions;
8 Grant water licenses and update of the water license register;
9 Develop educational activities to create public awareness of the need to conserve and preserve water and;
10 To promote the recognition of the economic and environmental value of water.

The main bodies of the National Water Authority are the Autoridad Administrativa del Agua – AAA (Water Management Authorities) and the Autoridad Local del Agua – ALA (Local Water Authority).

9.2 Autoridad Administrativa del Agua 'AAA' (Water Management Authorities)

The National Water Authority (ANA) is represented in the country through decentralized organizations defined as Water Management Authority (AAA). The Water Management Authority carries on the water management of its scope area according to the WRL and the National Water Authority regulations. One hundred and fifty nine river basins were delimited by ANA following the methodology detailed by the Ministerial Resolution N 033- 2008-AG, 96 are embedded in one regional government and 63 share more than one regional government. In order to manage these disparities the ANA joined together the basins, according to the legal administrative resolution N 546-2009-ANA, with geographic, biophysical, cultural, and socioeconomic matches to conform 14 hydrographic regions which are going to be the area of scope of the AAA (MINAG, 2009a).

The relevant tasks of Water Management Authorities (MINAG, 2010a) are:

1 To apply monitoring programs and surveillance activities on the water sources to ensure their conservation and sustainable use, with the faculty to apply penalties and sanctions in the case of a violation of the WRL;
2 Approval of studies and execution of projects on the water sources;
3 Perform studies, as well as inventories, to describe the characterization and situation of water resources while on the other hand to monitor the evolution of glaciers, Andean lagoons and groundwater, in coordination with the Ministry of Environment; grant water licenses and update the water license register;
4 Support the Basin Council in the preparation of Basin Water Resources plans, according to the law, with the validation of the basin councils in order to be submitted to the ANA for approval.

9.3 Autoridad Administrativa del Agua 'ALA' – (Local Water Authority)

The Local Water Authorities are local units of the National Water Authority, which manage the agricultural and non-agricultural water uses. The local water authority is bounded according to the administrative resolution 0546-2009-ANA, giving as result 73 offices around the country and located in strategic basins. They are hierarchically dependent of the Director of the Water Management Authority (AAA) and the most relevant tasks follow (MINAG, 2010a):

1 To approve water permits and authorizations with the approval of the Water Management Authority, approve, revoke and modify easements, conduct monitoring programs and surveillance activities in the water sources to ensure its conservation and sustainable use, with the power to apply penalties and sanctions with the approval of the Water Authority Management;
2 Submit any technical report required by the AAA to support its performance;
3 Solve water claims at the local level, except for those that have to be solved by the Water Management Authority;
4 Monitor, promote and evaluate the use and water management, the participation of water users, and the management of water infrastructure users in their areas of scope.

The creation of ANA and recently promulgated legal instruments allow, in theory, to the enforcement of conservation and protection of the country's water resources. However, in practice, tasks assigned to ANA, AAA and ALAs may not be fully enforced in the short term. Practical solutions are yet to be achieved and participation of the national, regional, and local governments, regional development agencies, users' associations, academic institutions and the civil society at large is necessary.

10 CONCLUSIONS AND RECOMMENDATIONS

The main intention of this chapter was to describe the current situation of the country and its relation to water resources. There are a number of issues that need to be urgently dealt with and represent immediate challenges for the current administration. The most significant conclusions are given below.

Peru is a very uneven country from the geographical, demographical and hydrologic viewpoint. The Pacific Basin is a very dry area, whereas the Jungle is a very humid area. The Sierra (Mountains) is a region with a wide variety of climates. The majority of the people live near the coastline, where water is most scarce.

The birth rate has decreased, especially in Lima, the largest metropolitan area. However, the population growth rate is one of the highest of the nation due to migration from the highlands and the jungle. A similar pattern is observed in other coastal cities. This creates pressure on water consumption in the Pacific Basin.

Economic growth has also created pressure on food consumption. It has been shown that as the annual per capita income has increased; the per capita consumption of chicken meat has also increased. A similar trend has been observed in other products.

Use of blue water is intense along the Peruvian Coast. Rice, a low cost product, is its higher consumer. In some areas, export products of high value are grown and they even use very water efficient irrigation methods. A more rational use of water must be encouraged or even enforced as the water resources are being exhausted along the Pacific Basin.

Population growth and economic growth have been also creating pressure on the demand for electric energy. Most likely, hydropower plants will be built to reduce the gap between the demand and supply. Alternative energy solutions are also being implemented. Ethanol facilities, which use sugar cane as raw material, have already built electric power plants. The use of alternative energies is expected to grow in the next years. For instance biodiesel might be made of vegetable oils. Crops markets are basically unregulated. Available land for raising crops is limited and food supply has to be given first priority. Provisions must be taken to avoid food scarcity for Peru's people.

Groundwater use is most intense along the Coast, which is the lower reaches of the Pacific Basin. An arid, highly variable hydrologic regime and the need of food production have led to an intensive use of groundwater resources. In a few valleys, where export products are grown, the National Water Authority has already limited the exploitation of new wells and setting limits on well extraction rates in a number of locations. Detailed studies are necessary to analyze the long-term trends of groundwater availability.

Uncontrolled and illegal economic activities, such as illegal gold mining, are destroying the environment and heavily damage Peru's water resources. In some cases, damage might be irreversible. It is necessary to enforce rules and regulations especially in the lower Jungle, where negligence from past authorities have allowed the formation of gold mining colonies that are predating the environment. Some measures such as the issuance of certificates of origin for gold products may be necessary in the long term.

Conflicts between agriculture, farming and other economic activities have occurred in the last years. These conflicts are related to the use and preservation of water. New avenues of communication are needed to avoid conflict which has already caused deaths and serious economic losses.

The recent creation of the Water National Authority, ANA, the Ministry of Environment and the publication of the new Water Law with new regulations have created a better framework for the integration of water resources management. However, Peru is a vast country and it is necessary to enforce the new rules and environmental regulations. There is an urgent need to form new capacities regarding the people in charge of natural resources management and to enforce laws and regulations.

REFERENCES

Agronoticias (2009) Revista para el desarrollo. *Agronoticias*. Edition 347. Year XXXI. Lima, Peru.

Alejos, R. (2011) *Proyecciones de la Matriz Energética al Largo Plazo*. Centro Nacional de Planeamiento Estratégico (CEPLAN). Documentos de Trabajo. Number: 12.

Álvarez, J., Sotero, V., Brack Egg, A. & Ipenza Peralta, C. (2011) *Minería Aurífera En Madre De Dios Y Contaminación Con Mercurio – Una Bomba De Tiempo*. Lima, Perú, Ministerio del Ambiente.

ANA, Autoridad Nacional del Agua (2009) *Política y Estrategia Nacional de Recursos Hídricos del Perú*. Ministerio de Agricultura, Autoridad Nacional del Agua. Lima, Perú.

ANA (2010) *Sistema de información de recursos hídricos 1970–2010*. Ministerio de Agricultura. Lima, Perú.

BCRP, Banco Central de Reserva del Perú (2010) *Memoria Anual 2010*. Banco Central de Reserva del Perú. Lima, Perú.

De Echave, J., Diez, A., Huber, L., Revesz, B., Lanata, X.R. & Tanaka, M. (2009) *Minería y Conflicto Social.* Instituto de Estudios Peruanos. Lima, Perú.

De Piérola, J.N. (2008) *El Valor y El Costo Del Agua en el Proceso De Producción Minera y Su Carácter Social En El Perú. Extemin.* Convención Minera. Arequipa, Perú.

Donoso, G., Jouravlev, A. & Zegarra Méndez, E. (2004) *Mercados (de derechos) de agua: Experiencias y propuestas en América del Sur.* United Nations, UNECLAC. Serie: Recursos Naturales e Infraestructura. Number: 80. Santiago de Chile, Chili.

El Peruano (2012) *Fatales consecuencias.* [Online] Newspaper article published on April 13, 2012. Lima, Peru. Available at: http://www.elperuano.pe/edicion/noticia-fatales-consecuencias–40878.aspx [Accessed November 2012].

FAO (2006) *Evapotranspiración del cultivo. Guías para la determinación de los requerimientos de agua de los cultivos.* Estudio FAO riego y drenaje, number: 56.

Gamio, P. & García, H. (2011) *Energía Renovables en el Planeamiento Estratégico del Mediano y Largo Plazo.* Centro Nacional de Planeamiento Estratégico (CEPLAN). Documentos de Trabajo. Number: 10.

Gestión (2011) *Perú, tercer país con mayor consumo de pollo en Latinoamérica.* [Online] Daily newspaper Gestión. June 23, 2011. Available from: http://gestion.pe/noticia/802197/peru-tercer-pais-mayor-consumo-pollo-latinoamerica [Accessed January 2012].

Gobierno Regional Lambayeque-Peru (2012) *PEOT: Proyecto Especial Olmos.* [Online] Available from: http://peot.regionlambayeque.gob.pe/ [Accessed September 2012].

Hendriks, J. (2005) *Legislación de aguas y gestión de sistemas hídricos en países de la región andina.* Derechos colectivos y políticas hídricas en la región Andina, 47–111.

INEI (2012) *Peru en Cifras* [Online] Available from: www.inei.gob.pe [Accessed October 2012].

INRENA (ed.) (1995) *Estudio de Reconocimiento del Uso del Recursos Hídrico por los Diferentes Sectores Productivos en el Perú.* Lima, Perú, Instituto Nacional de Recursos Naturales.

Instituto Geográfico Nacional- IGN (2012) *Cartographic Maps*- Scale 1: 100 000. Instituto Geográfico Nacional. Lima, Peru.

Kuroiwa Zevallos, J. (2012) *Diagnóstico de Aguas en Las Américas.* Red Interamericana de Academias de Ciencias, Foro Consultivo Científico y Tecnológico. Mexico.

Kuroiwa Zevallos, J. & Machuca Romero, F. (2008) Potencial Hidroeléctrico y Energías Renovables en el Perú. In: *Seguridad Energética en el Perú.* Consejo Departamental de Lima. Lima, Perú, Colegio de Ingenieros del Perú.

Kuroiwa, J.M., Castro, L.F., Lucen, M.N., Zafra, J., Montenegro, J.I. & Romero, F.M. a.o. (2013) *Estudio de Seguridad Hídrica y Alimentaria en el Perú.* Santander, España, Fundación Marcelo Botín.

MINAG. (2010) Ministerio de Agricultura. Series históricas de producción agrícola – Compendio estadístico 1950–2010.

MINAG, Ministerio de Agricultura (2009a) *Policy and National Strategy of the Water Resources in Perú.* Ministry of Agriculture.

MINAG (2010a) *Regulation of the water law.* [Online] Available from: http://www.ana.gob.pe/marconormativo.aspx [Accessed October 2012].

MINAG (2012) *Sector Pecuario en el Perú.* [Online] Available from: http://www.minag.gob.pe/portal/sector-agrario/pecuaria/situación-de-las-actividades-de-crianza-y-producción?limitstart = 0 [Accessed October 2012].

MEM, Ministerio de Energía y Minas (1968 & 1979) *Evaluación del Potencial Hidroeléctrico Nacional.* Lima, Perú, Ministerio de Energía y Minas.

Ordóñez, J.J. & Vera, H. (2012) *Balance Hídrico Superficial del Perú a Nivel Multianual.* Servicio Nacional de Meteorología e Hidrología (SENAMHI) and Dirección General de Hidrología y Recursos Hídricos.

Peña, F., Sánchez, M. & Lu, S. (2006) *Mapa Hidrogeológico De La Cuenca Del Río Ica.* Instituto Geológico, Minero y Metalúrgico. XIII. Congreso Peruano de Geología. Sociedad Geológica del Perú.

Rodríguez Zubiate, E. (2005) *Las Galerías Filtrantes de Nazca.* Revista Técnica, Number: 30 on Capítulo de Ingeniería Civil. Consejo Departamental de Lima. Colegio de Ingenieros del Perú.

Rojas Rubio, H.A. (2008) *Modelamiento para el Análisis de la Variación Hidrogeológica Espacial del Acuífero La Yarada – Tacna.* (Tesis Maestro en Ciencias con Mención en Ingeniería Hidráulica) Facultad de Ingeniería Civil, Universidad Nacional de Ingeniería. Lima, Perú.

Scotiabank (2008) *Peru en Cifras.* Negocios Internacionales. Monthly Magazine. October 2008. Sociedad de Comercio Exterior del Perú – COMEX. Lima, Peru. Available from:http://www.comexperu.org.pe/archivos/revista/octubre08/perspectivaseconomicas_134.pdf [Accessed October 2012].

World Bank (2012) *Data – Peru.* [Online] Available from: http://data.worldbank.org/country/peru [Accessed September 2012].

Chapter 12

Integrated water management in Chile

Guillermo Donoso
Departamento de Economía Agraria, Pontificia Universidad Católica de Chile, Santiago, Chile

ABSTRACT: The present chapter presents a general overview of Chile's water resources, its water institutionality and water legislation, and the obstacles that have prevented a decided adoption of an integrated water management framework. We identify key actions that are required to effectively advance towards an integrated water management framework in Chile. The most important of these are the following. Groundwater users must implement groundwater user associations and integrate these to the Juntas de Vigilancia so as to apply conjunct surface and groundwater management. Additionally, the Water Code of 1981 establishes that water use right owners are responsible for water management. However, it is imperative to strengthen all WUAs (Water User Associations) so that each one develops a strong rule of law, effective conflict resolution, and effective collective management. Finally, water users should implement Supra Organizations of Juntas de Vigilancia to integrate different river sections and aquifer hydrogeological sectors allowing for an integrated management of different river sections and aquifer hydrological sections. This does however not require a water legislation modification.

Keywords: Integrated water management, water institutionality in Chile, limitations to implement IWRM in Chile, key actions to implement IWRM

1 INTRODUCTION

Chile was one of the first countries in the world adopting market rules for the allocation of their water resources, and allowing the participation of the private sector in the management of the resource. The Chilean Water Code of 1981 treats water as an economic good based on the following principles: (i) water in not a factor of production only for agriculture, but for other sectors too, and must be transferable like any other economic input; (ii) separates the property rights from (mobile) water and (immobile) land resources; and (iii) establishes water use rights as any other property right, allowing for leases and sales between willing buyers and sellers.

Some of the benefits achieved by this approach are that water markets have been an effective reallocation mechanism of water under scarcity in expanding urban areas and have opened the opportunity to satisfy the demands from important social and economic activities, have played a key role in mitigating the negative impacts from droughts, and have promoted private investments to increase efficiency in resource utilization. Compared to the situation in most countries in Latin America and the Caribbean, Chile's water policies are unusually conducive to efficient resource use and development (Southgate & Figueroa, 2006).

Water markets have often been criticized for its potentially regressive impact on low-income agricultural producers and for deteriorating water resource distribution between the poor to the higher income (Brajer *et al.*, 1989; McEntire, 1989; Cummings & Nercissiantz, 1992; Metzger 1988; Syme, Nancarrow, & McCreddin, 1999; Molle, 2004). Limited empirical observations from operative markets in Australia and South Asia show that water markets can be beneficial to small farmers. Research in Chile on the impact of water markets on small farmers, however, has been limited and no reliable conclusions have been reached to date. Hadjigeorgalis (2008) is a notable exception: Her results indicate that water use rights markets have not been inequitable with respect to offer prices; resource-constrained farmers receive the same offer prices for their water and water rights as wealthier farmers.

However, the intensification of the demand for water has accentuated the recognition that water resources should be best analyzed and dealt with in an integral manner. Integrated water resource management has been emphasized as a means to incorporate the interest of the multiple users and uses of water in the planning process. Peña (1999) has pointed out that the unsolved challenge of the Chilean water policy is to implement an integrated water resources management. Initial proposed policy reform efforts in 1992 included proposals to create river basin administrative organizations. However, this proposal encountered significant opposition. Additionally, as Bauer (2004) points out, river basin organizations were poorly defined in this proposal. The 1992 proposal was eventually withdrawn and a less-ambitious proposal was finally approved by Congress in 2005. This water policy reform did not consider the necessary modifications to incorporate integrated water resource management.

Additionally, Chile's water policy approved in 1999, established as one of its goals the implementation of integrated water management. More specifically it states as one of its objectives the need to adopt an integrated water management that internalizes decisions of multiple water users, taking into account short and long-term externalities associated with traditional water management frameworks. Integrated water management should also strengthen and coordinate the actions of the multiple public services involved in water management. The main proposed actions were the creation of river basin organizations and the development of a water master plan based on the diagnosis of actual water management problems at the basin level. To date, advances have been achieved in the development of water master plans in a small number of watersheds. These are the result of a strong impulse during the years 2000–2001; however, there have been no major advances since then and, thus, only a few river basins count with water master plans. At the same time, the proposal of creating river basin organizations has not received the necessary support, which explains the inexistence of these organizations in Chile.

The present chapter presents a general overview of Chile's water resources, its water institutionality, and the obstacles that have prevented a decided adoption of an integrated water management framework. Finally, key actions that are required to effectively advance towards an integrated water management framework in Chile are proposed.

The structure of the chapter is as follows. Section 2 presents a general overview of Chile's water resources. The water sector institutionality and Chile's water legislation is summarized in the third section; analyzing the limitations it imposes to achieve an integrated water resource management. Section 4 concludes the chapter presenting the main actions that are required in order to implement an integrated water management approach.

2 GEOGRAPHICAL SETTING AND WATER RESOURCES

A long narrow strip of land (no more than 430 km wide) between the Andes and the Pacific Ocean, Chile stretches 4630 km from near latitude 18°S to Cape Horn (latitude 56°S). Chile's total land area is 743,800 km^2, of which 21.2% is agricultural land (157,687 km^2) and 21.8% is forest (162,148 km^2). Arable agricultural land is 1,294,000 hectares, which is 1.7% of the total land surface. Urban area covers approximately 0.06% of the total surface. Currently it is estimated that the area of wetlands in Chile is 4,498,060.7 ha, equivalent to 5.9% of Chile's total land area.

Chile is divided administratively into 15 regions, and has nearly 17 million inhabitants, of which 89% are concentrated in cities, mainly in the Metropolitan Region (RM) and Valparaiso, with 31% and 9% of the population of Chile, respectively. Population growth has declined in recent years from about 2% in the early 90's to 1% in 2009 (World Bank, 2010); estimated crude birth rate of Chile is 14.33 births/1000 population (INE, 2002).

In the last 30 years (1980–2010) Chile's real GDP has grown at an annual growth of 6.2%. The economy is based mainly on exports concentrated on natural resource-dependent production processes that are highly dependent on water, such as mining and agriculture (Central Bank of Chile, 2010). According to World Bank, Chile has a per capita GDP measured in purchasing power parity of U.S.$ 15,331 in 2010.

In 2005, the five classes of water-consuming activity with the largest share of GDP were manufacturing (12%), retail, restaurants and hotels (10%), mining (8%), agriculture and forestry (4%) and electricity, gas and water (3%), while in 2005 the contribution to merchandise exports were mining (57%), industrial (31%), and agriculture, forestry and fishing (7%) (World Bank, 2011).

Chile's unique geography provides a variety of climatic conditions and a number of short river valleys running from the Andes to the Pacific Ocean. Two primary mountain ranges, the Andes and the Coastal Mountains span the length of central Chile and provide the limits to the coastal plain and the central valley. Precipitation ranges from near zero in the north to an annual 2000 mm in the south. The rainy season is in winter, June to September, and much of the precipitation is stored in the snowpack in the Andes mountain range. Water flows in most basins have a mixed origin, since its waters come from winter precipitations and spring and summer snow melt, presenting highest flows in summer (November–February) with pronounced reductions in autumn and winter (April–June). Additionally, inter-annual rainfall fluctuations show greater variability in the arid and semi-arid north (between Arica-Parinacota Region and the Coquimbo Region). South of 37°S, rainfall becomes more uniform. Thus, the hydrological regime of Chile is irregular.

Within the global context, Chile as a whole may be considered privileged in terms of water resources. The average total runoff is on average equivalent to 53,000 m³/person/year (World Bank, 2011), a value considerably higher than the world average (6600 m³/person/year). However, there exist significant regional differences: From Santiago to the north, arid conditions prevail with average water availability below 800 m³/person/year, while south of Santiago the water availability is significantly higher reaching over 10,000 m³/person/year (see Figure 1).

Annual water withdrawals in Chile average approximately 4000 m³/second (World Bank, 2011). Of this total amount, almost 85% is used in non-consumptive hydroelectric generation. Consumptive water use in Chile is dominated by irrigation, accounting for 73% of the consumptive water use. Up to 12% of the total consumptive water withdrawals are destined for industrial use, and mining and potable water supply account for 9% and 6%, respectively.

In the northern Chile desert, approximately between 17° and 26° south latitude, the limited water resources sustain a few coastal cities, some specialized agriculture, and large mining operations such as the main copper mining area in Chile. In north central Chile, between 26° and 33° south latitude, there is an adequate supply of water in a few river valleys for canal irrigation. Water storage reservoirs have been constructed to support these irrigation systems, especially in the Limarí Valley where three reservoirs have a joint storage capacity of 990 Mm³. Central Chile, between 33° and 39° south latitude, contains the nation's major urban and industrial areas, including: Santiago with a population of 5,700,000. Irrigated crops include fruits, vineyards, basic grains, forage, and vegetables. Industrial products include processed food, pulp and paper, chemicals, plastics, and petroleum products. Also central Chile remains the region with the greatest hydroelectric generation capacity, especially in the Maule and Bío-Bío basins. Southern Chile, south of 39° south latitude, is humid, forested and scarcely populated. There is little irrigation in the area, which produces

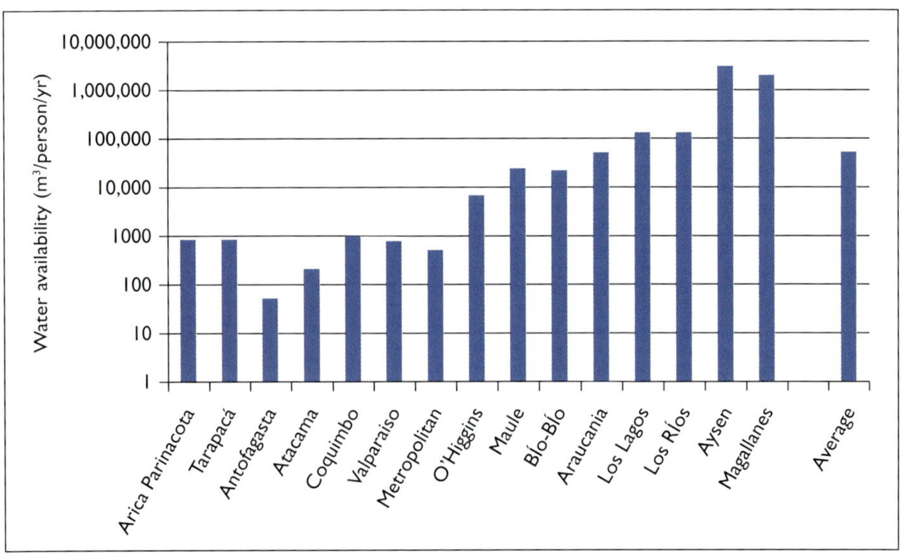

Figure 1 Average water availability per person per year (m³/person/year) (World Bank, 2011).

forest products, cereals, dairy and livestock, potatoes, and sugar beets. Because of its cool water, clear lakes, and coastal fjords, this area contains Chile's large aquaculture industry. In 2008 there were an estimated 493 marine and 185 freshwater-intensive salmon and trout farms in the region (ECLAC, 2011).

Chile has a high level of coverage of potable water and sewerage treatment systems. In 2010, 99.8% of the urban population (SISS, 2004 & 2011) and at least 72% of the rural population had access to improved potable water (Donoso & Melo, 2006). The national coverage of sewage treatment has significantly increased from 10% in 1990 to over 80% in 2010 (World Bank, 2011).

The situation of water resources over the past three decades in Chile has probably been more influenced by the country's development strategy than by the water sector itself (World Bank, 2011). The empowering role of the market and the promotion of an export-oriented economy based on products such as copper, fresh fruit, wood and pulp, salmon, and wine – all of which use water in their production process – have led to a significant increase in water use, particularly in relatively water-scarce water basins of the northern and central parts of the country. At the same time, water is increasingly becoming a limiting restriction to further economic development. As shown in Figure 2, most of the regions north of the Metropolitan Region have to deal with hydrological droughts, since water demand exceeds its supply.

Climate change impact projections indicate that water as a limiting factor will increase due to reduced water availability. Chile has developed important efforts to estimate the potential impacts of climate change (MMA, 2011; ECLAC, 2009 & 2012). These models project a general decrease in rainfall, mainly in the central and southern regions (30–42°S). This reduction becomes more important from the central valley towards the Andes mountain range area, showing reductions around 40% by the end of the century. Analyzing potential impacts on temperature, the models indicate that there will be an increase in all regions of the country between 2 and 4°C

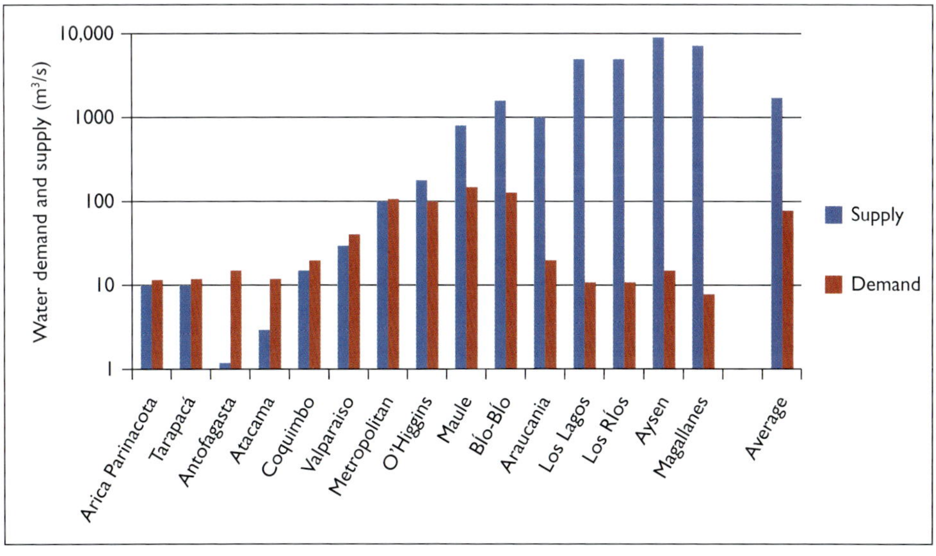

Figure 2 Water demand and supply per region (m^3/s) (World Bank, 2011).

by the end of the century. This temperature rise increases from the coast towards the Andes mountain range and decreases from the north of the country towards the south (ECLAC, 2012).

The availability of water resources is closely linked to climate variations, especially of precipitation and temperature. Thus, two effects are anticipated on what concerns water flows. In first place, there will be a reduction of total water flows at the basin level. This effect is due to the decrease in rainfall and a reduced snow accumulation in the Andes mountains; models project reductions that can reach up to 70% with respect to mean annual flows. In second place, there will be significant changes in the seasonality of water flows, due to an advancement of up to one month on the date of the centroid of water flow regarding the historical period (MMA, 2011).

In practice, the reduction in surface water availability will be dealt with by a proportional (re-) distribution of water according to each user's water use rights. However, in the medium and long term this approach may be insufficient to prevent significant reductions to private investment and increased conflicts between users. In the case of groundwater, on the other hand, the lack of user organizations in almost all of the aquifers of Chile prevents the establishment of specific actions to allocate water in a sustainable and peaceful manner. As a result of this trend, greater competition for water resources is foreseeable, as well as greater conflicts.

In addition to the projected changes in the supply of water resources, it is important to consider other processes such as economic and social, which are expected to change demand patterns, and therefore should be considered in the analysis of future water management challenges. The historical trend shows that all consumptive water uses have increased since 1990; total consumptive water use has increased 13% between 1990 and 2006. Industry is the sector with the highest consumptive water use increase (79%), followed by potable water and mining (48% and 46%, respectively). Furthermore, it is likely that many of these trends will continue in the short to medium term. Figure 3 shows the 2050 projected water demands by economic sector, indicating significant increases and thus greater hydrological droughts throughout the country.

Thus, it may be expected that the high level of conflicts between individual water users and between them and the Dirección General de Aguas (DGA) will increase in

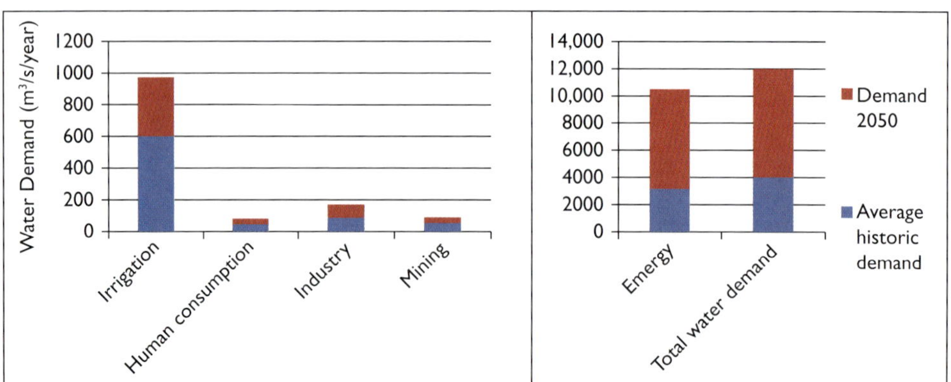

Figure 3 Average historic water demand and projected 2050 water demand ($m^3/s/year$) (Vicuña *et al.*, 2012).

the future due to the growing pressure on water resources. In order to minimize the negative impacts due to the increased water stress, it is essential to implement Integrated Water Management systems (IWRM). The feasibility of implementing IWRM in Chile, depends on the limitations imposed by its water legislation and institutionality to achieve an integrated water resource management. In the next section the water sector institutionality is being summarized.

3 INSTITUTIONALITY

Under the Water Code of 1981, the State reduced its intervention in water resources management to a minimum and increased the management powers of water use rights holders that are organized in Water User Associations (WUAs). The water resource management roles assigned to the State include:

- Measure and determine the availability of water resources and to generate the necessary data and information that allow for a well-informed management of water resources on the part of WUAs;
- Regulate the use of water resources avoiding third party effects and their overexploitation. For that purpose the State must analyze water resource availability and potential water use conflicts before granting new water use rights, authorizing water use right transfers and other authorizations such as changes in water distribution infrastructure;
- Conserve and protect water resources, by means of an environmental impact assessment of investment projects, establishment of minimum ecological flows and environmental policies.

However, multiple central authorities (ministries, departments, public agencies) are involved in water policy making and regulation at the central government level. In Chile the number of actors involved in water policy making is 15; one of the highest of OECD countries that were surveyed in an OECD (2011, 2012) study on water governance.

According to Vergara (2010), there is a distinction between centralized and decentralized institutions. Centralized organizations comprise the administrative bodies of the State. These centralized institutions include water quantity and quality management bodies and the judicial system that resolve most water use conflicts. Decentralized bodies are represented by user organizations, which are private organizations that are not part of the State administration.

The Dirección General de Aguas (DGA), part of the Ministerio de Obras Públicas (MOP), is the main public institution and is responsible for monitoring and enforcing water use rights. With its 15 regional offices, it collects and maintains hydrological data and Public Registry of Water use rights (Registro Público de Agua, RPA). As the leading government agency in water resources management it develops and enforces national water policy. In this role it has: Led efforts to amend the 1981 Water Code and developed a National Water Policy. In general, the DGA has maintained a limited role in accordance with the paradigm of limited state interference on which the Water Code of 1981 is inspired.

The Dirección de Obras Hidraulicas (DOH) is a unit of the MOP that plans, designs, coordinates and supervises the construction of major hydraulic public works, such as water dams. Its programs include not only infrastructure investment for water management, but also primary storm water infrastructure, flood control, and infrastructure for rural health services. A third important institution is the Comisión Nacional de Riego (CNR) of the Ministerio de Agricultura that establishes the policies and programs for the irrigation subsector. The CNR is headed by a Council of Ministers and an Executive Director. The Council of Ministers that includes the Ministers of Agriculture, Economy, Development and Reconstruction, Finance, Public Works, and Planning coordinates the institutions involved in irrigation and drainage, and an Executive Secretariat, which conducts research and implements programs and projects in order to submit proposals to the Council of Ministers. The Executive Director conducts studies and implements irrigation plans and projects. The CNR also administers a subsidy whose objective is to incentivize the adoption of water conservation technology by farmers (Ley 18,450).

Several institutions form part of the judicial system. Since neither the DGA nor any other governmental agency has authority to intervene in water conflicts, water use rights owners resolve their disputes either through voluntary negotiations, involving their respective WUAs, or through ordinary civil courts. Due to the lack of an effective conflict resolution mechanism, numerous conflicts end up in the hands of the courts (Bauer, 2004). Judges usually ask the DGA for expert opinions in any particular case, although they are not required to consult with experts. Broader conflicts over water, between different economic water consuming sectors and between water users and the DGA, are resolved in the Regional Court of Appeals. The decisions of the Court of Appeals may be appealed to the national Supreme Court. In addition, the Controlador General de la Republica controls the legality of various judicial and executive acts. This control includes the establishment restrictions and prohibitions on new water use rights concessions and on the declaration of scarcity, which allows for the intervention of the DGA in water management.

The Water Code of 1981 establishes that water use right owners are responsible for water management. Water user management has existed in Chile since the colonial era, and currently there are more than 4000 Water User Associations (WUAs) (Dourojeanni & Jouravlev, 1999). Three types of WUAs exist in Chile and are recognized by the Water Code of 1981: Water communities, channel user associations, and vigilance committees. Water communities are any formal group of users that share a common source of water. Channel user associations are formal associations with legal status that can enter into contracts; these associations operate on a distribution channel system. Vigilance committees are comprised of all the users and channel associations on any river, river section, or stream; they are responsible for administering water and allocating water to different channels. Some vigilance committees and channel user associations manage reservoirs for irrigation water and finance their operations with small hydroelectric plants.

The different user organizations have some common competences. In first place, their primary responsibility is the distribution of water resources between water users. Secondly, the management decisions are voted in general meetings by shareholders in proportion to their water use rights shares. Thirdly, under drought conditions, water is distributed proportional to the amount of water use rights each water user holds.

In 2010 the number of water communities was more than 10 times the number of channel associations; this is due to the fact that it is easier to form a water community than a channel association. Water communities and channel associations are responsible for both the management, maintenance and renovation of more than 40,000 km of primary and secondary channels, as well as dams built by the private sector or transferred to the user associations by the State (Verges, 2010). At present there is only one groundwater community in the country, in the region of Atacama. The Water Code of 1981 establishes that any aquifer that has been declared under restriction or protection must have a groundwater community. The compliance of this regulation is very low since several aquifers have been declared under restriction and protection and have not formed groundwater communities.

The vigilance committees are different from the other two types of WUAs, since all their competences and legal powers are over surface water before it's withdrawn. Since the Water Code reform in 2005, vigilance committees also must integrate groundwater into its jurisdiction, in an attempt to move towards a conjunct surface and groundwater management. Its main responsibilities are:

- Generate hydrological information in order to improve user's understanding of the water system;
- Manage surface- and groundwater withdrawals;
- Surveillance and monitoring of surface- and groundwater withdrawals;
- Water extraction enforcement;
- Application of sanctions to non-compliers.

Many of these WUAs have professional management (Hearne & Donoso, 2005). The effectiveness of some of these institutions in managing irrigation systems and reducing transactions costs for water market transactions has been noted (Hearne & Easter, 1995 & 1997). However, according to the DGA and the Dirección de Obras Hidraulicas (DOH), a large percentage of these institutions have not updated their capacity to meet new challenges. Many managers of these user organizations do not have a sufficient technical capacity and do not effectively communicate with their members. Additionally, Bauer (2004) points out that vigilance committees have not been effective in resolving inter-sectoral conflicts. To address some of these concerns, the CNR, DOH and DGA have implemented programs to train WUA managers and directors (Peña, 1999; Puig, 1998).

Brown (2004) has pointed out that the current water institutional structure has allowed for a multi-sectoral approach to environmental issues concerning water quality under the framework of Chile's environmental law (Ley Bases del Medio Ambiente). In Chile, most of the institutions related to water quality are separated from those that manage water quantity. In 2010 major changes were made to the Ley General de Bases del Medio Ambiente of 1994 (Law n° 19,300) taking into account recommendations of the OECD (OECD, 2005), as well as international experience. This institutional reform intended to (i) rationalize the competence of multiple agencies in the area of water quality management and to (ii) integrate and improve the effectiveness of Chile's environmental regulation (Library of Congress of Chile, 2010). The new Law (n° 20,417) created three new entities that replace the previous Comisión Nacional del Medio Ambiente (CONAMA): (i) the Ministerio del Medio Ambiente (MMA) with responsibilities in the formulation and implementation of water quality policies, plans, and programs also ensuring the protection and conservation of renewable natural resources such as water; (ii) the Servicio de Evaluación Ambiental (SEA) which is responsible, through its regional offices of environmental impact assessments, of managing, promoting, and facilitating public participation in the environmental evaluation of investment projects, and of presenting to the Council of Ministers a proposal of the environmental qualification resolution (Resolución de Calificación Ambiental, RCA); and (iii) the Superintendencia de Medio Ambiente (SMA) responsible for the monitoring of the compliance of the RCA, decontamination plans, and quality and emission standards. The Environmental Courts (Tribunales Ambientales) dependent of the Ministerio de Justicia were created in 2012 in Antofagasta, Santiago and Valdivia. The core competencies of these are related to environmental damage claims (Contreras, 2010).

However, Chile's water institutionality presents important limitations to effectively address integrated water resource management. In first place, Chile has sought to create institutional arrangements in which each economic sector has a defined regulatory framework, with appropriate incentives for the efficient management of resources in their particular area. This approach has not allowed for an effective management of the multiple interactions that arise between the public and private sectors present at a watershed level. Secondly, the fragmented water institutionality leads to the lack of a strong institution that identifies, formulates and implements national water policy as well as gives coherence to the actions of the various other institutions. In third place, OECD (2011, 2012) concludes that Chile's water institutionality presents obstacles to achieve an effective horizontal coordination between public agencies at the central level as well as a vertical coordination. The most important of these obstacles are the excessive fragmentation of Chile's water institutionality, the existence of overlapping and unclear allocations of responsibilities, competition of powers between ministries, lack of an adequate budget for public agencies, and the lack of citizens' concern for water policy.

Furthermore, the Instituto de Ingenieros (2011) points out that the current practice of managing water resources at the level of a river or aquifer section as if they were independent of other sections of the river basin presents limitations for the implementation of an integrated management. The Water Code of 1981 considers river sections and aquifer sections as independent bodies of water. Thus, each independent section has a WUA that optimizes water resources for its water users without considering downstream effects or impacts on groundwater users. For example, in the past three years that have been characterized by drought, several channel user associations have lined their channels so as to reduce water percolation and deliver more water to their surface water users. This is an optimal decision for surface water users, however, it significantly reduces groundwater recharge. What is more worrying is the fact that most of these investments have been subsidized by the CNR. Thus, government funded investments generate externalities on groundwater users.

Figure 4 depicts in a simplified diagram the negative impacts of this water legislation on downstream users when a river is separated in two sections. In the first section assume that there are two main water users: Agriculture and a potable water supply company. Agriculture extracts 80% of the available water flow while the potable water supply company extracts 20%. Suppose that agriculture's water use efficiency

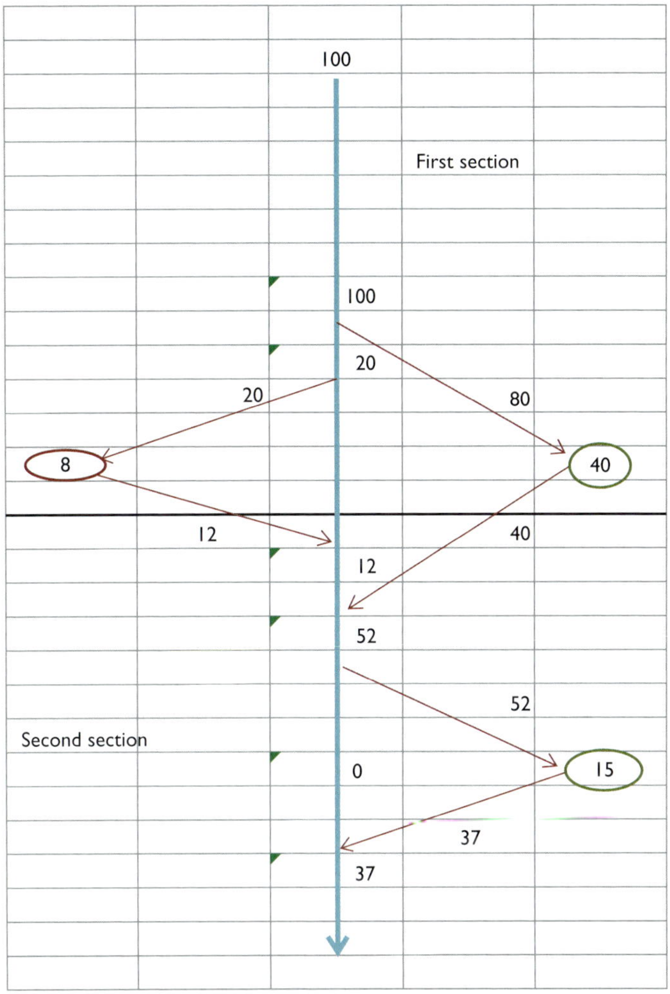

Figure 4 Simplified diagram of a river basin divided into two sections (own elaboration).

is on average 50% and that the potable water supply company has a water loss of 15% due to leaks and an inefficient distribution system and returns 25% in the form of treated water. The second section of the river receives water from the return flows from agriculture and the potable water supply company. Assume that this second section has only one user type that is agriculture, with an average water use efficiency of 29%. In this section agriculture consumes the total water available (52% of the original water flow) and returns 37%. Suppose that this final water flow availability satisfies ecological water flows at the final portion of the river basin which require 25% of the original water flow.

Now suppose that given a severe drought which has reduced water flows from 100 to 80 in the river basin, agricultural producers in both sections decide to increase their water use efficiency by 46%, taking advantage of the government's subsidy to adopt water conservation technology. This implies that the agricultural users in the first section reach a 73% water use efficiency, while agriculture in the second section achieves a 42.3% water use efficiency. At the same time, given population growth, the potable water supply company must increase its water extraction by 50% thus reducing the participation of agriculture in the first section to an extraction of 56%. However, given the increase in water use efficiency, the effective water consumption of the agricultural sector increases from 40 to 41 (Figure 5). The total return flow to the second section of the river basin is 43.4% lower than in the previous case (Figure 4). Note that the first section presents a 20% reduction while the second section sees their water flow reduced by more than twice that amount, due to the increased water

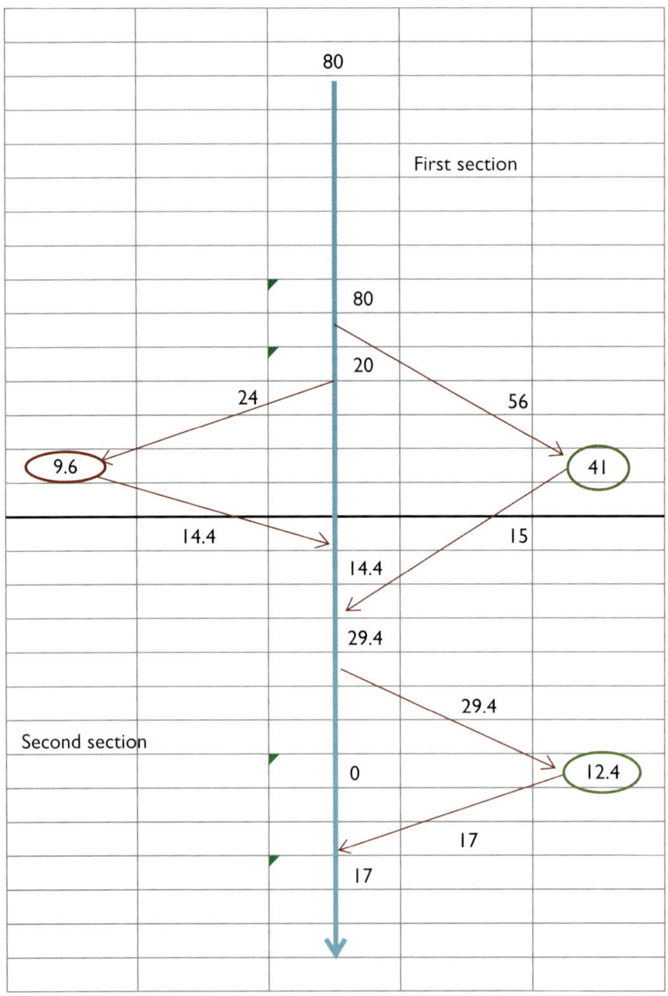

Figure 5 Simplified diagram of a river basin divided into two sections under drought and increased water demand and water use efficiency. (own elaboration).

Table 1 Water balance of each sector of the Copiapó Aquifer (DICTUC, 2010).

Sector	*Water Inflow (l/s)*	*Recharge (l/s)*	*Water use rights (N° Wells)*	*Water extraction (l/s)*	*Water outflows (l/s)*	*Hydrologic balance (l/s)*
1	19	1901	39	2187	513	–18
2	513	173	51	3380	66	–172
3	66	967	71	4107	118	–151
4	118	681	77	4115	227	–526
5	227	220	127	3895	112	–583
6	112	144	75	1938	0	–283

use efficiency in the first section. Thus agricultural users reduce their water extraction due to the drought and reduced return flow. The effective water consumption of farmers in the second section is 17% less than the previous case. The reduction is lower than the decrease in water availability due to their investment in water conservation technologies. Despite the increase in water use efficiency, agriculture in section two is clearly a loser. Finally, in this case, the second loser is the environment since final water flows are 17, well below the required amount of 25. This occurs at present in Chile's river basins from the Metropolitan Region to the north since minimum ecological flows were not set prior to the total allocation of available water flows.

Table 1 shows the hydrological connection between the six sections of the Copiapó aquifer, a division defined by Alamos and Peralta (1987) for practical reasons and following the natural narrowing of the Copiapó Valley. However, at present the DGA interprets these sections as Hydrogeological Sections which implies that each section must have a groundwater user association which takes water management decisions independently from others. Given the water flows between sections, four independent water management plans are clearly suboptimal to an integrated water management plan.

4 KEY ACTIONS REQUIRED TO ADVANCE TOWARDS AN INTEGRATED WATER MANAGEMENT IN CHILE

The interface of water resources and economic development is complex and includes many specific linkages. This wide array of important linkages that present many synergies, explains why pursuing each goal separately reduces the complex process of human and economic development to a series of conflicting, and unsustainable interventions. In addition, the interface of water resources and the achievement of sectoral development goals occurs at several different institutional levels, explaining why water resource policies in Chile have evolved in a fragmented and piecemeal fashion. Under this framework, policy objectives have been set without consideration of the implications for other water users and without consultation across sectoral and institutional boundaries. This traditional approach to water management has, in general, proven to be an ineffective policy strategy due to the fact that these problems fall outside of the normal purview of the agencies tasked with addressing them and, thus, require cooperation from multiple sectors. In order to optimize water resources for development, Chile must overcome constraints and thus generate an effective institutional coordination and make appropriate investments and management arrangements within broad planning and policy initiatives. In general, the solution can be found by greater institutional concentration since sectoral development goals occur at several different institutional levels.

In order to advance towards the implementation of IWRM Plans in Chile, it is necessary to implement the following key actions:

1 Increase the awareness of both the political leadership, water users and the society at large about the urgent need to move towards IWRM so as to involve stakeholders;
2 Implement groundwater user associations;
3 Integrate groundwater user associations to the Juntas de Vigilancia so as to implement conjunct surface and groundwater management;
4 Strengthen all WUAs so that each one develops a strong rule of law, effective conflict resolution, and effective collective management;
5 Implement Supra Organizations of Juntas de Vigilancia to integrate different river sections and aquifer hydrogeological sectors. This does not require a water legislation modification;
6 Implement efficient negotiation and conflict management since it will not be able to please everyone;
7 Work with the media to constantly inform society on the advances and short term wins so that society at large maintains its motivation towards the change process.

It is important to highlight, however, that there are no universal models that can be implemented. The specific solution and necessary action plan is country-specific. In addition, the experience indicates that IWRM Plans can be developed from scratch or be built on existing water plans. However, independent of the initial approach, it is clear that the strategies must go beyond the actions needed to solve current problems or to achieve immediate objectives; the implemented strategies should aim at promoting more strategic and coordinated decision-making on a dynamic basis so as to advance towards the development and implementation of IWRM Plans.

5 CONCLUSIONS

Lenton, Wright & Lewis (2005) remark that Integrated Water Resource Management (IWRM) builds on three basic pillars: (i) an enabling environment of proper water resources policies and legislation; (ii) an institutional framework of capable institutions at national, local, and river basin levels; and (iii) a set of management instruments for these institutions. Thus, IWRM allows for a more coordinated decision-making process across sectors and scales.

Hence, as Global Water Partnership (2000, 2004a) indicates, implementing an IWRM Plan is significantly different from the traditional approach used to develop water plans. Firstly, an IWRM Plan lays down a framework for a continuing and adaptive process of strategic and coordinated action and, thus, is dynamic rather than static. Secondly, implementing an IWRM Plan requires the involvement from multiple sectors. Traditional water plans tend to be concerned exclusively with water supply and demand issues, however, an IWRM Plan looks at water in relation to other ingredients needed to achieve economic development. Lastly, since an IWRM Plan allows for a coordinated decision-making process across sectors and scales, it requires more extensive stakeholder participation than traditional approaches.

Chile's current institutional framework makes it difficult to develop a multi-sectoral water management approach since it leads to single economic sector water planning. At the same time, Chile's water legislation favors the management of water resources exclusively for independent hydrological sectors, presenting legal obstacles to implement integrated water management. The problems that have been identified are not unique to water management in Chile. In general, they are present in many parts of the world and require a non-structural solution and a new approach to planning and management. In this context, Chile must overcome constraints through appropriate institutional reforms and legal modifications that allow different WUAs to integrate and develop an Integrated Water Resources Management (IWRM).

REFERENCES

Bauer, C. (2004) *Siren Song: Chilean Water Law as a Model for International Reform.* Resources for the Future. Washington D.C., Routledge.

Brajer, V., Church, A., Cummings R. & Farah P. (1989) The Strengths and Weaknesses of Water Markets as They Affect Water Scarcity and Sovereignty Interests in the West. *Nat. Res. J.*, 29, 489–509.

Brown, E. (2004) *Hacia un Plan Nacional de Gestión Integrada de Recursos Hídricos.* II Taller Nacional ECLAC, Santiago, Chile.

Central Bank of Chile (2010) *Síntesis de estadística de Chile 2005–2009.* Banco Central de Chile. Santiago, Chile.

Contreras, M. (2010) *Calidad de aguas y contaminación: Etapa diagnóstico.* Informe preparado para el diagnóstico de la gestión de los recursos hídricos. Dirección General de Aguas, Ministerio de Obras Públicas.

Cummings, R.G. & Nercissiantz, V. (1992) The Use of Water Pricing as a Means for Enhancing Water Use Efficiency in Irrigation: Case Studies in Mexico and the United States. *Nat. Res. J.*, 32, 731–55.

DICTUC (2010) *Análisis Integrado de Gestión en Cuenca del Río Copiapó.* Gobierno De Chile, Ministerio de Obras Públicas, Dirección General de Aguas.

Donoso, G. & Melo, O. (2006) Water Quality Management in Chile: Use of Economic Instruments. In: Biswas, A., Tortajada, C., Braga, B. & Rodríguez, D.J. (eds) *Water Quality Management in The Americas.* Berlin Heidelberg, Springer-Verlag. pp. 229–252.

Dourojeanni, A. & Jouravlev, A. (1999) El Código de Aguas de Chile: Entre la ideología y la realidad. *Debate Agrario,* 29, 138–185.

ECLAC (2009) *Economics of Climate Change in Latin America and the Caribbean. Summary 2009.* [Online] Available from: http://www.eclac.org/publicaciones/xml/3/38133/2009-851-Summary-Economics_climate_ change-WEB.pdf [Accessed October 2012].

ECLAC (2011) *Creciendo en base a los recursos naturales, "tragedias de los comunes" y el futuro de la industria salmonera chilena.* [Online] Available from: http://es.scribd.com/doc/55727880/10/CUADRO-2-DISTRIBUCION-DE-CONCESIONES [Accessed October 2012].

ECLAC (2012) *Agricultura y cambio climático: Del diagnóstico a la práctica.* [Online] Serie Seminarios y Conferencias. Report number: 71. Available from: http://www.eclac.org/publicaciones/xml/8/48088/Agriculturay CambioClimatico.pdf [Accessed October 2012].

GWP, Global Water Partnership (2000) *Integrated Water Resources Management.* [Online] TAC Background Papers. Number: 4. Available from: http://www.gwp.org/Global/GWP-CACENA_Files/en/pdf/tec04.pdf [Accessed November 2012].

GWP, Global Water Partnership (2004a) *Catalyzing Change: A Handbook for Developing Integrated Water Resources Management (IWRM) and Water Efficiency Strategies.* [Online] Available from: http://www.gwptoolbox.org/images/stories/gwplibrary/catalyzing%20change_english.pdf [Accessed November 2012].

GWP, Global Water Partnership (2004b) *Informal Stakeholder Baseline Survey: Current Status of National Efforts to Move Towards Sustainable Water Management Using an IWRM Approach.* Summary Report. Available from: http://protosh2o.act.be/VIRTUELE_BIB/Werken_in_het_Water/IWB-Integraal_WaterBeheer/W_IWB_E7_watermanagement_IWRL.pdf [Accessed November 2012].

Hadjigeorgalis, E. (2008) Distributional impacts of water markets on small farmers: Is there a safety net? *Water Resources Research*, 44(10).

Hearne, R. & Donoso, G. (2005) Water Institutional Reforms in Chile. *Water Policy*, 7, 53–69.

Hearne, R. & Easter, W. (1995) *Water Allocations and Water Markets: An analysis of gains from trade in Chile.* World Bank. Technical Paper number: 315.

Hearne, R. & Easter, K. (1997) The Economic and Financial Gains from Water Markets in Chile. *Agricultural Economics*, 15, 187–197.

INE (2002) Censo Chileno de 2002. [Online] Available from: http://www.ine.cl/cd2002/sintesiscensal.pdf [Accessed October 2012].

Instituto de Ingenieros (2011) *Temas Prioritarios para una Política Nacional de Recursos Hídricos.* Comisión de Agua.

Lenton, R, Wright, A. & Lewis, K. (2005) *Health, Dignity, and Development, What will it Take?* UN Millenium Project Task Force on Water and Sanitation, Final Report. London, Earthscan.

MMA (2011) *Segunda Comunicación Nacional de Chile ante la convención marco de las naciones unidas sobre cambio climático.* [Online] Available from: http://www.mma.gob.cl/1304/articles-50880_docomunicadoCambioClimatico.pdf [Accessed October 2012].

Metzger, P.C. (1988) Protecting Social Values in Western Water Transfers. *J. Amer. Water Works Assoc.* 80, 58–65.

McEntire, J. (1989) Water Farms and Transfer Conflicts in Arizona, USA: A Proposed Resolution Process. *Environ. Manage.*, 13, 287–295.

Molle, F. (2004) Defining Water Rights: By Prescription or Negotiation? *Water Policy*, 6, 207–227.

OECD (2005). *Environmental Performance Reviews: Chile.* Organization for Economic Co-operation and Development (OECD) and Economic Commission for Latin America and the Caribbean (ECLAC) of the United Nations.

OECD (2011) Water Governance in OECD Countries: A Multi-level Approach. [Online] OECD Studies on Water, OECD Publishing. Available from: http://dx.doi.org/10.1787/9789264119284-en [Accessed October 2012].

OECD (2012) Water Governance in Latin America and the Caribbean: A Multi-Level Approach. [Online] OECD Regional Development Working Papers, OECD Publishing. Available from: http://dx.doi.org/10.1787/5k9crzqk3ttj-en [Accessed December 2012].

Peña, H. (1999) *Política Nacional de Recursos Hídricos de Chile.* Ministerio De Obras Publicas, Dirección General de Agua.

Puig, A. (1998) *El fortalecimiento de las organizaciones de usuarios para una gestión integrada de los recursos hídricos.* [Online] Paper presented at the International Conference of Water and Sustainable Development, 19–21 March 1998, Paris, France. Available from: http://www.oieau.fr/ciedd/contributions/at2/contribution/aurora.htm [Accessed November 2012].

SISS, Superintencia de Servicios Sanitarios (2004) *Historia del Sector.* [Online] Available from: http://www.siss.gob.cl/577/w3-article-3681.html [Accessed October 2012].

SISS (2011) *Cobertura Nacional Agua Potable y Cobertura de Tratamiento de Aguas Servidas Nacional.* [Online] Available from: http://www.siss.gob.cl/577/w3-article-3683.html [Accessed October 2012].

Southgate, D. & Figueroa, E. (2006) Reforming Water Policies in Latin America: Some Lessons from Chile and Ecuador. In: K. Okonski (ed.) *The Water Revolution.* London, UK, International Policy Press. pp. 72–91.

Syme, G.J., Nancarrow, B.E. & McCreddin, J.A. (1999) Defining the Components of Fairness in the Allocation of Water to Environmental and Human Uses. *J. Environ. and Manage.*, 57, 51–70.

Vergara, A. (2010) *Diagnóstico de Problemas en la Gestión de Recursos Hídricos: Aspectos Institucionales para una Futura Propuesta de Modificaciones Legales, Reglamentarias y/o de Prácticas Administrativas. Informe preparado para el diagnóstico de la gestión de los recursos hídricos.* Dirección General de Aguas, Ministerio de Obras Públicas.

Verges, J. (2010) *Síntesis del diagnóstico de la gestión de los recursos hídricos: Primero borrador. Informe preparado para el Diagnóstico de la Gestión de Recursos Hídricos.* Dirección General de Aguas, Ministerio de Obras Públicas.

Vicuña, S., Meza, F., Bustos, E. & Poblete, D. (2012) *Los nuevos desafíos a la gestión de los recursos hídricos en Chile en el marco del Cambio Global.* Seminario Gestión de los recursos hídricos en Chile: Propuestas y desafíos, 6 de noviembre de 2012. Centro Políticas Públicas, Pontificia Universidad Católica de Chile.

World Bank (2011) *Chile: Diagnóstico de la gestión de los recursos hídricos.* [Online] Available from: http://www.dga.cl/eventos/Diagnostico gestion de recursos hidricos en Chile_Banco Mundial.pdf [Accessed July 2011].

World Bank (2010) *World Development Indicators 2010.* Washington D.C., USA.

Chapter 13

Towards IWRM in the upper Guadiana basin, Spain

José Ángel Rodríguez Cabellos
Head of the Hydrological Planning Office, Guadiana River Authority, Spain

ABSTRACT: The upper Guadiana basin provides an example of the difficulties involved in the practical implementation of IWRM and, more specifically, the need to deal with the burdens of the past. From the 1950s, wetlands were dried up in order to reclaim land for agricultural development. In the 1970s, based on the ancient private water abstraction rights from the XIX century water law, the advances in well drilling and pumping technologies contributed to the spread of intensive groundwater irrigation in the area. For decades, intensive groundwater use contributed to social and economic development. On the other hand, it also caused the water table to drop to the point that surface and groundwater bodies became disconnected. This had an adverse effect on riverine wetlands, as well as on Ramsar-protected wetlands such as the 'Las Tablas de Daimiel National Park' and, by and large, on UNESCO's 'Mancha Humeda Biosphere Reserve'. From 1987, a variety of plans, measures and laws have been implemented, including the EU Water Framework Directive. The last three decades have also witnessed dry and wet cycles, running parallel to continuous conflicts and negotiations between the main water users – irrigators – and water authorities. This chapter provides a description of the water management frameworks that have been in place since the 1950s, to show how social conflicts, economic interests and environmental protection play a part in integrated water resources management.

Keywords: upper Guadiana basin, groundwater, management plans, wetlands

1 INTRODUCTION: THE GUADIANA BASIN

The Guadiana river basin is located in the south west of the Iberian peninsula (Figure 1), and is shared by Spain (83% in surface) and Portugal (17%). The climate is Mediterranean-continental with the average rainfall being 522 mm/yr, ranging from 340 mm in La Mancha plain to 1100 mm in the upper Murtigas basin.

The average altitude of the basin is 550 meters above sea level. The landscape is most abrupt in the Bullaque-Tirteafuera system and in the middle basin, whereas the upper part, namely the Mancha Plain, is largely flat. This latter feature favours the existence of wetland ecosystems.

The upper Guadiana basin is known for its aquifers. These are fed by surface water and groundwater from adjoining groundwater systems. Groundwater emerges

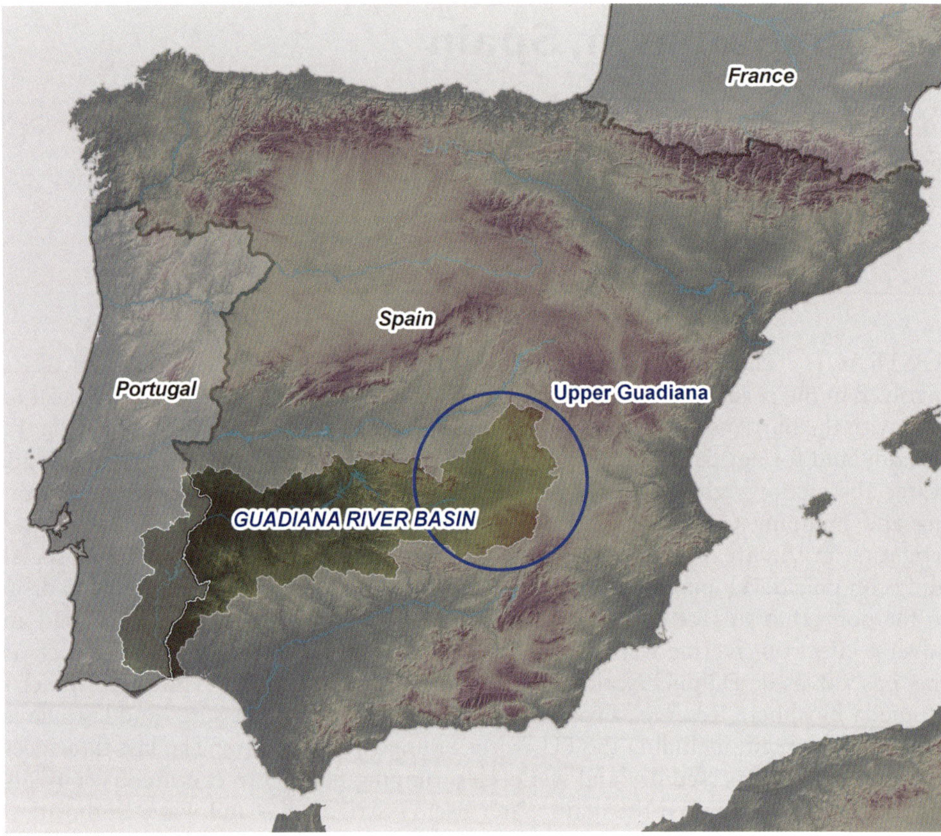

Figure 1 The Guadiana river basin (Source: Confederación Hidrográfica del Guadiana, n.d.).

locally in ground depressions, thus generating valuable ecosystems such as the Lagunas de Ruidera Natural Park and Las Tablas de Daimiel National Park, among others. Largely as a result, UNESCO declared the Mancha Humeda Biosfere Reserve in the 1980s. Some of the wetlands were also included in the Ramsar list and protected by national laws.

Throughout its upper reaches, located within the Campo de Montiel aquifer, the Guadiana river runs through a relatively abrupt topography. Limestone outcrops favor aquifer recharge through rainfall and discharge through a series of springs. All these factors contributed to create a complex of lagoons of high ecological value; the Lagunas de Ruidera. The Campo de Montiel aquifer discharges directly into the Mancha Occidental aquifer, also known as the Mancha Plain aquifer. This aquifer is also fed by other adjoining aquifers, including the Sierra de Altomira, Consuegra-Villacañas and Lillo-Quintanar units.

Another recharge mechanism is infiltration from rainfall and surface water courses. Take for instance the Guadiana, Zancara and Ciguela rivers, which gradually lose water to infiltration until they completely dry up. This, coupled with a flat landscape, gave rise to large extents of riverine wetlands. These exceeded 30,000 hectares in the 1950s.

Under natural conditions, the Mancha Occidental aquifer discharged through a series of springs and wetlands located towards its western end. These include the Ojos del Guadiana springs and the wetlands of Las Tablas de Daimiel National Park. Other significant wetlands ('lagunas') are the Lagunas de las Yeguas and Villafranca, Laguna de La Vega, Laguna del Alcahozo, Laguna del Prado, Laguna del Taray, Laguna del Hito or the Laguna de Manjavacas. All these provide a contrast against the dry landscape and are extremely vulnerable to environmental change.

2 LEGAL FRAMEWORK BEFORE AND AFTER 1985

Spain's 1879 Water Law, in place until 1985, established the rules for groundwater extraction. Under this provision, anyone who found groundwater within his or her property acquired the right to use it perpetually. This follows from the Roman law, which established that land owners are proprietors of all that is above and below their land ('from heaven to hell'). In fact, those who tapped groundwater within their properties needed not even inform the authorities until the 1973 update of the 1934 Mining Law, which established that they should register their wells.

The 1985 Water Law triggered an important change to groundwater management, as it declared groundwater as public property. Thereon, new groundwater users would be required to ask for a license prior to drilling. Existing wells however were allowed to remain operational. This was done so that the government could avoid the costs of compensating well owners for extirpating a pre-existing private right.

Under the 1985 law, all groundwater users – old and new – were asked to register their wells within three years. Owners of wells drilled prior to 1985 needed to present proof that their wells had been drilled before 1985, and then they were allowed to choose whether to register them in the Public Water Registry or the Catalogue of Private Waters. Choosing the latter meant that they could maintain private right over groundwater at the expense of forfeiting administrative protection. In contrast, choosing the former implied exchanging their right for a permit so as to obtain administrative protection.

The water authorities were swarmed by tens of thousands of applications following the enactment of the new law. The vast number of applications confirmed something which declining piezometric records had been anticipating since the late 1970s: That groundwater extractions exceeded the renewable aquifer resources by far.

3 EARLY EVOLUTION OF THE HYDROLOGICAL SYSTEM, FROM THE BEGINNING OF THE 20TH CENTURY UNTIL THE DECLARATIONS OF AQUIFER OVEREXPLOITATION

The 1939 law for the colonization of the land (Ley de Colonización de Grandes Zonas) and the 1956 law for sanitation and reclamation of wetlands (Ley de Saneamiento y Colonización de los Terrenos Pantanosos), were enacted to reclaim barren lands, including wetlands, for economic use. From the 1960s to the 1980s both laws underpinned a series of major changes to the local ecosystems. These included river channeling works, drainage of wetland areas and the transformation of reclaimed lands into agricultural surfaces. They were funded jointly by the public works and agricultural administrations.

In 1951, following studies by the Geological Survey of Spain and the National Institute for Colonization, the Government declared that transforming the Mancha plain into a major irrigation area was a project of high national interest. This declaration was furthered by groundwater research projects in the 1960s, which led to the delimitation of the regional aquifer systems.

Access to groundwater was cheap and easy for farmers, who had been extracting it by means of thousands of small waterwheels since the late 19th century. The technical developments of the 1970s, most notably the decrease in drilling costs and the invention of the submersible pump, led to the development of modern irrigation projects. For years, extraction rates exceeded aquifer recharge. In fact, it is estimated that extractions doubled the average recharge in the late 1980s. The water table was observed to drop as a result, to the point that the storage deficit in the system grew to 4000 Mm^3 by the mid-1990s.

In 1979, the Geological Survey of Spain pointed out the convenience of managing extractions. By then, thousands of individual farmers had taken control of groundwater extractions across the system. The water authorities did not have enough means to control groundwater development. Besides, the much-needed changes to the 1879 legal arrangements were systematically delayed.

About 130,000 ha were put under irrigation between the late 1970s and the early 1980s. This meant a four-fold increase on the then-existing figures. Traditional water-efficient crops such as vineyards, cereal and olive trees were gradually replaced by other water-intensive crops. These were favored by EU subsidies and included alfalfa, maize and sugar beet. All these changes provided a boost to the local agricultural economy, enhancing per capita income and creating an adequate environment for related industries. On the other hand, the water table dropped, and so did flows in the region's rivers. The Ojos del Guadiana springs dried up in the early 1980s. The wetlands of Las Tablas de Daimiel National Park also suffered a severe reduction in terms of water inflows. Agricultural intensification also triggered a decrease in groundwater quality.

The water and agriculture administrations, which had fostered intensive groundwater use, soon were surpassed by the situation. Poor planning and the lack of adequate coordination between them contributed to aggravate the region's environmental problems.

The Mancha Occidental aquifer was not the only one to experience the downside of groundwater-based development. Public policies also fostered the transformation of forest surfaces into agricultural areas in the Campo de Montiel aquifer. This had a negative effect on spring and stream flows in the southern part of the system. As a result, there were problems in the traditional irrigation schemes in Montiel and Villanueva de la Fuente, as well as in the urban water supply of Villahermosa and Montiel. Inflows to the Lagunas de Ruidera natural park and to Peñarroya dam were also reduced. Surface water farmers from Tomelloso and Argamasilla de Alba teamed up with environmental conservation groups to ask for pumping in the Campo de Montiel aquifer to be forbidden altogether. All these episodes led to public order problems in the area in the 1980s and early 1990s.

4 THE MANAGEMENT FRAMEWORK AFTER 1985

4.1 Declarations of overexploitation

The 1985 Water Law established a provision that allowed for aquifers to be declared overexploited. In turn, this allowed the water authorities to restrict extractions in heavily depleted aquifers in order to revert adverse situations such as the ones described above. On February 4, 1987, the Guadiana Water Authority passed a provisional declaration of overexploitation for the Mancha Occidental aquifer. This was made definite in November 1994. In parallel, the Campo de Montiel aquifer was declared provisionally and definitely overexploited in 1988 and 1989, respectively.

These declarations implied restrictions to pumping, implemented by means of yearly pumping plans. The maximum extraction rate was limited from the maximum right of 4278 m^3/ha/yr to as little as 2000 m^3/ha/yr. There was some degree of modulation depending on farm size (the larger the farm, the more stringent the restrictions). The declarations also placed a ban on new wells and forbade the administration to issue new permits.

Irrigation Communities were created to help the Water Authorities manage groundwater. Also, a Development Committee (Junta de Explotación) was implemented within the participation branch of the Water Authority.

Restrictions were barely enforced due to the absence of human, technical and economic means, as well as to strong social pressures. Farmers and irrigation communities claimed that restrictions caused large economic losses and contributed to unemployment, and called for due compensation. Besides, farmers were quick to realize that the water authorities did not have the means to enforce their own regulations. Hence, illegal drilling became rampant.

4.2 Compensation plans and the Vineyard Restructuring Plan

In 1992, the regional agricultural administration issued a plan to help recover the region's aquifers and wetlands. Pivotal to this plan were a series of economic incentives for farmers in the Mancha Occidental and Campo de Montiel aquifers to curtail water use (López-Sanz, 1996). The overall purpose of the plan was to balance the extractions with the renewable resources and limit the use of fertilizers and pesticides. The plan became widely known as the Compensation Plan (Plan de Compensación de Rentas).

The Compensation Plan was among the first to be implemented after the 1992 reforms of the EU Common Agricultural Policy and the introduction of regional agro-environmental programs. It was intended to last for five years, and endowed with 96 M€. Eventually, it was extended twice, lasting for ten years and with a total investment of 180 M€. The plan was funded partially by FEOGA (75%), Spain's Ministry for Agriculture, Fisheries and Food (12.5%) and the Castilla-La Mancha regional government (12.5%) (Rosell & Viladomiu, 1997).

Farmers took part voluntarily, choosing whether to cut down on water use (50% or 70%) or to stop pumping altogether (100%). This could be achieved by limiting the surface being irrigated or by switching to water-efficient crops. Compensatory payments were calculated based on aspects such as water consumption, productivity, irrigation alternatives and profitability, among others. From 1998 onwards, compensation was computed only based on water savings complementary to the yearly pumping plan.

The effects of the compensation plan were enhanced by a series of parallel initiatives, including the mentioned pumping plans, a program to foster the installation of water meters, programs to improve management of the national and natural parks, wetland restoration and (re-)forestation plans, as well as through the enactment of measures based on enhanced water supply from the Tajo-Guadiana transfer to a series of towns and water sanitation plans.

However, the most remarkable of the next initiatives was the Vineyard Restructuring Plan. In 2000, Castilla-La Mancha Autonomous Government (responsible for Agriculture) set off a Plan for restructuring the vineyard, with subsidies to restructure the vine farms in this region according the new Common Market Organization of the European Union. The main objective of this plan was the enhancement of vineyards to adapt wine production to the new national and international market. The effect of this Plan was an extraordinary contribution to change from high water-intensive crops (herbaceous crops such as maize and sugar beet, with a requirement exceeding 8000 m^3/ha) to vineyards (less than 1500 m^3/ha), consolidating the effect of the previous Income Compensation Plan (see Paragraph 5).

These sectorial Plans were not conceived as a means to provide a final solution to the problems. They are best described as an interim solution rather than a structural one, favoring an impressive switching from water-intensive to water-efficient crops, but with the persistent fear that this shift from water-intensive to less intensive crops will only last for so long as the subsidies lasted. Besides, it did not really contribute to create additional economic alternatives.

Other aspects of these Plans which were criticized:

- Insufficient coordination between all the administrations involved;
- Employment and economic activity regressed (although farmer income was maintained);
- Information to farmers on environmental matters was insufficient.

4.3 The 1998 Basin Plan

The Guadiana basin plan, approved by Royal Decree 1664/1998, July 24, was the last one within a long-standing tradition of basin management plans. Its purpose was to satisfy water demands while contributing to a balanced regional and sectoral development by increasing water availability and protecting water quality and the environment.

The 1998 plan emphasized the need to tackle the problems of the upper Guadiana basin. It highlighted the dramatic growth of irrigated agriculture in the Mancha Occidental aquifer, as well as the fact that pumping depths often exceeded 60 meters and that extractions had grown to 568 Mm^3 in 1988 (that is, twice the renewable resources). As a consequence, the plan referred to a gradual decrease in aquifer storage in the order of 2500 to 3000 Mm^3. It also mentioned the fact that this had reversed flow patterns in the system, causing the disappearance of ecologically-significant discharge areas such as Las Tablas de Daimiel national park. The plan also referred to the Campo de Montiel aquifer and to the fact that heavy pumping had considerably reduced inflows to the Lagunas de Ruidera natural park.

The plan incorporated the declarations of overexplotation as management instruments, and called for the monitoring and enforcement of the yearly pumping plans. It also mentioned the need to carry out studies in neighboring aquifers in order to inform eventual declarations of overexploitation. Besides, the plan reinforced the limitations to grant groundwater permits, while advocating the installation of metering devices and the establishment of measures to prevent and control contamination, as well as bringing into life protection perimeters and artificial aquifer recharge. All of these measures focused on reverting overexploitation and improving the overall quality of the ecosystems.

Finally, the plan acknowledged a strong deficit in terms of the relation between demands and available resources. This was mainly a consequence of the imbalance observed in the upper Guadiana basin (508 Mm^3 baseline, 461 Mm^3 for the 10–year horizon and 433 Mm^3 for the 20–year horizon, all of them greater under drought conditions). It also left the door open for the harmonizing provisions and alternative solutions of the then oncoming National Water Plan.

4.4 The National Water Plan

The National Water Plan (Law 10/2001, July 5), presented a series of coordinating efforts for water management at the national scale, including:

a Measures to coordinate the basin management plans;
b Solutions to the alternatives offered by these;
c Preview the need and conditions to transfer resources between basins and territories;
d Establish modifications to water planning that affect water supply and irrigation.

For the purpose of the Upper Guadiana basin, the National Water Plan did not establish the need to authorize the implementation of an inter-basin transfer. Rather, it advocated internal management and restriction measures to restore the hydrological balance. Its fourth additional disposition established the need to develop a special water plan for the upper part of the basin.

4.5 Water Framework Directive – National Water Plan

On October 23, 2000, the European Parliament approved Directive 2000/60/CE, better known as the Water Framework Directive (WFD). The WFD established a common water policy framework across the EU, and intended to modernize water management practices as well as to protect (or restore) all continental, marine and transitional waters.

Public participation in the planning process is a pivotal aspect to the WFD. The status of water bodies is defined first, in order to establish the baseline conditions for negotiation and the environmental objectives. Then specific programs are devised in order to tackle the deficiencies and avoid potential adverse effects. Planning takes place at the water district scale (demarcaciones hidrográficas), seeking consensus with stakeholders and the general public and establishing economic instruments to reach planning objectives.

After being transposed into Spanish law, the WFD inspired all water planning processes in the country.

4.6 The Upper Guadiana Water Plan

As explained above, the 2001 National Water Plan established the need to elaborate a specific water plan for the upper Guadiana basin. Broadly speaking, the purpose of such a plan would be to ensure a sustainable use of the basin's water resources. Its instruments included a reorganization of water rights, modifications in pumping plans and the possibility to issue groundwater permits under exceptional circumstances, among others.

The WFD establishes that all surface and groundwater bodies must attain a good status by 2015 at the latest. In the case of groundwater bodies, it establishes that they should attain a good status both in quantitative terms and in terms of water quality.

The Upper Guadiana Water Plan was intended to be instrumental in meeting these objectives. It should also be incorporated in the measures of the National Water Plan. The Upper Guadiana Water Plan was approved by Royal Decree 13/2008, January 11, following an intense public participation process that generated a strong social consensus among the groups with an economic interest, the environmental conservation organizations, the public administrations and the civil society.

The resulting plan aimed at being:

- An integral solution that contemplated water issues from all different perspectives (hydrological, agricultural, economic, social and environmental);
- A final solution to obtain lasting structural transformations in the basin (i.e. it transcended the provisional dimension of some previous plans);
- A plan endowed with enough means and carried out in a coordinated manner by all national and regional administrations with a stake in water, environment and socio-economic development;
- A plan devised with strong input from public participation, taking care to foster environmental education and divulgation.

The goals of the Upper Guadiana Water Plan included:

- Attaining a good quantitative and qualitative status of groundwater bodies and associated aquatic ecosystems, restoring connectivity between surface and groundwater.
- Correcting the existing deficit between demands and resources, following the principles of sustainable development for agricultural uses and other socio-economic activities.

Measures to attain these goals included:

- Transformation of private groundwater rights (1879 and 1985 laws) into permits, in an attempt to strike a balance between legal rights and available water resources. Private rights would be exchanged into permits for an equal amount of water except in the case of overexploited aquifers, where transformed rights would be minorated according to the available resources;
- Legal contracts to reassign water rights;
- The Water Authority would be funded with 810 M€ and enabled to purchase land and water rights. In turn, these would be devoted preferentially (70%) to restoring degraded water bodies. The remaining 30% would be transferred to the regional government for reassignation among users based on different criteria (social needs, inclusion of illegal users in the system and so on);
- A 432 M€ hydrological program, based on management and control measures such as information systems, vigilance, installation and monitoring of water meters, etc.;
- A 34 M€ support program for irrigation communities;
- A 1669 M€ environmental program to reduce diffuse pollution, restore the public water domain, foster forestation and restore hydraulic heritage;
- A 55 M€ program to propagate information and environmental awareness.

Overall, the plan was endowed with 3000 M€, to be funded entirely by the Spanish government. Complementary programs, funded by other sources, amounted to 2000 M€ and included:

- A water supply and sanitation program, funded jointly by the Spanish government and the Castilla-La Mancha regional government;
- A program for rural development and modernization. This program aimed at steering the agricultural sector to more productive and water-efficient crops. This would be funded by the Castilla-La Mancha regional government;
- A socio-economic development program, funded by the Castilla-La Mancha regional government. This program was intended to foster socio-economic development in sectors other than irrigated agriculture.

In practice, the effects of the Upper Guadiana Water Plan were limited. This is largely due to budgetary cuts following Spain's 2010 economic crisis. Within this context, the plan gave priority to the following actions:

- Transformation of water rights;
- Purchase of water rights (about 14 Mm^3). The recovered rights were issued to the regional government to be reassigned to water-efficient crops (vineyards);
- Installation of water meters;
- Agreements with irrigation communities;
- Information and environmental awareness raising.

There were no practical implementations of the modernization and rural development program or the socio-economic development program.

This plan has been heavily criticized. From the viewpoint of socio-economic interests, many consider it unfulfilled (particularly, regarding the purchase of water rights on the part of the administration). In this regard, it has been termed 'overly ambitious' given the current context of economic crisis. From the environmental standpoint, its application is considered partial and not sufficiently transparent. It is argued that the purchase of water rights did not focus on priority areas, and that the principle of purchasing land with a history of recent irrigation was not respected. Another criticism is that recovered water rights have been used exclusively to incorporate illegal users to the system, and that none of the savings have been destined to restore the aquifer. Finally, little was done in regard to structural changes.

As of 2012, there are talks on how the plan needs to be reformed. It is argued that it should only focus on management measures and pumping restrictions, avoiding initiatives such as the purchase of water rights, forestation and the environmental program; or, in other words, getting back to the mid-1990s situation. This is currently the source of bitter disputes among the region's main political parties.

4.7 The 2009–2015 Basin Plan

Following in the tradition of Spanish water law, the Basin Plan for the Guadiana district must provide answers to water demands for all socio-economic activities. It must also cater for the environmental objectives of the WFD, which establish the need to restore the upper Guadiana aquifers to a good quantitative and qualitative status.

The Upper Guadiana Water Plan, provisioned by the National Water Plan and approved by Royal Decree, constituted an essential reference for the Basin Plan that was submitted to public consultation. The public consultation process was highly participative. There were disputes as to how to modify the Upper Guadiana Plan. These were accompanied by strong criticisms to the fact that it has been thoroughly unfulfilled due to the economic crisis (see previous paragraph), and that some of it – including the management Consortium and the costlier programs – should be suppressed. In addition, many of the allegations called for the implementation of inter-basin transfers as a means to alleviate the basin's long-standing problems. Such comments came from social, environmental and professional organizations (irrigators), as well as from the new regional government.

As a result of this process, the projected Basin Plan advocates that the Upper Guadiana Plan needs to be reformulated. It also proposes to the National Water Plan that the solution to the upper Guadiana water woes should not only be a matter of internal adjustments and restrictions, arguing that external resources should also be taken into consideration.

These aspects, together with the directives of the General Directorate for Water for the elaboration of a budget for the Basin Plan Project, led to reviewing of the initial provisions of the plan, including its timing and funding. The purchase of water rights and forestation were suppressed together with other minor items. Following from this, and based on new and existing laws (Royal Decree 17/2012-Law 11/2012 and Royal Decree 9/2006) the current Basin Plan proposal presents the following highlights:

- Transformation of private rights into permits;
- Implementation of a system to reassign water rights by means of contracts, fully compliant with the existing laws, as well as of legal instruments to reduce

groundwater allocation to water users in order to balance extractions and demands (a water market);
- Declaration of groundwater bodies at risk or not complying with the WFD requirements, in order to establish specific management measures (including variable pumping patterns to cater for wet and dry years);
- Implementation of a center for the exchange of water rights.

The Basin Plan has triggered technical and scientific advances, contributing to enhance the existing knowledge about the area's groundwater bodies. Models have also been developed to underpin management practices. Finally, the plan defines the available resource in each case so as to attain the WFD objectives and establishes appropriate management measures. As of late 2012, the Basin Plan is pending on the approval of the Spanish Government.

5 EVOLUTION OF ABSTRACTIONS & IRRIGATED CROP AREA

It is remarkable that the groundwater abstraction has been well monitored through Satellite Remote Sensing (Global Monitoring Earth System – GMES nowadays), already since the mid 80's of the last century in this area – which was the first time this happened in Europe – (see next images). This monitoring system has allowed a good understanding of what was happening in this huge area of 5000 km^2 in the last thirty years, and made it possible to establish a punishment and penalty scheme according to Spanish Water Law. This monitoring system has been complemented with a huge effort to install flow-meter sets in each well (as described in the paragraphs before).

In the following images from the Satellite Remote Sensing (GMES) monitoring system, the change from herbaceous crops (intense red areas) to vineyards (less intense red-ochre color) may be observed along the years. The reduction of the area covered with herbaceous crops and the increase of vineyards may also be seen from the graphics.

Finally, the result of all of these processes was an impressive reduction of groundwater abstraction in the main central aquifers from 640 hm^3/yr in the mid 80's to a current 240 hm^3/yr. Nowhere in the world, a reduction of this magnitude in groundwater abstraction or such a shift in crops' surface has occurred because of environmental reasons.

6 QUANTITATIVE EVOLUTION OF THE HYDROLOGICAL SYSTEM FROM THE DECLARATION OF OVEREXPLOITATION TO 2012

Two distinct water table drops can be observed in the Mancha Occidental aquifer between the 1987 declaration of overexploitation and the present (Figure 2 and 3). These were followed by partial recoveries. Drops can be correlated to dry sequences, which may last for up to five years at a time. During these, rainfall is scarce and pumping exceeds the renewable resources. In contrast, recoveries are associated to

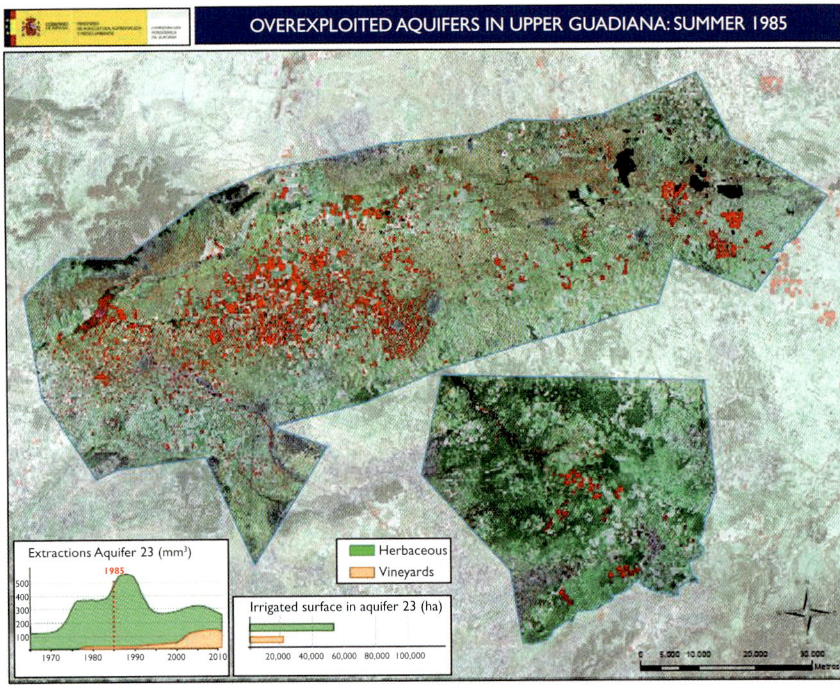

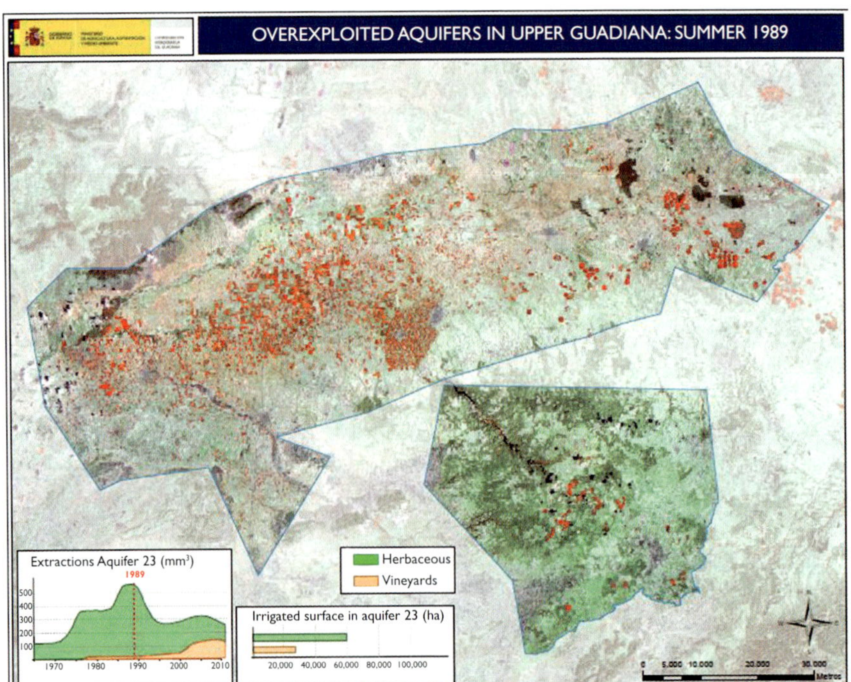

Figure 2 Continued.

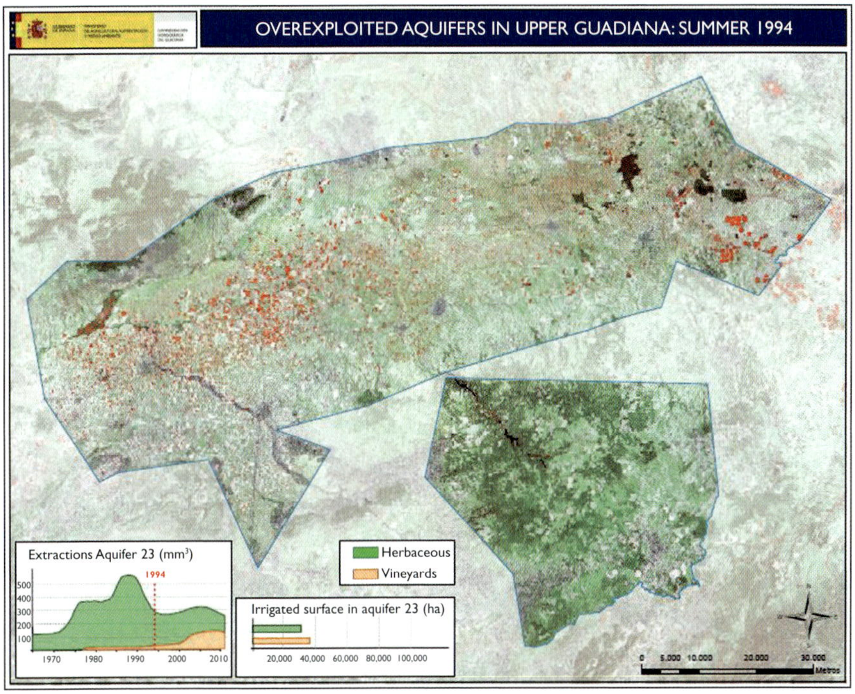

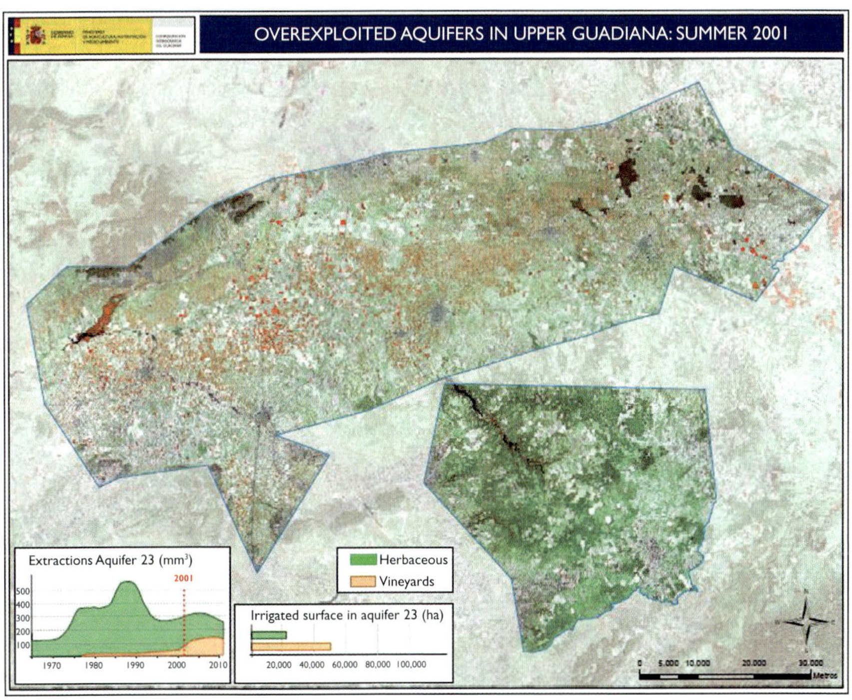

Figure 2 Continued.

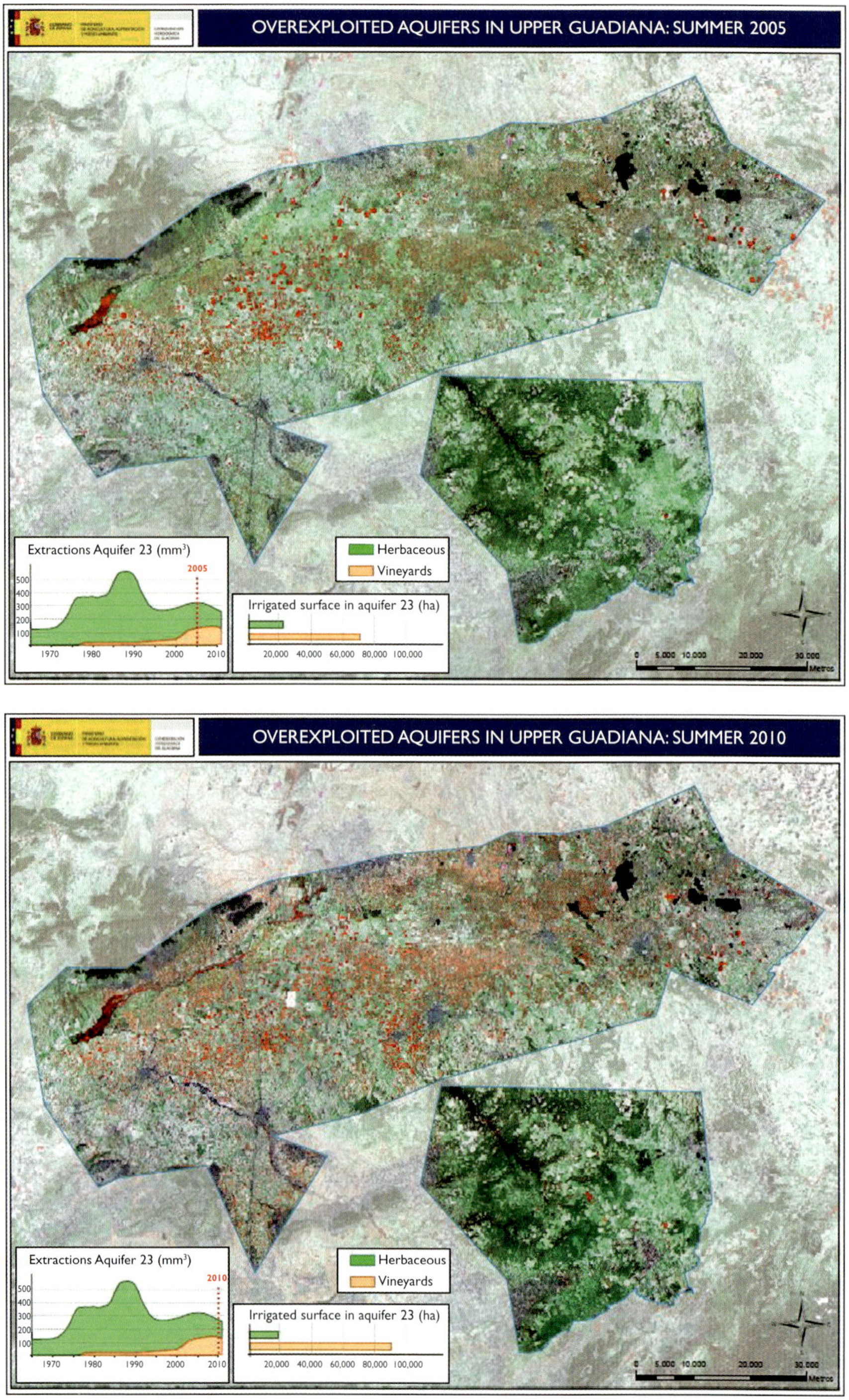

Figure 2 Evolution of crops surface and water abstraction in central La Mancha Aquifers (1985–2010) (Elaborated by SM GEODIM for Confederación Hidrográfica del Guadiana. Copyright ESA).

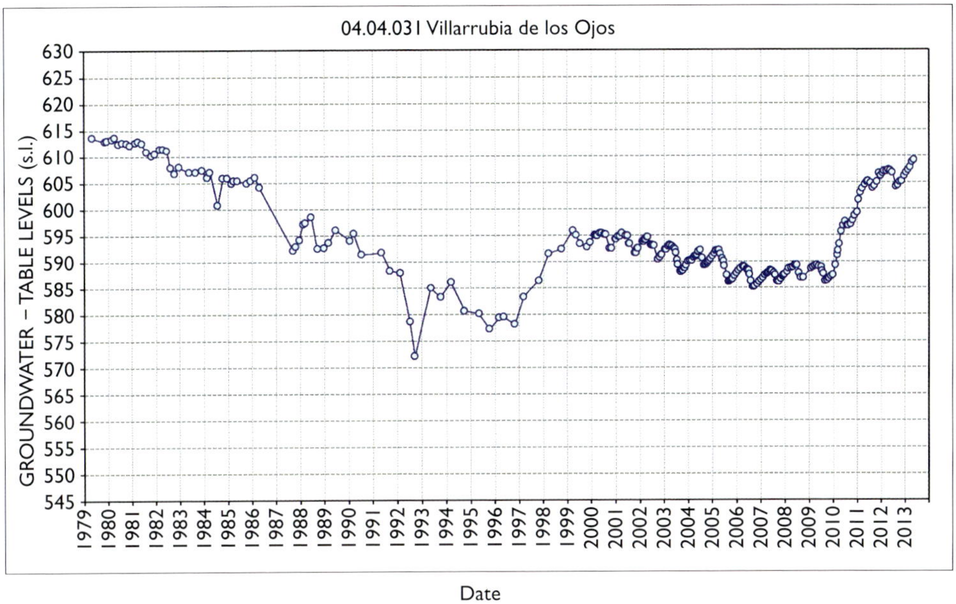

Figure 3 Evolution of the water table in piezometer 04.04.031 (Villarubia de los Ojos, Mancha Occidental aquifer).

shorter wet periods, up to two years, where rainfall far exceeds the average rate and pumping is significantly reduced. Variable periods can be observed between. Rainfall during these is generally below the long-term average. This implies that recharge is far from stable over time.

In the 1979–1993 period, a piezometer located near the Ojos del Guadiana springs registered drawdowns in the order of 42 meters. The downward trend was aggravated between 1990 and 1995 due to the most severe drought of the last decades. In fact, a 20 meter drawdown can be attributed to the 1990–1993 period alone. This aggravated the hydrological status of the whole basin but it was particularly harmful to the ecosystems of the upper Guadiana aquifers.

A 20 meter recovery is observed between 1996 and 1999. This is mostly attributed to an exceptionally humid period, although the above described management measures undoubtedly played a part in this.

The second downward period is registered between 1999 and 2005. A 10 m drawdown was accumulated over these years, less intense than the rates observed in the 1980s and 1990s.

The stable groundwater table trend from 2006 to 2010 is very remarkable due to the fact that this was a dry period. Probably the application of management measures played a key role in mitigating drawdowns, which is encouraging news to prove the effectiveness of the implied measures.

From then on, there was a 21 meter recovery (7 meter per year), which far exceeds the intensity of recovery from the 1996–1998 period. As a result, the system is currently only six meters below the 1979 reference (although a further five meters would be needed to restore it to an undisturbed state).

7 CONCLUSIONS

The last three decades have witnessed intense droughts as well as exceptionally wet periods, improvements in the knowledge of the water cycle, a series of water management plans and measures, new laws (most notably the WFD) and conflicts and negotiations between farmers and water authorities.

Overall, pumping restrictions and incentives to save water, together with other supporting measures, have contributed to shift from water-intensive to less water-intensive crops and reduce groundwater consumption. As a result, there is a reorientation of irrigated agriculture towards more water-efficient practices.

The quantitative status of water bodies has impressively improved to reach a nearly-good quantitative status of groundwater bodies and related surface water ecosystems. Therefore, the new Management Plan (which has set management measures such as risk declarations, water rights exchange system, limits to abstraction, etc.), and parallel actions, allow optimism in achieving a good status of groundwater bodies over the period 2015–2021.

However, and notwithstanding a recent recovery in the aquifers, governance problems often prevent the resolution of social conflicts. Public participation is generally limited, socio-and economic development is compromised.

ACKNOWLEDGEMENTS

The author would like to thank the staff from the Oficina de Planificación Hidrológica of the Confederación Hidrográfica del Guadiana and Pedro Martínez Santos for all his help.

REFERENCES

Confederación Hidrográfica del Guadiana (n.d.) *Internal Studies*. Documentación interna.

López Sanz, G. (1996) *El programa de compensación de rentas agrarias en los acuíferos de Mancha Occidental y el Campo de Montiel: Consideraciones sociales y ambientales*. X Reunión Anual ASEPELT-España, Albacete el 20–21 de junio 1996.

Rosell, J. & Viladomiu, L. (1997) El Programa de Compensación de Rentas por reducción de regadíos en Mancha Occidental y Campo de Montiel. *Economía Agraria*, 179, 331–350.

Chapter 14

Water resource vulnerability & adaptation management to climate change & human activity in North China

Jun Xia
State Key Laboratory of Water Resources & Hydropower Engineering, Wuhan University, Wuhan, China
Key Laboratory of Water Cycle and Related Land Surface Processes, Institute of Geographic Sciences and Natural Resources Research, Chinese Academy of Sciences, Beijing, China

ABSTRACT: The impact of climate change and human activity on water resources vulnerability is a challenging issue with widespread concern globally. It is also the key of water security evaluation and adaptation management issues in the national sustainable development of China. Especially in North China, water shortages and related environmental degradation are major issues the country is facing. This study proposed a method applied to water pressure evaluation to represent water vulnerability as a function of people per flow unit of one million cubic meters per year, water use to availability ratio, and per capita water use. The levels of water vulnerability for five main river basins in North China are assessed and the impact of climate change and human activity are considered for Hai River Basin in which the most serious water vulnerability occurs. The situation of water vulnerability is very serious under climate change and socio-economic development to 2050 in Hai river. The results show that rising water demands greatly outweigh greenhouse warming in defining the state of global water systems in 2050. Consideration of direct human impacts on global water supply remains a poorly articulated but potentially important facet of the larger global change question.

Keywords: water security, water vulnerability evaluation, synthetic analysis, North China

1 INTRODUCTION

Water is a key issue for social & economic sustainable development. China is one of the thirteen water-poor countries all around the world; particularly the Eastern China monsoonal area with its dense population has witnessed a more serious imbalance of water resources between supply and demand. The spatial-temporal variability and uncertainty of water cycle components and water resources distribution under climate change is still the main challenge to solve water problems. In addition, drought and water logging problems frequently occur in the East China monsoon area. The national average for water resources per capita is only 2173 m^3, no more than 1-fourth of the world's average, and per unit area and irrigated area this figure is 29.9 × 104 m^3/km^2 and 21,600 m^3/ha, respectively, corresponding to about half of the world's average.

North China, covering an area of more than 1.5 million square kilometers, is one of China's six administration regions, and plays a vital economic role. It was shown in the case of Haihe River Basin, the major river basin in the region, that among the total river length of 10,000 km, 4000 km of them has been turned into seasonal rivers. Comparing with the beginning of the 1950s, the wetland area within the basin decreased from 10,000 km^2 to 1000 km^2 at present. Over-extraction of groundwater in this area covers nearly 90,000 km^2, 70% of the plain areas. Compared to the end of the 1950s, the accumulated over-extracted groundwater volume is 90 billion m^3; what means an average drop of 10 m. Two-thirds of the mountainous area, or 110,000 km^2, is affected by water overextraction and soil loss. The sandstorms induced by desertification endanger Beijing and other cities. Thus, the problems of water shortage and related eco-environmental issues in North China, that is the political, cultural, and economic center of China, have become the most critical issues to impact with the sustainable development. In the next 40 years, this will be one of the crucial problems in restricting the continual development of the Chinese economy (Xia *et al.*, 2007).

The problems have received considerable attention from the international and Chinese governments, leading to a notable progress (Liu, 2003; Zhang & Wang, 2007). At present, the Ministry of Science and Technology of China (MOST) has started the '973-National Basic Research Program of China' (2010–2015) and the research project 'The impact of climate change on terrestrial water cycle, regional water resources security and the adaptation strategy for east monsoon area of China' in a preceding stage in the realm of science. Also, the MOST would like to positively encourage the joint development of educational institutes through the project and to bring into full play of its advantage, respectively. Studying this together with the safety problems of water resources in northern China, would provide a scientific foundation for continual development of social economy, continued availability of water resources and strategic planning of environmental protection.

2 CLIMATE CHANGE IMPACT

As humanity enters the 21st century, it is faced with an increasingly warming world that threatens human survival. Chief amongst the threats brought about by global warming is its negative effects on water resources. The impact of climate change on water resources security is a challenging issue with globally widespread concern. It is as well the great strategic issue in the national sustainable development of China. Water resources management, both from the sciences and humanity, is confronted by current and emerging new challenges as nations worldwide face serious problems of water availability due to global climate change. It was shown (Lv & Wang, 2004) that air temperature in North China was increased with 0.6°C–3.2°C during the last 50 years, with an above-national rate, which could result in the enhancement of water evaporation over land. The trend of precipitation change in North China is continually decreasing over the last 50 years. According to the precipitation data set and the variation of droughts and floods during the past 500 years in North China, the impact of climate change on precipitation and runoff is rather complex.

Regarding the issue of air temperature change in the future, some integrated research bodies from the Ministry of Water Resources (MWR), the Chinese Academy of Science (CAS) and others have implemented several related projects, such as the China Research Project on Climate Change. Using four GCMs (DKPZ, NCAR (CCMOA), GFDL, and UKMO), it was shown that the simulated annual air temperature change could reach 1.68°C in 2020 and 2.22°C in 2050 (Lv & Wang, 2004). For scenario 2 × CO_2 (a doubling in CO_2 levels), the averaged air temperature change would be 2.94°C. Since 1991, MWR cooperated with other water organizations to complete the research projects on the impact of climate change on water resources in China. For instance, the Hydro-Information Center of the MWR (Xie *et al.*, 2003) used the VIC Model, coupled with a regional climate model (PRECIS, 50 km × 50 km–SRES A2–B2) to predict the average runoff change over 2061–2090 in China under the A2 scenario (population continually increasing using the present trend in terms of baseline years 1961–1990 and the prediction years 2061–2090), and B2 as the other scenario (growth of population is controlled, and regional development is sustainable). Preliminary results indicated that runoff will significantly decrease in North China, and increase in South China for scenario A2. The risk of flooding in South China (such as the Yangtze River) and droughts in North China (Yellow and Haihe Rivers) will increase due to climate change. For scenario B2, the water conflict will be reduced compared to scenario A2. In both the scenarios, extreme hydrological events, particularly in North China, will see an increase due to climate change (Zhang, 1997; Xie *et al.*, 2003; Lv & Wang, 2004).

Under the circumstances of climate change; drought aggravation in the northern region, ecological water quality deterioration, and increasingly extreme floods in the southern region severely restricted the sustainable development of the economy and society during the past 30 years. The future climate change will have great influence on the existing pattern of 'northern drought and southern flooding' in China and the water resources distribution in the near future, and consequently exert some unexpected influence on the effects of major engineering projects in China, such as the food increasing project in North and Northeast China, the water transfer project, flood control system planning of southern rivers etc. This study will focus on the major river basins in the eastern monsoonal region of China, and investigate the mechanism of the impact of climate change and human activities on water pressure and vulnerability and the relevant adaptation strategies. The study aims to meet the major strategic demand of enhancing the water resources security for China. Trying to understand and address the above key scientific issues, this study puts forward a new theory and method to assess the expected changes in water resources vulnerability under climate change, to make an important contribution to develop adaptive strategies for climate change.

3 WATER STRESS INDICATORS

In general, the water resources stress is concerned with the integrated socio-economic capacity, and with the science-technical and managerial aspects of water resource organization. It is a vital index for assessment of the degree of water security in the region. However, the main difficulty remains in assuring that the represented indicators assess the pressure of climate change and human activities on the water resources system. Charles *et al.* (2000) take the relative water usage, i.e. the ratio of water withdrawal or water use (WD) to discharge (Q), WD/Q, to assess the water stress. Falkenmark (1989) demonstrated two dimensions of water scarcity and proposed a graphic instrument of water stress evaluation through three indicators. This framework provides a rather concise but reliable indicator system and explicit linkage between water use efficiency, water inherency, and the pressure they exert on the water resources system. Considering WD and Q are only related to 'blue' water, therefore the green water is not considered in these indicators. Also water withdrawals are usually higher than just the consumptive use which is usually considered in water footprint analyses. Thus, indicators should be further improved.

Based on Falkenmark & Molden's suggestions, the water scarcity demonstrates two dimensions, i.e., Demand-Driven (WD) water stress (high usage compared to the availability of water) and Population-driven (P) water shortage (many people dependent on the availability of water). The link between population-driven water shortage (water crowding, people per flow unit of one million cubic meters per year, P/Q), the water-use-driven mobilization level (use-to-availability, percent of water availability, r), and per capita water use (withdrawals in cubic meters per capita year, WD/P) are given in Figure 1.

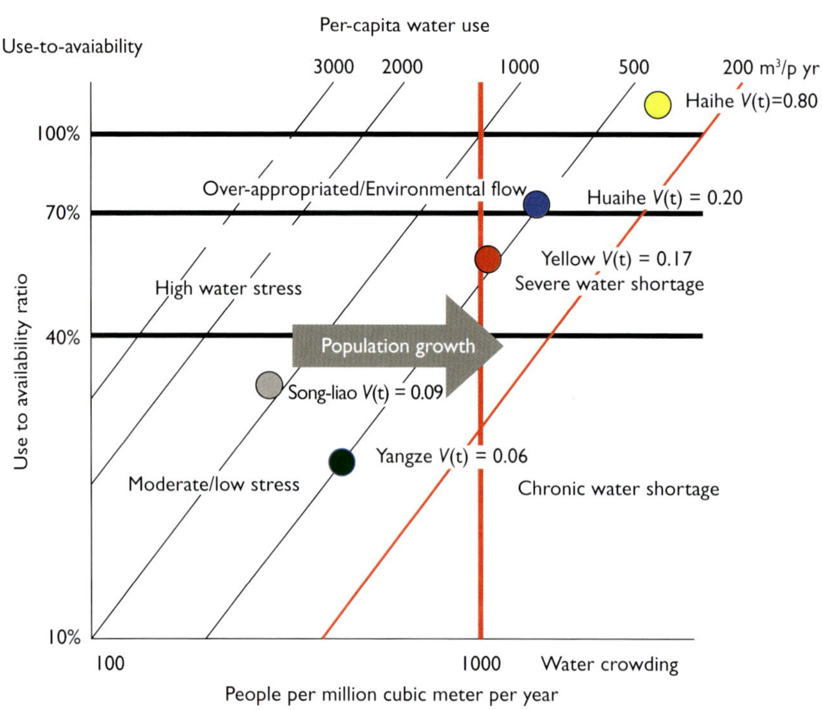

Figure 1 The link between people per million cubic meters per year, use to availability, and per capita water use. The dots show the level of water stress of five main river basins in North China.

Basically, the environmental flow requirements are seen as limiting the appropriation potential, i.e. possible use to 70% of availability, leaving the remaining 30% as reserve for aquatic ecosystems (Falkenmark & Molden, 2008). In reality, urban and agricultural uses are however very different: The former are of through-flow character, the latter of consumptive use character, depleting the river. Urban water might therefore be recirculated after wastewater treatment, whereas irrigation water will be literally consumed/evaporated, the more the higher the irrigation efficiency. Rather than competing, the two uses are in other words complementary, provided that the through-flow-based use precedes the consumptive use.

High water stress occurs when there is intensive usage compared to the amount of water available. Based on the global statistic (Falkenmark & Molden, 2008), the situation of more than 1000 people having to share each flow unit (1 million cubic meters of water per year) is a critical value for water stress. The higher the water crowding, i.e., water resources bearing capacity (Falkenmark & Molden, 2008), the more serious the water stress situation and water management and governance measures required will need to be more sophisticated. The level of water stress was divided into 5 levels: Moderate low stress ($r < 40\%$; $P/Q < 1000$), chronic water shortage ($r > 40\%$; $P/Q > 1000$), high water stress ($40\% < r < 70\%$; $P/Q < 1000$), middle water shortage ($40\% < r < 70\%$; $P/Q > 1000$), and over-appropriate/environmental flow ($r > 70\%$). Based on Falkenmark's framework, the results for water stress evaluation of five main river basins in North China are calculated. Further, if the relationship between vulnerability, i.e., V(t), and water stress indicators (r, P/Q, WD/P) can be defined, we can analyze vulnerability of water resources under climate change conditions. Part of these results are shown in Table 1 and Figure 1. The contemporary condition is represented by 2000, the year that is most compatible with the time span represented by the runoff climatology and historical water use statistics. It suggested that the Haihe River is undergoing serious water resources stress and water withdrawal is exceeding appropriate environmental flow. Especially, the value of water crowding is beyond the water barrier of water crowding of 2000 people per unit water in Haihe River (Falkenmark, 1989). The water stress in Huaihe and Yellow River is relatively high and they will face more serious water shortages in the future. For the Yangze and Song-Liao River Basin, the situation is more alleviated as they deal with a more moderate/low water stress.

Table 1 Contemporary evaluation results of water-related indicators and vulnerability of five main river basins in North China.

Catchments	*People (billion)*	*Water crowding (p/Mm³/yr p)*	*Per capita water use (m³/p yr)*	*Use-to-available ratio (%)*	*V(t)*
Huaihe	0.167	2021	253	51	0.20
Haihe	0.137	3711	252	94	0.80
Yellow	0.111	1556	292	45	0.17
Yangze	0.438	443	333	15	0.06
Song-Liao	0.059	354	737	26	0.09

4 VULNERABILITY OF WATER RESOURCES SYSTEM WITH CLIMATE CHANGE AND HUMAN ACTIVITY

Climate change is an everlasting ongoing process. Despite the arguing of whether global warming is due to the increasing concentration of CO_2, it has been shown that the contribution of climate change and land use/cover change (LUCC) combined, accelerates the occurrence of water scarcity through decreasing water availability and increasing its very uneven distribution in time and space (Xia & Zhang, 2008). On the other hand, population increase and economic development continue to enhance the water demand. The combination of the impact of global warming and socio-economic water demand increase on water resources is the main issue treated in studies to investigate climate change impact on water security. Even without the anthropogenic factors, the influence of natural variation makes that climate change can't be omitted from water security studies. In pace with the growing population, agriculture-industrial development and urban economic development, the poor management of water resources may lead to serious water safety problems. So, it is necessary to give effective countermeasures of scientific water resources management through the study of water resources vulnerability.

In general, the vulnerability of water resources systems affected by climate change and human activities may be described in terms of both sensitivity and adaptability. The former is related to the natural features of the hydrological cycle in the region, while the latter is concerned with the integrated socio-economic capacity and the scientific technical and management levels. However, most measures that reduce sensitivity are closely linked with increasing the systems adaptability to climate change or human activity and reach the final target of vulnerability reduction of water resources system (Liu, 2003). Thereafter, numerous studies have been proposed to assess the vulnerability of a water resources system in consistence with the evaluation of water adaptability which is the reciprocal of water stress (Liu, 2003; Zhang & Wang, 2007). In that case, the relation between water vulnerability, $V(t)$, and adaptability, $C(t)$, can be described as

$$V(t) = \frac{1}{C(t)} \tag{1}$$

where t is time. The evolution of water resource vulnerability is a dynamic process as social – and economic development and the enhancement of adaptive capacity are dynamic processes too.

Reduction of vulnerability is consistent with increasing the adaptability of water resources and decreasing the water stress. As described, balancing demand to the available water resources is the key factor for regional water security and measurement of water stress. In that view, water adaptability is closely correlated with two dimensionless water use characteristics, i.e. the ratio of water use to runoff discharge and the relationship between water supply and water demand. It is also possible to estimate the water vulnerability from the water resources supply and demand statistics. This study selects the above two dimension variables as the key descriptors for the relationship between water supply and demand in a basin to determine the water resource adaptability as follows:

$$C(t) = C\left\{r \cdot \frac{Q}{W_D}\right\} = f_1(r) f_2\left(\frac{Q}{W_D}\right) \tag{2}$$

where f_1 and f_2 are functions to be determined.

The relationship between water supply (Q) and water demand (W_D) can be described as:

$$\frac{W_D}{Q} = \frac{P}{Q} \cdot \frac{W_D}{P} \tag{3}$$

where P is the population. Substitution of equation (3) in equation (2) gives

$$C(t) = f_1(r) \cdot f_2\left(1/\left(\frac{P}{Q} \cdot \frac{W_D}{P}\right)\right) \tag{4}$$

It should be noted that the equation is consist with Falkenmark's (2008) graphic instrument for seeing the link between water resources stress determined by water crowding (P/Q), use to availability (r), and per capita water use (W_D/P).

On the basis of phenomenological considerations the boundary conditions are determined as:

$$\begin{aligned} &f_1(r) \to 0, \ \ i.e., \ \ C(t) \to 0, when \ \ r \to \infty \\ &f_1(r) \to 1, \ \ when \ \ r \to 0 \\ &f_2(P/Q) \to 0, \ \ i.e., \ \ C(t) \to 0, when \ \ P/Q \to \infty \\ &f_2(W_D/P) \to 0, \ \ i.e., \ \ C(t) \to 0, when \ \ W_D/P \to \infty \end{aligned} \tag{5}$$

Considering the correlations shown in Figure 1 and the above boundary conditions, the functional forms of f_1 and f_2 are selected and thus equation (4) is obtained as:

$$C(t) = C\left\{r \cdot \frac{Q}{W_D}\right\} = \exp_1(-r \cdot k) \exp\left(-\frac{P}{Q} \cdot \frac{W_D}{P}\right) \tag{6}$$

where the value of coefficient k should be larger than zero and its value can be estimated by regression analysis for the critical value of 40% water use ratio. Through calibration with the critical value 0.4 it is found that $k = 2.3$. Substitution of equation (6) into equation (1) and standardized by the critical value with $r = 1$ and $\frac{P}{Q} \cdot \frac{W_D}{P} = 1$ gives the assessment of water vulnerability. Three indicators are systematically assessed against a set of 5 categories of different water vulnerability levels with the critical value of 0.05, 0.1, 0.2 or 0.4 points, reflecting the relevance of those factors to increase sensitivity to water vulnerability (Table 2).

This study applies water resources vulnerability equation (1) to estimate the water vulnerability of five main river basins in North China. Indicated by the results shown in Table 1 and Figure 1, the assessed water vulnerability level is consistent with Falkenmark's categories of water pressure. For the Haihe River Basin, water is being

Table 2 Categories of water resource vulnerability.

No vulnerability	*Low vulnerability*	*Moderate vulnerability*	*High vulnerability*	*Serious vulnerability*
<0.05	0.05–0.1	0.1–0.2	0.2–0.4	>0.4

over-exploited, which places the basin under the category of serious vulnerability and over-appropriated environmental flow. Evaluation of the Huaihe river leads to a value of 0.2 being a critical value between high and moderate water vulnerability, while the value for the Yellow river with 0.17 belongs to moderate vulnerability. The Yangze and Song-Liao rivers both have a low vulnerability level with values of 0.06 and 0.09, respectively.

We formulated three scenarios to quantify the contribution of climate change and socio-economic development pressure to the degree of relative water demand in 2050 for the Haihe River Basin. The first scenario (Sc1) varied climate but fixed the magnitude and spatial distribution of human population and water withdrawals at the year 2000 levels. Sc2 applied projected water demands for 2050 but used runoff and discharge based on the contemporary climate. Sc3 changed both climate and water demand (Charles *et al.*, 2000).

The water resource prognosis was made by the National Climate Center based on a China multi-model ensemble. It has shown that the temperature is expected to increase by 1.5°C–2.2°C and the precipitation to increase by 5%–16% in 2050 for the Haihe River Basin. Generally, mean runoff varied in response to changes in precipitation and temperature. For the Haihe River Basin, a 5% precipitation increase results in a more than 20% increase in streamflow but this would be reduced to a 13% and 6% increase if the temperature increases with 1.0°C and 2.0°C, respectively. A 1°C decrease in the model results in a 8% decrease in streamflow and a 10% precipitation decrease results in a 26% decrease in streamflow. However, both a 1°C decrease and 10% decrease in precipitation results in a 30%–35% decrease in streamflow (Hao *et al.*, 2009; Liu *et al.*, 2004). In this study, we take the 10% increase of streamflow as the 2050 scenario of water resource availability in the Haihe River Basin. Domestic and industrial water demand and withdrawals are determined by population and per capita use statistics (Zhang & Wang, 2007). These index values are projected according to the prediction of socio-economic driving factors under climate change (Shen *et al.*, 2008). The results are shown in Table 3.

For the Haihe River Basin, climate change under Sc1 decreased the V(t) value by 0.18. In contrast, rising water demand alone (Sc2) increased V(t) by 1.15, whereas Sc3 combing both climate and development effects produced a relative increase of 0.57. Considering the fact that the estimated use-to-available ratio exceeds 1, the water resource vulnerability surpasses 1 under Sc2 and Sc3. Despite the projected improvements in water resources, most likely a sustained and severe pressure on water supplies will remain in this basin. Water transfer into the Haihe River Basin is necessary to release the serious water shortage situation. Contemporary conditions along this

Table 3 Water vulnerability evaluation for 2050 under climate change and human activities for the Haihe River Basin.

Scenario	*Total population (billion)*	*Level of water crowding (p/Mm³/yr)*	*Per capital water use (m³/p yr)*	*Use to available ratio (%)*	*V(t)*
2000	0.137	3711	252	94	0.80
Sc1	0.137	3373	252	85	0.62
Sc2	0.169	4579	262	120	1.95
Sc3	0.169	4163	262	109	1.37

river are already more severe than indicated, because of the rapid increase in water use and decrease in discharge into the 2005 benchmark.

5 CONCLUSION

This chapter proposed a mathematical model that offers insight into catchment-scale water resources stress and vulnerability assessment. The water vulnerability in five main river basin in North China was evaluated and emergent issues for water security were addressed. The cause of these problems, mainly resulting from climate change and human activity, are analyzed by the case study of the Haihe River Basin. The major conclusion of the present study is given as: The systematic analysis of water pressure by the proposed model is consistent with Falkenmark's method, in which the water situation for the Haihe river is most serious with water crowding over 3500 people for each unit water, for the Huaihe and Yellow river a moderate to high level of vulnerability, while for Yangze and Song-Liao the situation is much better. Despite the fact that water resources availability is expected to increase due to climate change for the Haihe River Basin, however, socio-economic development causes water demand acceleration, aggravating water safety problems. Besides the improvements in the water use efficiency and scientific water resource management of basin, it is necessary to look for new extra water resources to relief the serious situation expected in 2050 by for example water transfers from other basins, re-use of treated water, and desalination. The results show that rising water demands greatly outweigh greenhouse warming in defining the state of global water systems in 2050. Consideration of direct human impacts on global water supply remains a poorly articulated but potentially important facet of the larger global change question.

ACKNOWLEDGMENTS

This study was supported by the External Cooperation Program of the Chinese Academy of Sciences (No. GJHZ1016), and CAS-CSIRO Cooperative Research Program, Grant (No. GJHZ1223).

REFERENCES

Charles, J.V., Pamela, G., Joseph, S. & Richard, B.L. (2000) Global water resources: Vulnerability from climate change and population growth. Science, 289, 284–288.

Falkenmark, M. (1989) The massive water scarcity now threatening Africa-why isn't it being addressed? Ambio, 18(2), 112–118.

Falkenmark, M. & Molden, D. (2008) Wake Up to Realities of River Basin Closure. *International Journal of Water Resources Development*, 24(2), 201–215.

Hao, L., Yao, X. & Zhi, D. (2009) Correlation between climate change and surface water resources in Haihe River Basin. *Haihe Hydrology* (10), 1–10. (in Chinese).

Liu, C. (2003) The vulnerability of water resources in Northwest China. *Journal of Glaciology and Geocryology*, 25(3), 309–314.

Liu, C., Liu, Z. & Xie, Z. (2004) Study of trends in runoff for the Haihe River Basin in recent 50 years. *Journal of Applied Meteorological Science*, 15(4), 385–393. (in Chinese).

Lv, X.D. & Wang, W.Y. (2004) *Global Climate Change Research: Development and Perspective*. Beijing, China, China Meteorology Press. (in Chinese).

Shen, Y., Oki, T., Utsumi, N., Kanae, S. & Hanasaki, N. (2008) Projection of future world water resources under SRES scenarios: Water withdrawal. *Hydrological Sciences Journal*, 53(1), 11–33.

Xia, J., Zhang, L., Liu, C. & Yu, J. (2007) Towards better water security in North China. *Water Resour Manage*, 21, 233–247.

Xia, J. & Zhang, Y. (2008) Water security in North China and countermeasures to climate change and human activity. *Physics and Chemistry of the Earth*, 33, 359–363.

Xie, Z., Su, F., Liang, X., Zeng, Q., Hao, Z. & Guo, Y. (2003) Application of a surface runoff model with Horton and Dunne runoff for VIC. *Adv. Atmos. Sci.*, 20(2), 165–172.

Zhang, H.L. (1997) *Flood and Drought Hazards in China*. Beijing, China, Chinese Water Resources and Hydropower Press. (in Chinese).

Zhang, J. & Wang, G. (2007) *The impact of climate change on water resources*. Beijing, Science Press. pp. 138–181. (in Chinese).

Chapter 15

Blue water transfer versus virtual water transfer in China – with a focus on the South-North Water Transfer Project

H. Yang

Systems Analysis, Integrated Assessment and Modeling, Swiss Federal Institute for Aquatic Science and Technology (Eawag), Duebendorf, Switzerland
Department of Environmental Sciences, Basel University, Switzerland

Y. Zhou

Syngenta Foundation for Sustainable Agriculture, Basel, Switzerland

J.G. Liu

School of Nature Conservation Beijing Forestry University, Beijing, China

ABSTRACT: Alongside its astonishing economic growth during the past decades, there has been increasing water stress in many areas in China. Water diversion has been one of the measures in dealing with the problem. The South-North Water Transfer Project currently under construction is the largest project as such in the world, which aims to transfer water from the Yangtze River to primarily the North China Plain to alleviate the water stress in the region. Water diversion projects play an important role in supporting the continuous economic growth and safeguarding food production in the country. However, they also bring about many negative impacts concerning the environmental and ecosystem sustainability, as well as socio-economic development, both in the source and destination regions of diversions. One question arising is whether a virtual water transfer, primarily in the form of agricultural products, would be one of the tools economically and environmental advantageous over transferring massive amounts of water to water deficit regions. This chapter presents an overview of China's water and land endowments and uses across regions, and the spatial distribution of food production. Based on this, the extent to which the virtual water strategy may be useful in dealing with the water stress in northern China as well as its limitations will be discussed. The focus will be on the South-North Water Transfer Project.

Keywords: China, South-North Water Transfer, virtual water, water scarcity, policy support

1 INTRODUCTION

China is a large country with significant regional disparities in natural conditions, resources endowments and economic development. In terms of water resources, the distribution is highly uneven. The bulk of the water resources concentrates in the southern part of the country. The northern part is generally short of water. Over the years, China's water policies have relied heavily on the construction of massive water diversion projects. These projects brought tremendous benefits in the economic development, food production and water scarcity alleviation. With the increasing water stress in the northern regions, diverting water from the water rich regions has become increasingly important in the water management strategies in China. The South-North Water Transfer Project (SNWTP) which is currently under construction is the largest and the most strategically important water diversion project in China. Apart from many other economic and ecological objectives of the project, supporting the agricultural production in the north, which relies highly on irrigation, is an important objective of many such diversion projects. But massive water diversion generally also poses ecological, environmental, economic, and social problems, such as losses of biodiversity, degradation of aquatic and terrestrial ecosystems, soil and river erosion, and resettlement. A question arising is whether China should rely on water diversion to alleviate water stress in the water-scarce regions or it should import virtual water to serve the purpose. The SNWTP has been in the center of the debate.

By and large, the virtual water trade strategy has been useful for: 1) alleviating local water scarcity by importing water-intensive products; 2) conserving water by producing water-intensive products in water abundant and high water productivity areas and exporting them to water-scarce areas; 3) enhancing sustainable water use by promoting regional structural adjustment and environmentally friendly trade (Faramarzi *et al.*, 2010; Yang *et al.*, 2006; Aldaya *et al.*, 2010; Hoekstra & Mekonnen, 2012).

The North China Plain region is one of the major granaries of China. While lacking water resources, it is currently exporting food, especially wheat and maize to the south (Yang & Zehnder, 2005; Lin *et al.*, 2012). There is much debate on whether bringing the water from south to north through the SNWTP is worth its environmental consequences. Some have considered that from a water resources point of view, this does not make sense to do so (Ma *et al.*, 2006). Interestingly, the negative view on the SNWTP is mostly prevalent outside of China. The project has been mostly portrayed as a negative case against the virtual water strategy. Within China, the view is generally less critical, although the concerns on environmental impacts are high. Hence, apart from water resource endowments and water uses per se, there must be other decisive factors to justify the water transfer strategy and to shape the contrast opinion. A broader, integrated assessment is necessary to gain a more comprehensive understanding of the roles of the SNWTP and the virtual water transfer in the integrated water resources management in China.

The purpose of this chapter is to provide an overview of water resources and use status across river basins and provinces in China. Regional disparities will be elaborated with respect to water and land endowments, agricultural production, crop distribution, irrigation requirement and crop water productivity. Factors that play important roles in the decision on the SNWTP are highlighted. The importance of incorporating the virtual water strategy in the integrated water resources management is addressed.

2 AN OVERVIEW OF WATER AND LAND RESOURCE ENDOWMENTS

2.1 Water resources distribution

According to the official statistics, China's average annual total available water resources amount to 2812.4 billion m³. Dividing the figure by the total population of 1.34 billion in 2010 (NBSC, 2011), the average water availability per person is approximately 2100 m³/capita. This figure is roughly 1/4 of the world average and 1/6 of the figure for the United States. Thus, China as a whole can be said a water-scarce country by the world standards.

The spatial distribution of China's water resources is uneven (Figure 1). Of the nine major watersheds, the Haihe (the Hai River) watershed has the lowest water availability on a per capita basis: Slightly over 300 m³/capita/year. In the adjacent Huanghe (the Yellow River) and Huaihe (the Huai River) watersheds, per capita water availability is also considerably low. The area covered by the Haihe, Huaihe and Huanghe is commonly called the HHH region in China. The lower reaches of the three rivers in the east form the famous North China Plain, the major breadbasket of China. The SNWTP will divert water from the Yangtze basin to the HHH region, particularly the North China Plain.

Table 1 presents numerical indicators of internal water resources in each province. As the provincial borders are inconsistent with the river basin boundaries, the provinces encompassed in the respective river basins shown in the table are indicative of the location of their major territories and the economic centers. Provinces with extremely low water resources availability include Beijing, Tianjin, Hebei, Shanxi, Shandong, Henan, Ningxia, Jiangsu and Shanghai. Except Shanghai and the southern part of Jiangsu, the rest of the provinces are located in the Haihe, Huaihe and

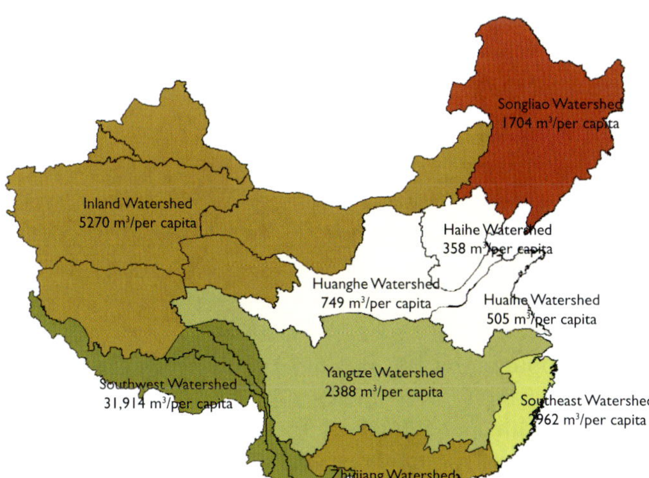

Figure 1 China's nine watersheds and water resources (m³/per capita) (WR, 2010).

Table 1 Indicators of internal water resources by provinces in the respective river basins, 2010.

River basin	*Province*	*Total internal water resources* (Billion m³)*	*Per capita water resource (m³/capita)*
Haihe	Beijing	2.31	124
	Tianjin	0.92	73
	Hebei	13.89	195
Huaihe	Shandong	30.91	324
Huanghe	Shanxi	9.15	262
	InnerMongolia	38.85	1576
	Henan	53.49	566
	Ningxia	0.93	148
	Shaanxi	50.75	1360
Inland	Qinghai	21.52	842
	Gansu	74.11	13,225
	Xinjiang	111.31	5125
		0.00	
Song-Liao	Liaoning	60.67	1392
	Jilin	68.67	2503
	Heilongjiang	85.35	2229
Yangtze	Shanghai	3.68	163
	Jiangsu	38.35	489
	Zhejiang	139.86	2609
	Anhui	92.28	1527
	Jiangxi	227.55	5117
	Hubei	126.87	2217
	Hunan	190.66	2939
	Chongqing	46.43	1617
	Sichuan	257.53	3174
	Guizhou	95.65	2727
Southeast	Fujian	165.27	4492
Zhujiang (South)	Guangdong	199.88	1943
	Guangxi	182.36	3853
	Hainan	47.98	5539
Southwest	Yunnan	194.14	4233
	Tibet	459.30	153,682
National average		2812.41	2150

Note: * Internal water resources refer to the renewable water resources generated within the province (NBSC, 2010 and MWR, 2010).

Huanghe watersheds. Topographically, Beijing, Tianjin, Hebei, Shandong and Henan are located in the North China Plain, the legendary region that has been commonly referred to in addressing China's water crisis (Yang & Zehnder, 2001; Ma *et al.*, 2006; Lin *et al.*, 2012). Shanghai and the southern part of Jiangsu are located in the Yangtze River basin. Although the volume of water engendered within their territories is small, abundant external water flows into these areas before draining into the sea.

In contrast to the extremely low water resources availability in the North China Plain provinces, water is relatively abundant in the southern regions. It is, however, noticeable that the water resources availability in per capita terms is relatively small in Anhui, Hubei and Chongqing provinces in the Yangtze river basin. In the Inland basin, Xinjiang and Gansu have rather high per capita water resources. However, some of the water is not accessible or has to be released to the downstream areas. These provinces are suffering water stress partly because of the large water use for irrigation.

2.2 Land endowments and irrigation development

With less than 9% of the water resources of the country, the HHH region possesses about 22% of the country's total cultivated land. The monsoon climate dominates the region and over 70% of the annual rainfall is concentrated to the period between June and September, leaving the rest of the months relatively or very dry (Yang & Zehnder, 2001). Irrigation is essential for the practice of multiple-cropping. Given the importance of irrigation in increasing output per unit of land, expanding irrigated areas has been generally linked to the strategy of food security of the country.

Compared with the water resources, China's land endowment is perhaps more unfavorable. Table 2 shows that the per capita arable land is only 0.091 ha. In the provinces in the Yangtze basin, Southeast and South regions, the arable land per capita is mostly below the national average. In contrast, many provinces in the HHH region and other northern regions have the arable land per capita above the national average. This indicates that land endowments are relatively better in the north than in the south.

The ratio of irrigated land to the arable land varies largely across provinces. In the northern provinces, irrigation, together with high fertilizer input, is important for obtaining high crop yields. In some areas, irrigation is essential for agriculture. However, the expansion of irrigated areas is generally constrained by water availability. In the southern regions, water resources are less a constraint to expanding irrigation. With the relatively high annual precipitation, mostly over 1000 mm/year, irrigation is not always needed for some crops (Chen *et al.*, 1995). The marginal return of irrigation with respect to crop yields is not as high as in the north. The fertilizer application rate is also lower.

The per capita food production is an important factor in determining the role of a province in the interregional as well as international food trade. Table 2 shows that the provinces with food possession above the national average of approximately 400 kg/capita are mostly located in the Song-Liao, Huanghe and Inland basins. These provinces generally have grain surplus and hence are the exporters of grain. The grain possession is particularly large in Inner Mongolia, Ningxia, Henan (the Huanghe basin) and Jilin and Heilongjiang (the Song-Liao basin), as well as Xinjiang (Inland). This has formed the current food grains flow patterns which are dominantly from the north regions to the south. Such trade is associated with a large amount of virtual water transferring from the water-scarce north to water relatively abundant south.

2.3 Basin and sectoral water use

The continued increase in water demand has put growing pressure on the water resources, particularly in the North China Plain (Table 3).

Currently, the ratios of water withdrawal to water resources availability in the Haihe, Huanghe and Huaihe basins are excessively high compared with the internationally recommended sustainable ratio of 40%, indicating a severe water

Table 2 Land endowments, crop areas and irrigation ratio (NBSC, 2010).

River basin	*Province*	*Per capita arable land (ha/capita)*	*Ratio of irrigated land to arable land (%)*	*Multi-croping index* (%)*	*Fertilizer application (kg/ha)*	*Grain production per capita (kg/capita)*
Haihe	Beijing	0.012	91.3	136.9	431	59
	Tianjin	0.034	78.1	104.1	556	123
	Hebei	0.088	72.0	138.0	370	414
Huaihe	Shandong	0.078	65.9	143.9	439	452
Huanghe	Shanxi	0.113	31.4	92.8	293	304
	InnerMongolia	0.289	42.4	98.0	253	873
	Henan	0.084	64.1	179.8	460	578
	Ningxia	0.175	42.0	112.7	304	563
	Shaanxi	0.108	31.7	103.3	470	312
Inland	Qinghai	0.096	46.4	100.8	160	181
	Gansu	0.182	27.4	85.8	213	374
	Xinjiang	0.189	90.2	115.4	352	536
Song-Liao	Liaoning	0.093	37.6	99.7	344	404
	Jilin	0.202	31.2	94.3	350	1035
	Heilongjiang	0.309	32.8	102.8	177	1308
Yangtze	Shanghai	0.011	82.4	164.5	295	51
	Jiangsu	0.061	80.2	159.9	448	411
	Zhejiang	0.035	75.5	129.4	371	141
	Anhui	0.096	61.4	158.0	353	517
	Jiangxi	0.063	65.5	193.1	252	438
	Hubei	0.081	51.0	171.5	439	404
	Hunan	0.058	72.3	216.8	288	433
	Sichuan	0.074	42.9	159.4	262	401
	Guizhou	0.129	25.2	109.0	177	320
	Chongqing	0.078	30.6	150.2	273	401
Southeast	Fujian	0.036	72.7	170.7	533	179
Zhujiang	Guangdong	0.027	66.1	159.8	524	126
	Guangxi	0.091	36.1	139.8	402	306
	Hainan	0.084	33.5	114.6	557	208
Southwest	Yunnan	0.132	26.2	106.0	287	333
	Tibet	0.120	65.5	66.4	197	303
National average		0.091	49.6	132.0	346	408

Note: * Multi copping is the practice of growing two or more crops in the same land area during a year. Multi cropping index is the ratio of total area cropped in a year to the land area cultivated.

stress in these basins. The Haihe basin has the ratio of 124%, meaning that the basin is using more water than it has. This is possible mainly due to the exploitation of non-renewable deep aquifers in the North China Plain, supplemented with a small amount of desalinated sea water. The result has been a drop of groundwater table at an alarming rate and a depletion of water resources in the region. It is estimated that the accumulated overdraft of groundwater during the last two decades in the North China Plain has exceeded 90 billion m^3 (Yang & Zehnder, 2005). The depletion of

Table 3 Blue water resources availability and water withdrawal in the major river basins, average of 2006–2010 (MWR, 2006 & 2010).

Basins	*Per capita water resources availability (m³/capita)*	*Water resources availability (billion m³)*	*Water withdrawal (billion m³)*	*Ratio of withdrawal to availability (%)*
Song-Liao	1704	172.07	64.38	37.42
Haihe	358	29.75*	37.00	124.37
Huanghe	749	61.06	46.67**	76.43
Huaihe	505	89.06*	63.97	71.83
Yangtze	2388	839.60	197.04	23.47
Zhujiang	3327	453.62	87.68	19.33
Southeast	2962	173.52	34.36	19.80
Inland	5270	132.34	64.13	48.46
Southwest	31,914	594.44	11.18	1.88
Nation	2100	2475.52	596.52	24.10

Notes:
* Including the water transfer into the Haihe river basin and the Huaihe river basin.
** Including the water transfer out of the Yellow river basin.

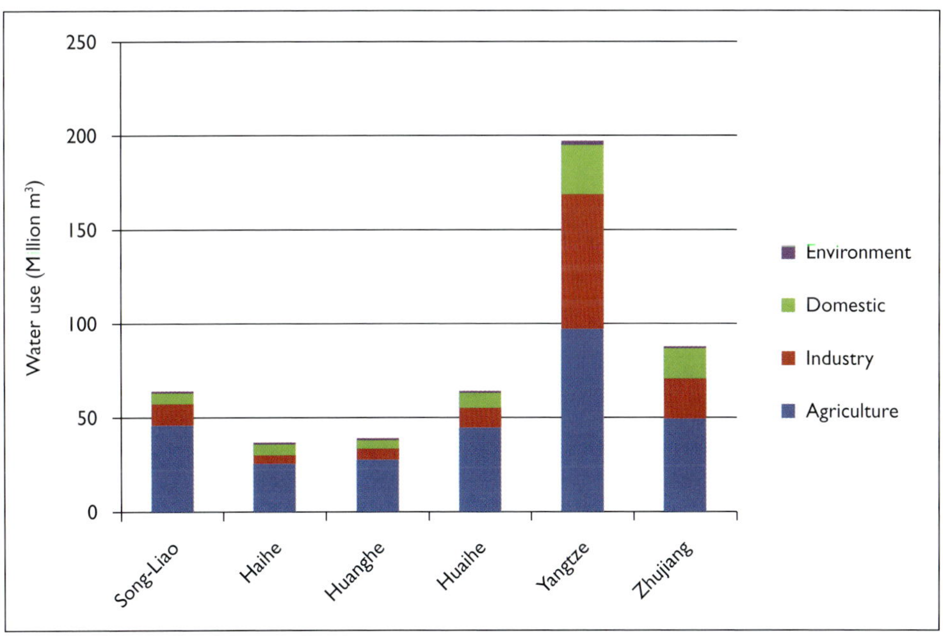

Figure 2 Sectoral water use, average of 2006–2010 (MWR, 2006 & 2010).

groundwater has serious consequences, including land subsidence, sea water intrusion, and loss of ecosystem functions. With the excessively high level of water withdrawal, many rivers and their tributaries in the north have been in extremely poor ecological status. The Yellow River had once become a seasonal river and sent little or no water to the sea most of the time in the late 1990s (MWR, 2002; Yang &

Jia, 2008). In contrast, the water withdrawal ratio is much lower in the Yangtze and Zhujiang basins, only 23% and 19%, indicating the relatively low pressure on their water resources.

Agriculture, mainly irrigation, is the biggest user of withdrawn water. At the national level, the agricultural water use accounts for about 50% of the total water withdrawn. For the individual river basins, the ratio varies. For the Haihe, Huanghe and Huaihe basins, the ratios are respectively 68%, 60% and 70%. The severe water stress has put increasing threat on the irrigated agriculture in the HHH region, particularly the North China Plain. Given the importance of the region in the national food production, the continuation of the water stress can have a significant impact on the food security in China. The ratio of water use to water availability is relatively low in the southern regions, although the total quantity of water use is large, particularly in the Yangtze basin.

3 THE SOUTH-NORTH WATER TRANSFER PROJECT VS. THE VIRTUAL WATER TRANSFER IN CHINA

3.1 An overview of the SNWTP

The increasing water stress in the north and the continued increase in food demand have heightened the aspersion of the Chinese government to implement water diversion projects. The SNWTP is a key project in the national water development strategy. It is the longest water diversion project in China and the world. The idea of transferring water from the Yangtze to the north was first proposed in 1952 by then Chairman Mao. In 2002, after much research and debate on the feasibility of the project, the SNWTP was approved and the construction began. The first phase of the project is expected to be completed in 2014 (Yang & Zehnder, 2005; Liu *et al.*, 2013). The general layout of the SNWTP includes three routes: Eastern, Middle, and Western (Figure 3).

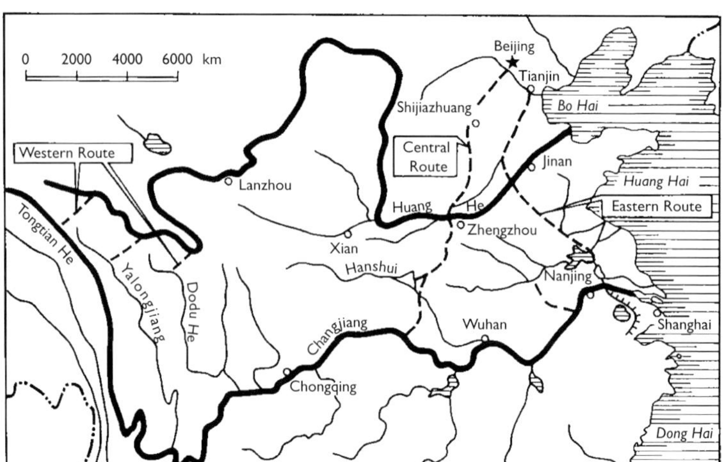

Figure 3 The three routes of the South-North Water Transfer Project (Yang & Zehnder, 2005).

The Eastern Route is 1467 km long, and 14.8 billion m^3 of water will be transported along this route annually. The diverted water will pass through and impound a chain of lakes and link up with the northern section of the 600-year-old Beijing-Hangzhou Grand Canal to provide water to Tianjin and Shandong. The initially planned total investment for the Eastern Route is 42 billion Yuan, and its construction began first. The length of the Middle Route is 1432 km, and a total of 13 billion m^3 of water will be transported to Beijing and Tianjin cities in the north through Henan and Hebei provinces. The water conveyance system is a canal from the Danjiangkou Reservoir to Beijing. The water is diverted from relatively high terrain and will hence flow by gravity. The planned investment is 92 billion Yuan, and its construction began in 2003. The West Route is currently under evaluation. The proposed route passes through the Qinghai-Tibet Plateau. The terrain is mountainous, and the proposed design involves construction of high dams, tunnels over 100 km long, and elevated canals that are extremely costly and difficult to construct. The proposed route is highly controversial, largely due to potentially huge environmental damages. The investment is expected to be much higher than the Eastern and Central Routes in part due to its technical complexity.

The main purpose of the SNWTP is to alleviate water scarcity problems in the north, especially the North China Plain. A total of 44.8 billion m^3 of water per year is planned to be transferred from the Yangtze basin to the north upon the completion of the three routes. The Eastern Route will bring the water to help environmental conservation, groundwater conservation, and agricultural production in these areas in the eastern part of the North China Plain and Shandong peninsula. The Middle Route is proposed to bring water to solve the water shortage crisis and to boost economic development in the route of the transfer. In addition, it is expected to control floods in the regions surrounding Danjiangkou Reservoir, the northern Jianghan Plain, and Wuhan City. The large amount of water transferred by the Middle Route is also expected to have potential ecological benefits, such as groundwater recovery, creation of artificial niches, and water allocation for ecological purposes (Chen *et al.*, 2008; Lin *et al.*, 2012). The Western Route will bring water from three tributaries of the Yangtze River to the upper reach of the Yellow River to ease water scarcity and significantly boost development in the northwest areas of China by increasing irrigated cropland by 2 million ha and supplying 9 billion m^3 of water for domestic and industrial purposes (Liu *et al.*, 2013).

The SNWTP links four major rivers in China, and affects almost one-third of China's land mass. Many studies have conducted the assessments on the economic and ecologically gains and losses of the project. The conclusions are mixed and the uncertainties are high. For example, Lin *et al.* (2012) used the life cycle impact assessment modeling, including regionalized water consumption impact assessment and an interregional input-output model, to examine the potential of the SNWTP to reduce the impact of water consumption embodied in final demand. The results show that the SNWTP increases the environmental impact of water consumption in southern China and reduces it in northern China. In total, the SNWTP will lead to a 5.74% net reduction in the environmental impact of water consumption embodied in the final demand from both southern and northern China. While the benefits are huge, there are serious concerns about the negative impacts on the water source areas, surroundings of the water transfer routes, and water-receiving areas. Two major environmental threats have been commonly concerned.

The first one is the reduction of sediment supplies to downstream regions, particularly the coastal regions in the Yangtze River. This will slow down or even reverse the formation of the Yangtze delta, reduce land reclamation, and threaten coastal environments and Shanghai's socio-economic development (Zhang, 2009). The second threat is saltwater intrusion. The total amount of water diversion accounts for 4%–5% of the average annual discharge of the Yangtze River, but this percentage may reach over 20% in dry seasons, accelerating saltwater intrusion into the Yangtze estuary (Yang *et al.*, 2003).

3.2 Substituting real water diversion with virtual water transfer?

According to the estimation by Ma *et al.* (2006), the HHH region currently transfers a large amount of virtual water (about 52 billion m^3/year) to southern China in the form of agricultural commodities, particularly wheat and maize. The rationale of the virtual water and real water transfers between the south and the north has been an issue of debate in the political arena and the scientific community both in China and abroad. One of the major criticisms is that the transfer patterns are inconsistent with the virtual water strategy which advocates the import of water-intensive food products as a means to alleviate local water scarcity.

The distribution of crop production is directly affected by the climate conditions. In China, over 60% of wheat production is from the HHH region (Figure 4). Meanwhile, almost 50% of the maize production is from this region. In contrast, rice production is dominantly in the southern regions, accounting for over 80% of the total rice production. This spatial pattern is closely related to the climate conditions in the respective regions. In general, the climate condition in the northern regions is suitable for wheat and maize production, whereas rice is more suitable to be grown in the south. The bio-physical requirements of different crops limit spatial expansion of their

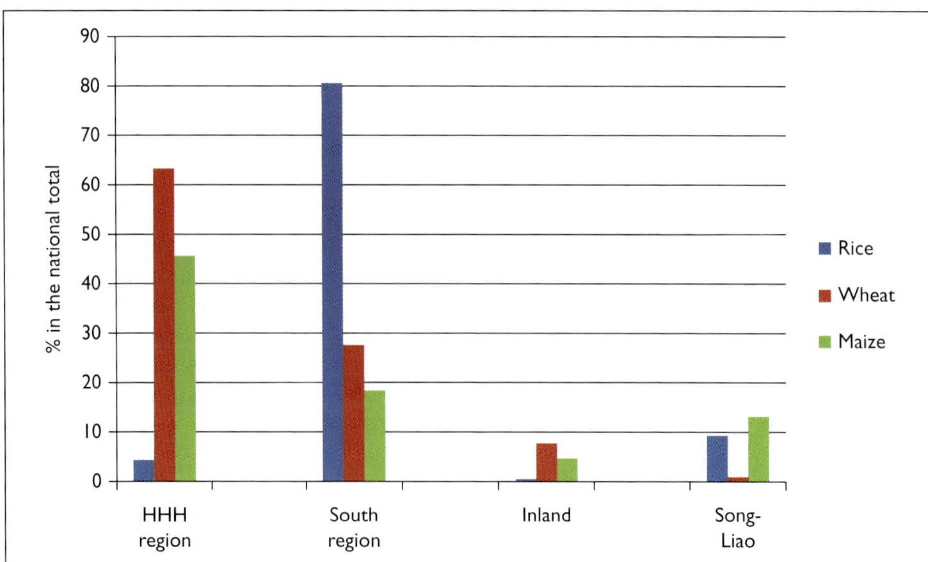

Figure 4 Crop production in regions (% in the national total) (NBSC, 2010).

production. Wheat is a crop whose optimal temperature during the growing season is 15–20°C. It is hence not practically feasible to shift wheat production to the south. The increasing demand for wheat will still have to be met primarily by the north, where the natural condition is suitable for its production. Given the severe water stress in the north and the increasing demand for wheat, the SNWTP would have been a necessity to safeguard the national food production.

In addition to water, land is also an important factor for agricultural production. In the southern regions, although water resources are relatively abundant, the arable land availability is generally scarce. Meanwhile, the land-use intensity, indicated by the multi-cropping index (Table 2), in major food producing provinces in the south, such as Hubei, Hunan and Jiangxi, is already high. The possibility for further increasing the multi-cropping index is limited. In addition, the rapid economic development in the south in the last decades has led to increases in labor cost. Many rural labors have left the land to work in factories in cities and their hometowns. All these factors have limited the potential to expand the food production in the southern regions, despite the relatively abundant water resources. In contrast, the northern regions, -especially the Huanghe basin, Song-Liao and Inland-, have a relatively larger land availability than the south. The economic development level there is much lower than in the Yangtze basin. Labor cost for agriculture is also relatively lower. The potential for expanding food production is relatively high provided the availability of irrigation. This is particularly so given the situation that the agricultural water price is currently much lower than the shadow price. Farmers have little incentive to shift to high value added products.

3.3 Wheat water productivity

Crop Water Productivity (CWP) in kg/m^3 is a commonly used indicator for water use efficiency. It can be measured through field experiments, and through bio-physical crop modeling. With the support of GIS techniques, the application of crop models for simulating crop water use, yield and water productivity has flourished in the last two decades. Figure 5 and Figure 6 show the wheat water productivity in its producing areas simulated by GIS-based EPIC model (GEPIC) (Liu *et al.*, 2007; 2009).

CWP can be calculated based on the yield and ET for rainfed and irrigated winter wheat, respectively. Figure 5 shows that under rainfed conditions, winter wheat had CWP values generally in the range of 0.20 to 1.20 kg/m^3. Relatively higher CWP values were found in the high-yielding rainfed winter wheat belt, mainly Shandong, Hubei, Chongqing and Sichuan provinces. This belt had the most favorable climate conditions for rainfed winter wheat in China (Liu *et al.*, 2007). High precipitation was an important reason for the high CWP values under the latter.

In the north of the high-yielding rainfed winter wheat belt, low precipitation amounts seriously limit rainfed winter wheat production while south of this belt, high temperature pose constraints to the productivity. In addition, high temperature together with plentiful precipitation leads to high evapotranspiration, while yield is low. The lowest values of CWP were seen in some Southwest provinces (Figure 5) such as in Guangxi and Guizhou. The soil and terrain slope were not suitable for winter wheat, leading to a relatively low yield and high ET.

The effects of supplemental irrigation on CWP varied significantly among regions (Figure 6). The North China Plain provinces stood out to be the region with the most

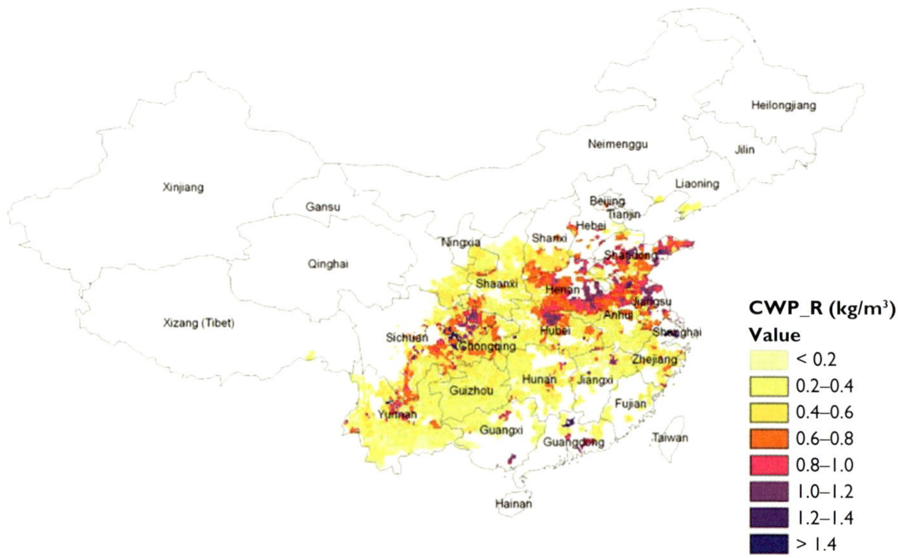

Figure 5 Simulated crop water productivity for rainfed winter wheat in 2000 (kg/m³).

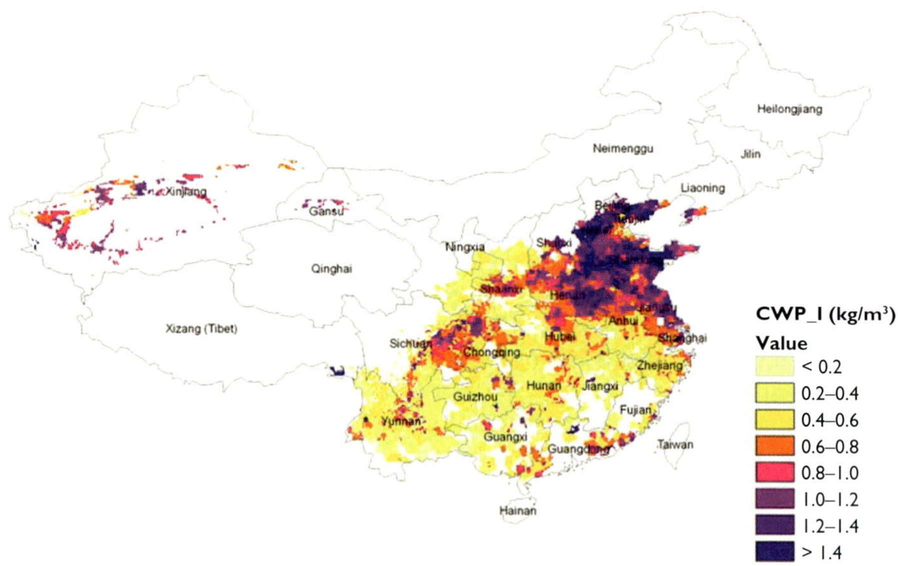

Figure 6 Simulated crop water productivity for irrigated winter wheat in 2000 (kg/m³).

significant improvement on CWP under irrigated conditions. On average, CWP under irrigated conditions is appropriately 56% higher than that under rainfed conditions (this is also partially attributable to the higher fertilizer input in the irrigated land than that in the rainfed land). The Northwest provinces and the Southwest provinces can improve the CWP by 33% and 27%, respectively. In the Southeast provinces, CWP for irrigated winter wheat is only 10% higher on average than for rainfed winter wheat. These provinces generally received high precipitation during the winter wheat growing period. The irrigation does not increase crop yield much, but the actual ET increases substantially (Liu *et al.*, 2007).

In the precipitation-rich Southwest and Southeast provinces, no irrigation or only little irrigation is required for winter wheat. In addition, irrigation could not increase crop yield significantly. On the contrary, in the precipitation-poor North China Plain provinces and Northwest provinces, winter wheat strongly depends on irrigation to achieve high crop yields and high CWP. However, in water-scarce provinces, competition exists among agriculture and other water uses (economic sectors and the environment). The opportunity cost of irrigation is high. Efficient allocation of the precious water among different sectors is of importance for both regional economic development and food production. Therefore, achieving high values of CWP needs to be considered an important objective for agricultural water management.

At the national level, the current pattern of transferring wheat from the north to the south generates an overall reduction in water use. This is because of the generally high water productivity in the major producing provinces, such as Hebei, Shandong and Henan, compared with the southern regions. When a product is produced in areas of high water productivity (thus low virtual water content) and exported to the areas of low water productivity (thus high virtual water content), less water is used for the production of the product than if there was no trade, holding other factors constant.

4 INCORPORATING THE VIRTUAL WATER STRATEGY IN INTEGRATED WATER RESOURCES MANAGEMENT

Water resources management and trade decisions have multifaceted dimensions. They require consideration of multi-objectives and trade-offs of different options. The virtual water concept represents one important dimension. However, as illustrated by the SNWTP, alone it cannot determine the optimal water resource allocation in importing and exporting countries and regions. Figure 7 depicts integrated measures in addressing regional water scarcity, especially for the HHH region, and the socio-economic and environmental implications.

Water scarcity problems can be addressed by different means, i.e., improving water use efficiency locally, transferring water from outside into the region to increase the total water supply, and transferring virtual water into the region to reduce the local water demand or reducing the virtual water export to reduce the water outflow. These measures are not contradictory to each other, but can be combined to form the integrated approach in addressing water scarcity problems. Concerning the HHH regions, the transferring of real water through the SNWTP does not mean that the region should not improve water use efficiency (Yang *et al.*, 2010). Meanwhile, the total amount of water transfer through the SNWTP is limited and the water scarcity is a long-term challenge to the socio-economic development and the environmental sustainability of the region. Hence, regional crop adjustment and reducing virtual water export from the HHH are integral part of the strategies in dealing with the water scarcity.

Improving water use efficiency, including more 'crop-per-drop', is at the core of the integrated approach to addressing water scarcity. Efficiency gains can be achieved at the household level through awareness raising, water conservation campaigns as well as water pricing measures. More water efficient technologies and processes can be put in place in a

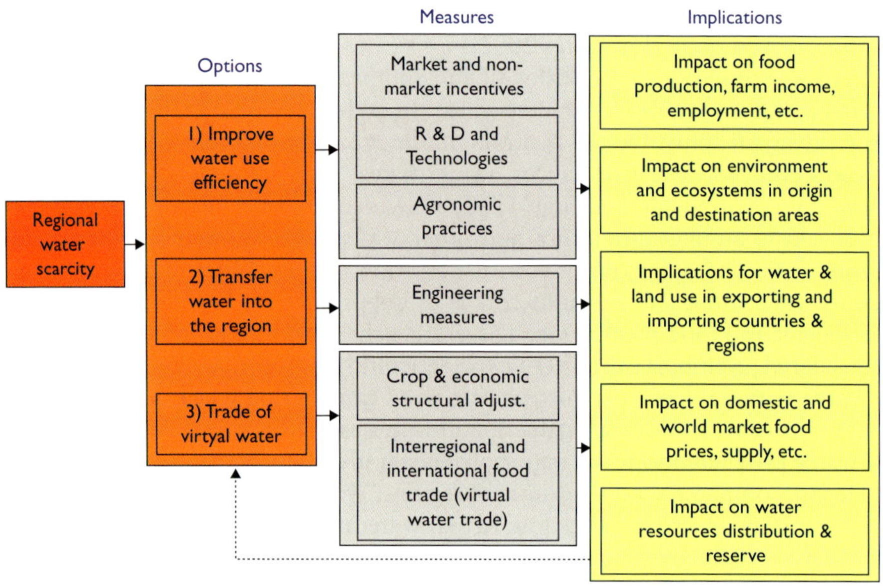

Figure 7 Integrated approach to addressing regional water scarcity (own elaboration).

variety of industries to reduce their water footprint. In agriculture, there is huge potential to reduce water wastage and increase 'crop-per-drop' through adopting more advanced irrigation technologies such as drip and sprinklers. Increasing water use efficiencies in various sectors can go a long way in balancing local water demand and supply.

In areas where demand management is insufficient to tackle the water stress problem, supply-oriented solutions should be taken into consideration including long-distance water transfer projects like SNWTP. Water transfers usually involve altering of natural hydrological regimes, landscape and relocation of people, and thus should be assessed carefully with adequate considerations on ecological, environmental and human impacts beyond engineering feasibility. Water transferring should be combined with demand management measures in water-recipient regions whenever appropriate to improve overall water efficiencies in the long term.

Trade of virtual water is another important component of integrated strategies in tackling water scarcity. The essence is that countries/regions should undertake economic activities (incl. agriculture) in which they have a comparative advantage. In some cases, in water-scarce areas (e.g. arid, semi-arid, saline), water can be used for other high-value alternative uses rather than being used for agriculture. Hence it might be advantageous to import certain food instead of growing locally. The virtual water aspect should be considered as part of the solution, especially across regions within a country, while at the same time considering the potential externalities of this option. Virtual water strategies could potentially improve overall water use efficiencies in agriculture by adjusting crop structure and importing most water-intensive crops, thereby easing the level of water stress.

The situation witnessed in Beijing since the late 1990s reflected to some extent its efforts in the integrated water resources management, particularly in dealing with its water stress, with the multiple-approach strategies. Between 1997 and 2007, the total water footprint of Beijing increased largely, from 4342 million m^3 to 5748 million m^3 (Zhang *et al.*, 2012). However, the increase was almost entirely attributable to the external sources, i.e., the import of virtual water, whereas the internal water footprint was rather stable. Of the total water footprint increase of 1406 million m^3, 1380 million m^3 was external, only 26 million m^3 was internal. As a result, the total use of its own water resources in Beijing did not increase (Beijing Water Authority, 1997 & 2007), despite the rapid economic development and relatively large expansion of population during the period.

Economic structure adjustment has been one of the measures for Beijing to control its total water use and deal with its water scarcity. For agriculture and the related down-stream sectors, such as food processing and textile, the internal water footprint witnessed a large decrease. These sectors generally have high water use intensity. The large reduction in the internal water footprint (i.e., the consumption of water originates from the local source), and substantial increase in the external water footprint, (i.e., the consumption of water originates from other regions) in these sectors indicate a shift of Beijing's water consumption to rely more on the water resources outside, due to the increases in virtual water import into Beijing.

5 CONCLUSIONS

This chapter provides an overview on water and land endowments and uses in China from river basin and provincial perspectives. The spatial distribution of the production of major crops and the crop water productivity across provinces and regions is also investigated. The rationality of the SNWTP is discussed against the water and land resources endowments and crop physiological requirements.

The analysis concludes that the relative water endowments in the north and the south cannot be a sole criterion to judge the rationale of the real water and virtual water transfer patterns between them. Land resources availability, among others, is also an important factor, if not more crucial. Compared with the north, southern China is scarcer in arable land. Meanwhile, the rapid expansion in urban sectors in the south has led to a rise in opportunity costs of agricultural inputs, particularly land and labor. A comprehensive assessment of trade-offs taking into consideration the natural and socio-economic conditions is necessary to gain a better understanding of the cons and pros of the SNWTP. The information on the virtual water embodied in the trade between the north and the south is important for such an assessment.

To achieve sustainable water management and food security in China, integrated assessments of different water demand and supply management options and their socio-economic and ecological implications, taking the virtual water concept into account, are needed based on transparent data. Water is posing a serious constraint on future food supply, and thus should be looked at judiciously to ensure an efficient and effective use of this resource.

REFERENCES

Aldaya, M.M., Martinez-Santos, P. & Llamas, M.R. (2010) Incorporating the water footprint and virtual water into policy: Reflections from the Mancha Occidental Region, Spain. *Water Resources Management*, 24(5), 941–958.

Beijing Water Authority (1997, 2007) Beijing Water Resources Bulletin. Beijing, Standards Press of China.

Chen, H. & Du, P. (2008) Potential ecological benefits of the middle route for South-North Water Transfer project. *Tsinghua Science and Technology*, 13, 715–719.

Chen, Y.M., Guo, G.S., Wang, G.X., Kang, S.Z., Luo, H.B. & Zhang, D.Z. (1995) Main Crop Water Requirement and Irrigation of China. *Hydraulic and Hydroelectric Press*, 73–102.

Faramarzi, M., Yang, H., Mousavi, J., Schulin, R., Binder, C. & Abbaspour, K. (2010) Analysis of Intra-country Virtual Water Trade Strategy to Alleviate Water Scarcity in Iran. *Hydrology and Earth Systems Sciences*, 14, 1417–1433.

Hoekstra, A.Y. & Mekonnen, M.M. (2012) From water footprint assessment to policy. *Proceedings of the National Academy of Sciences*, 109(22), E1425.

Lin, C., Suh, S. & Pfister, S. (2012) Does South-to-North Water Transfer Reduce the Environmental Impact of Water Consumption in China? *Journal of Industrial Ecology*, 16(4), 647–654.

Liu, J.G., Liu, J.G., Yang, H. & You, L.Z. (2013) Water conservancy projects in China: Achievements, challenges and way forward. *Global Environmental Change*. (In press).

Liu, J.G., Wiberg, D., Zehnder, A.J.B. & Yang, H. (2007) Modeling the role of irrigation in winter wheat yield and crop water productivity in China. *Irrigation Science*, 26, 22–33.

Liu, J.G., Zehnder, A.J.B. & Yang, H. (2009) Global crop water use and virtual water trade: The importance of green water. *Water Resources Research*. [Online] 45. Available from: doi.10.1029/2007 WR006051 [Accessed 10 August 2010].

Ma. J., Hoekstra, A.Y., Wang, H., Chapagain, A. & Wang, D. (2006) Virtual versus real water transfers within China. *Philosophical Transactions of the Royal Society*, 361(1469), 835–842.

MWR, Ministry of Water Resources (2002, 2006, 2010) *Water Resources Bulletin*. [Online] Ministry of Water Resources. Available from: http://www.chinawater.net.cn [Accessed 2 July 2012].

NBSC, National Bureau of Statistical of China (2000–2011) *China Statistical Yearbook*. Beijing, China Statistical Press.

World Bank (2001) *China: Agenda for Water Sector Strategy for North China*. World Bank. Report number: 22040-CHA. Available from: www.worldbank.org [Accessed 15 January 2003].

Yang, H. & Jia, S.F. (2008) Meeting the basin closure of the Yellow River in China. *International Journal on Water Resources Development*, 24(2), 265–274.

Yang, H., Wang, L. Zehnder, A.J.B. & Abbaspour, K.C. (2006) Virtual water trade: An assessment of water use efficiency in the international food trade. *Hydrology and Earth System Sciences*, 10, 443–454.

Yang, H & Zehnder A.J.B. (2001) China's regional water scarcity and implications for grain supply and trade. *Environment and Planning A*, 33, 79–95.

Yang, H. & Zehnder, A.J.B. (2005) The South-North Water Transfer Project in China: An analysis of water demand uncertainty and environmental objective in decision making. *Water International*, 30(3), 339–349.

Yang, S.L., Belkin, I.M., Belkina, A.I., Zhao, Q.Y., Zhu, J. & Ding, P.X. (2003) Delta response to decline in sediment supply from the Yangtze River: Evidence of the recent four decades and expectations for the next half-century. *Estuarine, Coastal and Shelf Science*, 57, 689–699.

Yang, Y.M., Yang, Y.H., Moiwo, J.P. & Hu, Y.K. (2010) Estimation of irrigation requirement for sustainable water resources reallocation in North China. *Agricultural Water Management*, 97(11), 1711–1721.

Zhang, Q. (2009) The South-to-North Water Transfer Project of China: Environmental Implications and Monitoring Strategy1. *Journal of the American Water Resources Association*, 45, 1238–1247.

Zhang, Z.Y., Shi, M.J. & Yang, H. (2012). Understanding Beijing's Water Challenge: A Decomposition Analysis of Changes in Beijing's Water Footprint between 1997 and 2007. *Environmental Sciences & Technology*. 46(22), 12373–12380.

Chapter 16

The institutional organization of irrigation in Spain and other Mediterranean countries

Laurens A.D.R. Thuy
Hydrogeology, Earth Sciences, Utrecht University, The Netherlands

Juan Valero de Palma
General Secretary of the Spanish Federation of Irrigators Communities, Madrid, Spain

Elena López-Gunn
Water Observatory, Botín Foundation, Universidad Complutense de Madrid, Spain

ABSTRACT: This chapter will be focusing on the topic of agricultural irrigation in Spain and other Mediterranean countries. Spain has a long and rich history concerning the tradition of irrigation and the formation of users' communities. Due to their long history, it is worth considering the organizational structure and functioning of these communities. The chapter is structured in two major sections: The first part undertakes a comprehensive overview of water endowment in Spain, understood as both its natural water resource assets and its institutional assets, looking at Spanish water use, and the role of the agricultural sector therein. In order to better understand the evolution and recent changes that have taken place in the irrigation sector and potential further improvements that could be made in the near future, the properties of the different irrigation methods and techniques are pointed out, as well as their change in use over time. The second part of the chapter analyses the incentives which lead to the creation of these irrigation communities, some of the oldest and best studied self-governance systems, their results and the main factors for their success. Special attention is given to the history, functioning and achievements of irrigators' communities, in particular analyzing the pivotal role played by the Spanish National Federation of Irrigators Communities (FENACORE) and its European equivalent, the Euro-Mediterranean Irrigators Community (EIC). The structuring and formalization of collective action into these institutional organizations is a key element for robust water management. The special case of a self-governed independent judicial body: The Valencia Water Court, which has succeeded in keeping peaceful communal water management in the region through its unique functioning, is also briefly discussed.

Keywords: irrigation, Spain, Mediterranean, surface water, groundwater, agriculture, drought, Irrigators communities, river basin communities, collective action

1 INTRODUCTION

Spain has an average annual precipitation of 436 mm and is characterized by a Mediterranean climate. The average inter-annual variation of precipitation is significant at 20%, where – except in the Northwest – summer precipitation over the summer period (June, July and August) is well below 100 mm in all regions, implying the need for supplementary irrigation. In addition, the dry period shows a large inter-annual variability with some very dry summers while others are merely wet. This variability in rainfall, together with other factors like unfavorable soil properties, a hilly topography and a climate characterized by hot summers, makes agriculture in Spain more complex than in other (wetter) regions. This is reflected by an agricultural sector which is more vulnerable compared to other EU regions, as can be seen for example in the annual crop production which can vary by up to 20% not only temporally, but also spatially (Iglesias *et al.*, 2000). In Spain, variations in rainfall are large: For example the Southern and Central Mediterranean river basins (Guadiana, Guadalquivir, Sur, Segura and Júcar), the Balearics and the Canaries form 44% of the total Spanish territory where 40% of the population lives, yet they only receive 20% of total water resources. In contrast, the rest of the Spanish territory (56%) has 80% of water resources at its disposal, and concentrates 60% of the population (EU, 2004). These inequalities in water endowments point to a need for irrigation to guarantee a level of stability: Firstly to ensure a good crop yield during dry spells, secondly to increase average productivity, and thirdly to enable crop diversification (Iglesias *et al.*, 2000).

Water resources are experiencing an increased pressure due to current demand, which is in some cases outstripping (sustainable) supply, together with the need for higher quality water. The difficulties to meet these demands are causing greater competition for available water resources between different water use sectors such as agriculture, industry and urban supply. Also competitiveness is growing internally, which can lead to tensions and disputes at times of drought and the occurrence of local water scarcity. This competition could place constraints on the economic development of some areas. Moreover, the increasing scarcity of water resources has, due to competing demand, led to some inter-regional tensions for available water resources (Moreno-Perez & Roldan-Cañas, 2013). Concerns are increasing, since studies suggest that over some regions, such as for example in Spain, precipitation amounts decreased in the second part of the 20th century, and projections point to a further decline and a greater irregularity in the future (Silvestre & Clar, 2010).

Agriculture is the main water consumer worldwide, accounting for up to 70% of water consumption – mainly through irrigation – next to 20% for industry and 10% for domestic use (FAO, 2013). This is also mirrored in Spain where irrigation accounts for 68% of water consumption. At present, the use of traditional or historical irrigation techniques is still in practice in some areas like along the Mediterranean coast of Spain, which is remarkable taking into account these practices are several centuries old. Irrigation entails the management and regulation of the natural flow of surface water from rivers and aquifers, other open water bodies, or systems of wells, through a network of channels and ditches to the field. The increase in the total irrigated area in Spain in recent decades however has been made possible through the use of artificial regulation, and by making use of modern technology such as hydraulic pumps.

Due to the large quantities of water needed for most irrigation activities, it seems logical that one of the key elements to overcoming water scarcity problems lies in reducing water consumption in the agricultural sector. This could be achieved through a wide range of measures, from technological adjustments to a shift in growing different crop species (less water-intensive and more productive). According to data from the Federación Nacional de Comunidades de Regantes (National Federation of Irrigator Communities) (FENACORE, 2012), water consumption from irrigation in Spain has dropped from 80% to 63% thanks to the on-going efforts of farmers and official bodies towards better irrigation practices and greater investment in more water-efficient technologies and equipment.

Generally, as recognized by the White Paper on Water (MIMAM, 2000), water-related issues are deeply interconnected to distinctly institutional aspects to best optimize water use. Therefore, water has been subjected to regulation, to a greater extent than other resources and assets, especially in those areas where availability was limited or had a strong variability, since sound integrated management is fundamental. In Spain, the legal framework for water uses has a long tradition (Valero de Palma, 1994). The modern State as owner of water and as manager of the public water domain, has been a fundamental actor to define water rights. The Water Act of 1879, and its predecessor, the Water Act of 1866, were the basis of water law in Spain. These declared the public character of surface waters with use regulated by a concession procedure. The 1985 Water Act extended procedures to include groundwater into the public water domain, except for groundwater wells that were in use before 1985, which ones remained as private property (Pérez Picazo, n.d.). It also introduced other aspects to improve water planning and thereby rationalized and systematized water policy decisions. This will be explained in more detail later on, since it also had provisos for institutions (Valero de Palma, 1994). The 1985 Water Act has reformed the practice of irrigation in Spain on a local scale by formalizing those already existing and mandating the formation of different organizations to unite irrigators with the aim of extending benefits to the whole community. This decision was partly based on the view that through the organization of water users in groups and organizations, these water user groups could take over the financial and management responsibilities from the government (Ghazouani *et al.*, 2012).

2 NATURAL RESOURCE ENDOWMENTS AND IRRIGATION IN SPAIN

The mainland Spain has a surface of approximately 500,000 km² and generates, with an average annual precipitation of 436 mm, an annual runoff of about 105,000 million m³ ($1.05 \cdot 10^{11}$ m³). Surface water and groundwater comprise respectively 90,000 and 15,000 m³ of this figure. MIMAM (2000), notes that the most important agricultural use of water is for irrigation, which basically includes water volumes required for crop evapotranspiration and associated bio-assimilation. Water is required for plant development, and its application increases productivity and enables crop diversification (Lopez-Gunn *et al.*, 2012a). In Spain, the great irregularity of rainfall does not always allow applying water when needed, and therefore requires the provision of water through irrigation. According to the Spanish Irrigation Plan (PNR), presented in 1998 the number of irrigated farms in Spain at that time was over one million, with the total irrigable area in Spain being 3.76 million ha. The term 'irrigable area' is defined as the area with irrigation infrastructure and that has been irrigated at least once. The real irrigated area in a normal season – important for hydrological planning purposes – is about 3.34 million hectares on average. This represents 14.5% of the Useful Agricultural Area (UAA) and 6% of the total Spanish territory (Figure. 1).

So-called historical or traditional irrigation dates back centuries, through the regulation of surface water from rivers and groundwater inputs from aquifers through a network of irrigation channels to fields and crops. Traditional irrigation was most often adapted to a smaller scale with a surface area of tens to several hundreds of hectares and sustained by local communities without significant governmental support (FAO, 2012). Artificial regulation and government policies however, have allowed the growth in irrigated area in Spain over recent decades. Plots operated thanks to artificial regulation cover an extensive area and are generally funded by the government or by private organizations and often make use of more advanced or modern techniques such as hydraulic pressurized systems.

In terms of irrigation in relation to demand and use in 2005, Spain's total water demand amounted to 35,323 hm³/yr. A breakdown between different uses is presented in Figure 2, where irrigation demand represents 68% of the total, whereas urban demand accounts for only 13% of the total. Spanish forecasts indicate an increase in total demand to around 44,000 hm³/year for the year 2015 (Ruiz Cañete & Menéndez, 2009). Spain

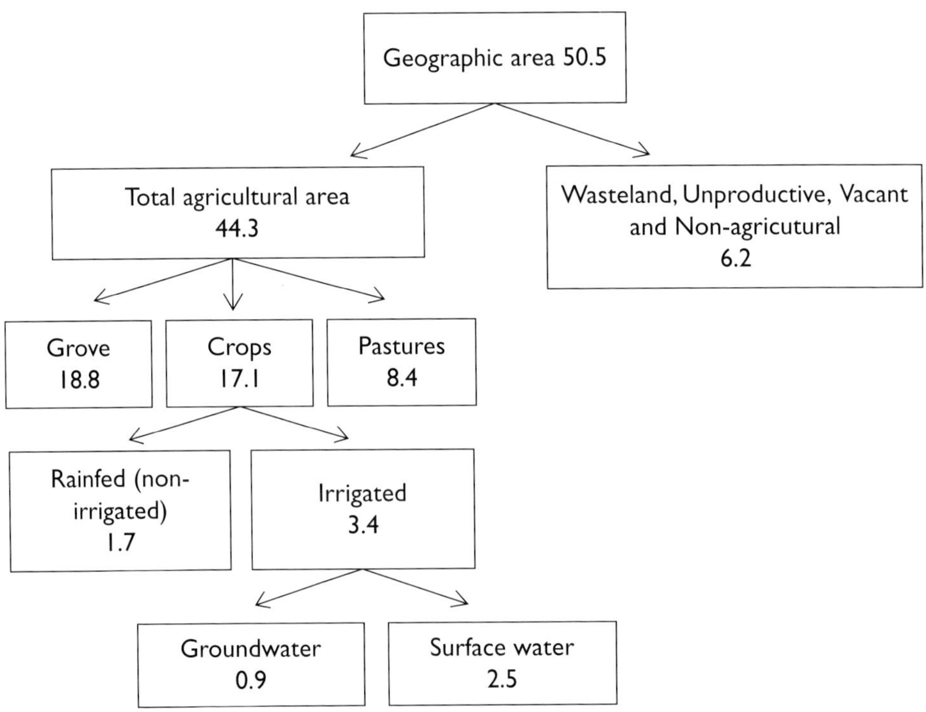

Figure 1 Land use in Spain, agriculture and irrigation. Figures are expressed in 10^6 ha. (own elaboration from MAGRAMA, 2012).

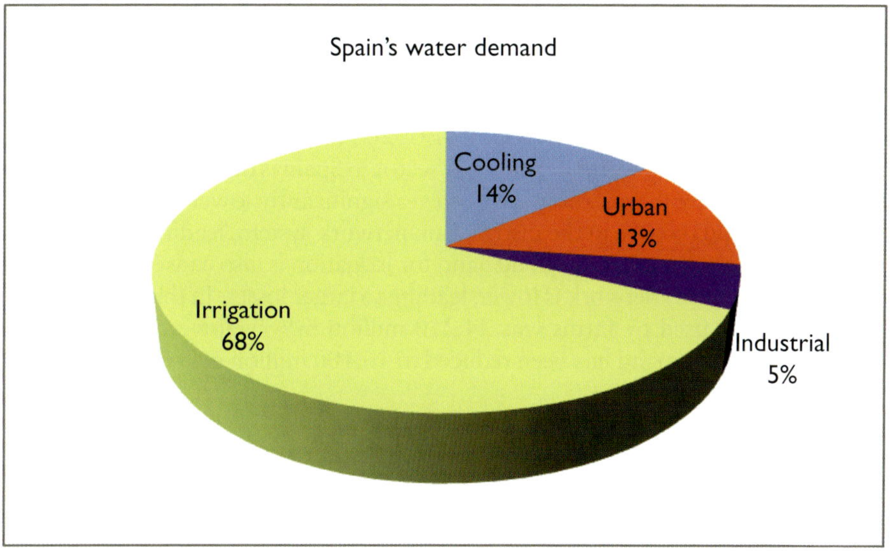

Figure 2 Water demand in Spain (2010) (own elaboration from INE, 2011).

with an annual average of about 2000 m³/capita is among the countries with world's highest water footprint (Waterfootprint, 2013). However, water use of the consuming sectors does not match their direct economic value. Irrigation is a key part of the agricultural sector since irrigated fields account for 60% of the total agricultural production (i.e. 13,000 out of an estimated 20,500 Million Dollars), and even 80% of total farmer exports. Yet it represents 14% of the total agricultural area, with $3.5{\cdot}10^6$ ha of the total Agricultural Used Area (Barbero, 2008). The difference between the average production of a hectare of dryland compared to irrigated land is 1 to 6.5, i.e. on average 1 ha of irrigated land produces over 6 times more than a hectare of rainfed agriculture (Camacho Poyato, 2005; Maestu & Gómez, 2010). This in turn generates an income four times higher than in rainfed agriculture and in addition generates more employment per hectare (Camacho Poyato, 2005; Lopez-Gunn *et al.*, 2012a). Irrigation generates 30% of wages in the agricultural sector, which represents about 600,000 jobs in irrigation. A study with data from the province of Huesca for the late 1980s period showed that yields from cereals on irrigated land were up to 4.4 times higher compared to those from non-irrigated areas. In the case of fruits, the yield on irrigated land in the provinces of Huesca and Lérida was up to almost six times greater than on non-irrigated plots (Silvestre & Clar, 2010).

Yet precisely because it is the main water-consuming sector, changes in the irrigation sector and in its productivity in terms of either resource use (more efficient use of resources) or in terms of e.g. euros per drop are fundamental for strategic Spanish water policy. The economic significance of changes in irrigation are very important.

The vast increase in water use over the last decades has been enabled by an increase in the number of dams, and regulation of water volumes. A boost in water consumption by irrigation was the result of irrigation projects by the government in the 1950s to 1960s. Currently, a smoother water supply is ensured by about 1170 large dams which collectively have a capacity of 56,000 hm^3 (EEA, 2013). Water storage capacity in total is about 48% of the natural water runoff flows. In addition, about one million groundwater wells, 5000 km of water distribution pipes, and 10,000 km of irrigation canals have been installed. Thanks to these extensive hydraulic projects, five times the natural amount of water is now available (EU, 2004).

In Table 1, the annual water demand per sector in Spain is outlined. The actual water amounts consumed by the end-user are however significantly lower than reported. This is due to losses of up to 20% in the distribution network system, leading to return flows. Next to this high consumption, high demand for irrigation is also caused by the fact that the quality of the supply network is lower, leading to larger losses. In the 90's, the volume of irrigation water used by farms was 24,250 million m^3/yr. However, with irrigation modernization, this demand has been reduced to 16,100 million m^3/yr (INE, 2010). The

Table 1 Annual water demand in Spain per sector (own elaboration from MIMAM, 2000).

Annual water demand in Spain		
Uses	*(hm^3/year)*	*%*
Urban	4667	13.2
Industrial	1647	4.6
Irrigation	24,094	68.2
Water cooling	4915	14
Total	35,323	100

Table 2 Gross irrigation water supply and consumption in existing irrigation plans per water basin (own elaboration from MAGRAMA, 2008).

Basin	*Gross supply (m^3/ha)*	*Irrigated surface (ha)*	*Gross supply (hm^3)*	*Water return flow (hm^3)*	*Water consumption (hm^3)*
Galicia Costa	8337	26,371	220	44	176
Norte	7734	74,032	573	63	510
Duero	6801	447,576	3044	322	2722
Tajo	8262	201,336	1663	230	1433
Guadiana	6657	335,590	2234	236	1998
Guadalquivir	6635	602,966	4000	505	3495
Sur	5620	142,457	801	75	726
Segura	6240	276,316	1724	157	1567
Júcar	6122	384,802	2356	184	2172
Ebro	8033	738,662	5934	962	4972
Cataluña CI	5962	67,774	404	36	368
Baleares	7804	17,376	136	25	111
Canarias	7147	29,379	210	27	183
Total	6965	3,344,637	23,298	2866	20,432

water volume for irrigation used by farms in 2010 reached almost 16,118 million m^3, representing an increase of 1.3% over the previous year and indicating an upward trend in water demand if water use efficiency is not increased significantly. The table below shows water demands in River Basin Management Plans, expressing their values in absolute terms and as a percentage of total irrigation demand in Spain.

As shown, the total irrigation water supply in the whole of Spain was around 23,000 million m^3 in 2008, with more than half consumed in large river basins like the Ebro, Duero and Guadalquivir. Average allocation ranges from 5962 m^3/ha/year in the Catalonia river basin, to 7147 m^3/ha/year in the Canary Islands, with an overall mean value of around 7000. Figure 3 presents a graphical display of these results.

Data concerning the origin of the Spanish irrigation water is not sufficiently contrasted with direct inventories. The Government adopted different estimations (MAPA and MOPTMA), setting percentage distributions with the same order of magnitude, as shown in the following table.

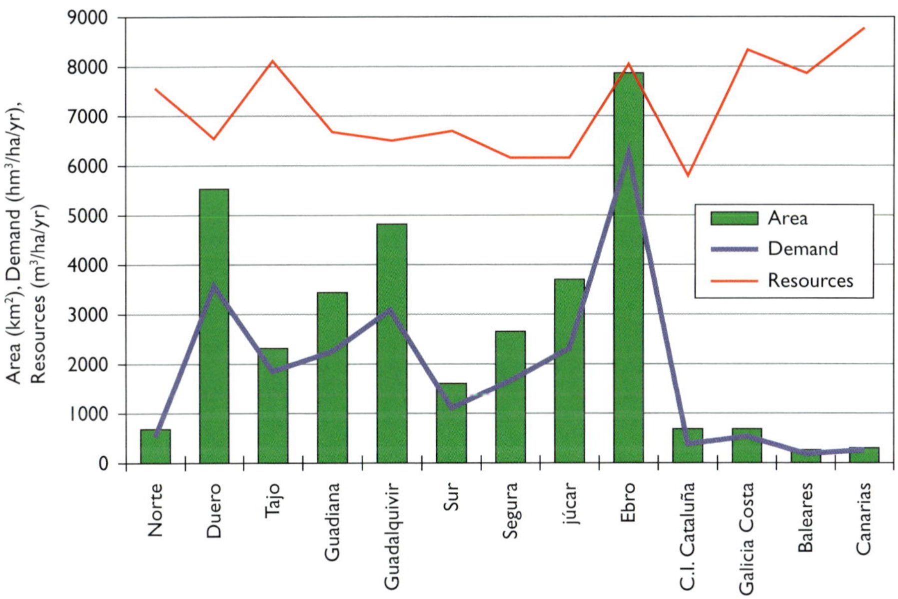

Figure 3 Surface area, demand, and allocation of planning areas (MIMAM, 2000).

Table 3 Origin of irrigation water according to the two different ecological territory maps of Spain MAPA and MOPTMA (own elaboration from MIMAM, 2000).

	Fraction of irrigated area (%)	
Origin of water	*MAPA*	*MOPTMA*
Surface	68	67
Subsurface	28	23
Mixed and other	4	10

The Spanish Statistic Institute (or Instituto Nacional de Estadistica, INE) gave the following figures on volume (m³) extracted for 2010 (Table 4). Surface water is the main source, followed by groundwater and finally with a small but significant role due to its importance for high-value crops, desalination and treated water with 1.2%.

Today, the mixed use of surface and groundwater is common (Lopez-Gunn *et al.*, 2012c), since groundwater wells may provide an additional security for water availability in times of low surface water availability, (e.g. during drought) and the use of mixed resources instead of using only groundwater may prevent or reduce aquifer overexploitation. The geographic location of areas irrigated by surface, ground or mixed source water is shown in Figure 4.

Finally, in terms of irrigation methods and systems, 30% of Spain's irrigated area is equipped with gravity-driven, or surface irrigation systems (basin, border, furrow or surge), 22% with spray systems (by sprinklers), and 48% by micro-irrigation (drip) technology.

Table 4 Origin of irrigation water (own elaboration from INE, 2011).

2010	*Volume (*1000 m³)*	%
Surface waters	14,474,666	81.1
Ground waters	3,168,585	17.7
Desalinated, treated...	208,710	1.2
Total	17,851,961	100.0

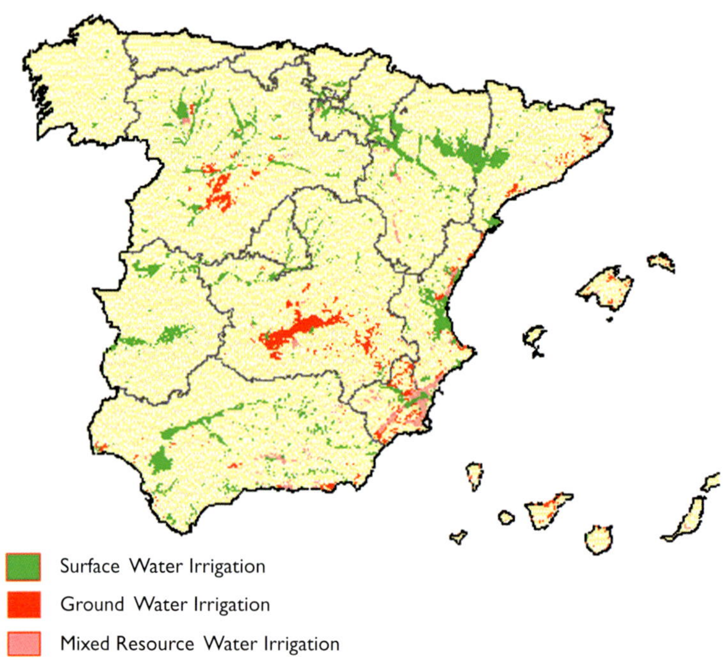

Figure 4 Irrigated areas in Spain and their water sources (own elaboration from Moneo & Iglesias, 2004).

Table 5 Application Efficiency of Irrigation Systems (own elaboration from Rogers D.H. *et al.*, 1997).

System type	*Application efficiency Range (%)*
Surface Irrigation	
Basin	60–95
Border	60–90
Furrow	50–90
Surge	60–90
Sprinkler Irrigation	
Handmove	65–80
Traveling Gun	60–70
Center Pivot & Linear	70–95
Solid Set	70–85
Micro-Irrigation	
Point source emitters	75–95
Line source emitters	70–95

The above table evaluates the irrigation technique with the aid of the Application Efficiency (E_a), which is the percentage of water delivered to the field that is effectively used by the crop, in which W_c is the water available for use by the crop, and W_f the water delivered to the field:

$$E_a = 100 \cdot \frac{W_c}{W_f} \tag{1}$$

This water application efficiency indicates how well an irrigation system performs in getting water to the plant roots (Rogers *et al.*, 1997). The average application efficiency range for micro-irrigation (drip) is better than for surface and sprinkler irrigation techniques. However, there are different ways to calculate irrigation efficiency. Another one is to consider the volume of irrigation water beneficially used, thus including beneficial effects such as plant absorption, salt leaching, frost protection, etc., rather than just the amount of water reaching the roots. The resulting efficiencies however show similar trends for different techniques, only yielding lower results for gravity irrigation: Surface irrigation 20–80%, sprinkler 60–80% and micro-irrigation 70–90% (Manero, 2008).

Significant work has been done in the past few years on irrigation improvement and modernization of irrigation techniques (Table 6). The area operated under gravity irrigation has significantly decreased from 59 to 30%, while a strong increase has occurred for the area under drip or micro-irrigation: From 17 to 48%. The end result of these efforts has been a decrease in the total water consumption of the irrigated agricultural sector over the last decade.

Table 6 Evolution in the Spanish irrigation systems between 2000 and 2011 (own elaboration from MAGRAMA, 2012).

	Before 2000		*2011*	
Irrigation system	*Hectares*	*%*	*Hectares*	*%*
Flood irrigation (gravity)	1,973,336	59	1,031,669	30
Sprinkler and others	802,712	24	783,487	22
Drip irrigation	568,588	17	1,658,317	48
Total	3,344,636	100	3,473,473	100

3 INSTITUTIONAL ENDOWMENTS AND IRRIGATION IN SPAIN

The Spanish river basin organisations (or RBOs) are some of the earliest formal water governmental authorities, created in 1926, providing Spain with one of the longest histories together with the USA, in developing these kind of authorities at a river basin scale. Currently, water resource management in Spain can be characterized as decentralized (Lopez-Gunn & De Stefano, in press). Decentralization in water management is one of the key principles in the IWRM paradigm (GWP, 2000) which can result into a better water resource management that takes all local resources and basin-specific resources (land-use, water, etc.) into account. Another complementary aspect is public participation, which – if effective, rather than symbolic – can lead to better future management decisions (Blomquist *et al.*, 2005). Ultimately decentralization in the water sector can lead to the creation of autonomous and financially self-dependent organizations (Saleth & Dinar, 2000).

The history of irrigation in Spain goes far back in history. The first foundation of irrigation systems were laid out by the Romans, but fell into disuse after their departure at the end of the 4th century (Glick, 1970; Del Campo García, 2008). During the Arab period in the 8th century, these systems were further developed and became wide-spread. After 1490, the Spanish crown started to promote and recognize existing irrigation organizations. A legal framework was established in which, according to law, water judges were appointed to deal with water-related problems or conflicts resulting in a more centralized approach (Giménez & Palermo, 2007).

Since Spain's first Water Law in 1866, most surface waters have been considered part of the public domain, while groundwater was considered to be of private ownership. A result of the laws of 1866 and 1879 was the separation of irrigation system management and river management when the formation of irrigator communities became mandatory, while this was not the case for water management organizations (Llamas & Garrido, 2007). In 1926, River Basin Organizations (RBOs) or 'Confederaciones Hidrograficas' in Spanish (CHs) were established by the central government, and in 1955 the National Federation of Irrigators Communities (FENACORE) was established. The aforementioned territorial bodies are not autonomous; RBOs exercise the power of state water policy at the river basin scale. RBOs are directed by the national government in the case of inter-regional river basins, and by the regional government for intra-regional basin (Blomquist *et al.*, 2005; Lopez-Gunn and De Stefano, in press). The national water law (1985) recognized 13 RBOs: 9 in inter-regional and 4 in intra-regional basins. These are responsible for water works (dam construction and operation), water planning (physical conditions of the basin and its water use), monitoring of water resource conditions (to anticipate on water events), water licensing (i.e. processing water concessions), water transfers and finally the enforcement of water regulations from Spain or the EU (Maestu, 2003).

The 1978 Constitution allocates powers among the Central Government, Regional Governments, as well as Local Governments, like municipalities. This institutional set-up requires participation and coordination between the various authorities since some Regional Governments initiatives, like changes in irrigation, can have a major impact on water resources management and determine decisions falling within the State's powers, and *vice versa*.

The 1985 reform of Spain's Water Law placed groundwater under the category of public property, although only for wells drilled after 1985 (Llamas & Garrido, 2007). All waters (including also desalinated for example) (Lopez-Gunn *et al.*, 2012b) became part of the public domain and thus the law allowed users to use specific water volumes, as well as make discharges into water, after the necessary licensing and permits were obtained from the RBOs. The only exception was for groundwater resources which were abstracted before 1985. At that time, an attempt was made to register existing private wells, which was largely unsuccessful. The White Book of Groundwater (1994) contained a program of action (ARYCA) for groundwater management, aimed at making an inventory of all existing wells, which was only partly successful. In 2001 with the new national water plan, another attempt was made to make an inventory of all existing wells in order to solve the legal situation (ALBERCA), indicating the presence of 25,000 illegal wells (Fornes *et al.*, 2007).

Decentralization has now unravelled and in what concerns the agricultural sector, the task of the national government lies in checking whether regional performance is complying with EU Regulations and arranging inter-regional initiatives. Since Spain is part of the European Union (EU), water policy is strongly influenced by EU's agricultural and environmental policies (Saleth & Dinar, 2000). The National Government still has a strong influence on the operation of River basin organisations, which ones receive national funding to design hydrological plans and evaluate the status of public water resources and infrastructure (Oñate & Peco, 2005). Thus, while the main regulatory and policy tasks are in the hands of the MAGRAMA, the executive power is in hands of the RBO's.

Changes in the institutional structure of the water sector have been suggested due to both internal drivers (such as water scarcity, water conflicts, performance improvements, and financial autonomy) as well as external drivers (economic development, political reforms, changes in demand, meteorological extreme events, natural disasters, and technological changes) (Saleth & Dinar, 2000). Changes in demand through demographic growth or extensive shortages due to droughts have a large impact on institutional changes, since the net benefit from the reforms is a direct function of water scarcity.

The history and functioning of Spanish irrigation organisations represent one of its richest, more complex and oldest institutional endowments. While on an inter-regional to regional scale, water management is done by the RBOs, on a more local scale the physical and social issues related to water management, are dealt with by irrigators' communities. The current River Basin Organisations must act as conflict resolution or mediation bodies to reconcile the various interests of the local users, the Autonomous Regions, and the State. Therefore, they must be composed of approximately one third representatives of the State, one third of the Autonomous Regions and one third of the Users. Water management is therefore increasingly a shared responsibility that requires the effort and cooperation of all stakeholders. Moreover, developed countries with an adequate democratic political system require user participation of citizens and society in its attempts to tackle major issues concerning water, environment, education, or health. These users organizations have made numerous achievements, but have also encountered problems. Because of this, it is worth pausing to study these Spanish organisations, which have succeeded up to now to maintain a collectively negotiated system, which provides fair and convenient sharing and distributing of water among its users.

In Spain the institutional set-up in terms of self-governing institutions is diverse and sophisticated. This chapter only concentrates on surface water institutions yet there is an equally interesting phenomena in the emergence of groundwater user groups, and more recently desalinated and even recycled – and recharged water user communities (Lopez-Gunn *et al.*, 2012; Lopez-Gunn *et al.*, 2013). Most of the 7196 irrigation communities in Spain belong to an umbrella organisation, in a federated structure. There is also a parallel institution, the AEUAS, which focuses solely on groundwater user communities. The Spanish Federation of Irrigators Communities (FENACORE) in Spain is a non-profit association, founded in 1955, which gathers organizations dedicated to the administration of irrigation by surface – and groundwater. During the period when it was founded, in the 1950s, the existing political regime tried to control all aspects of public life in the country. Irrigators Communities associated together to oppose this intrusion, to be able to maintain the freedom and independence they had always had. Nowadays, FENACORE in Spain is fully recognised and integrated into the public organisational structure of Spain, seeking to safeguard the interests and rights of users, mainly irrigators, and harmonizing the main use increasingly with other users like municipalities and industry. At present, there are Irrigators Communities in all Spanish provinces which collectively represent an area of about 2 million hectares, or over 60% of the Spanish irrigated area. This National Federation argues for political independence, working with (and on occasions questioning) governments of very different ideologies, with beneficial results for the members. The National federation has become an influential lobby and key group in Spain's water policy-making.

The Spanish Federation is one of the main representative bodies of the irrigation sector at the national level, through sheer volume of affiliates and also due to its influence. In this sense, the Public Administration has recognized FENACORE as a main actor in numerous occasions, like the preparation of the Water Act and its Regulations, the elaboration of the National Hydrological Plan, the collaboration in the preparation of the National Irrigation Plan, the White Paper on Water, the Water Framework Directive, etc. The Spanish Federation carries out a wide range of activities among which the ones displayed in Box 1. Its internal organization is structured

Box 1: Main roles internally and externally of the national umbrella irrigation association (FENACORE)

Internal roles and functions:

- Advice to all kinds of consultations (juridical, practical, technical).
- Inform, trying to keep the associated members informed about those topics users are interested in (legislation, technical reports, sentences, communications, interesting articles, etc.) through brochures issued periodically.
- Training, organising courses for the farmer-irrigator, technicians and secretaries of the Irrigators Communities.

External roles and functions:

- Member of the Permanent Commission of the Irrigators Communities National Congresses.
- (OM MOP 12.10.1972).
- Consultant Body of the Ministry of Works (Ministerio de Obras Públicas) (OM MOP 12.01.1978) nowadays called Ministry of Environment.
- Natural member of the Water National Council.

by a General Board, a Direction Board, a Permanent Commission, a General Secretary, and a President.

At a lower level and as full members of the Federation, are irrigators communities. The institutional character of irrigation associations has become strongly prevalent as a result of the acknowledgement of the public status of the water resource domain. Since the first Water Law in 1879 it was mandatory for users of the same outlet or concession, to establish a Community of Users, meaning a strong support for each individual self-administered Spanish farmer. These entities have the character of Public Law Corporations which are affiliated to the River Basin Organisation and should monitor the proper use of water. The legal nature of Irrigators Communities has been described by Toledo & Arrieta (1987) as being 'Public Law Corporations mandated by the Act under its autonomous system function to distribute and manage waters granted, subject to the sanctioned rules by the Administration and developed by users themselves.'

In the Water Act of 1985, Article 82.1, a legal definition was given for user communities: 'User communities have the character of Public Law Corporations and are assigned to a River Basin Organisation, which will ensure compliance with its statutes or ordinances and good order of use. These statutes and ordinances shall comply with the procedures laid down in this Act, in accordance with Law 30/1992 of 26 November, on the Legal Regime of Public Administrations and the Common Administrative procedure' (Del Campo García, 2008).

Irrigators Communities are institutions steeped in history, key to good water distribution and organization of Spanish irrigation itself, constituting entities deeply rooted in the popular consciousness. The public character of these communities is determined by the aim to achieve better administration and management of public water, distributing flows, resolving conflicts between irrigators and exercising police functions (Valero de Palma, 1996). In addition to this, the character of Public Law Corporation of the Irrigators Communities, with all the consequences that this entails, is compatible with private management of the Community that is expected to be more efficient and effective.

Their organizational structure has been considered the most appropriate by the Spanish legislature to group and integrate users. Because of this, Irrigators Communities are prominent in the water institutional organization of Spain and their structure has been transferred: Article 82 of the Water Act transposes the schemes of Irrigators Communities to all users calling them 'User Communities' (Rojas Calderón, 2007; Del Campo García, 2008). The success of this concept is thought to be due to the fact that community-managed irrigation systems are empowered with 'human and social capital'. This term is referring to the social bonds and interconnectivity between the members of these communities which are at the base of a high level of social trust and respect of norms, enabling the community to solve common problems. Due to the successful history in Spain, this model has been taken up in many foreign laws such as in the countries of Argentina, Mexico, and others. This has been possible because of the versatility and flexibility of its structure and functioning, its ability to adapt to different irrigation methods (traditional or new, using ground- or surface water, with abundance or scarcity of resources, etc.), to the local customs and country-specific conditions (climate, hydrology, economics). Box 2 summarizes the main advantages of Irrigators Communities.

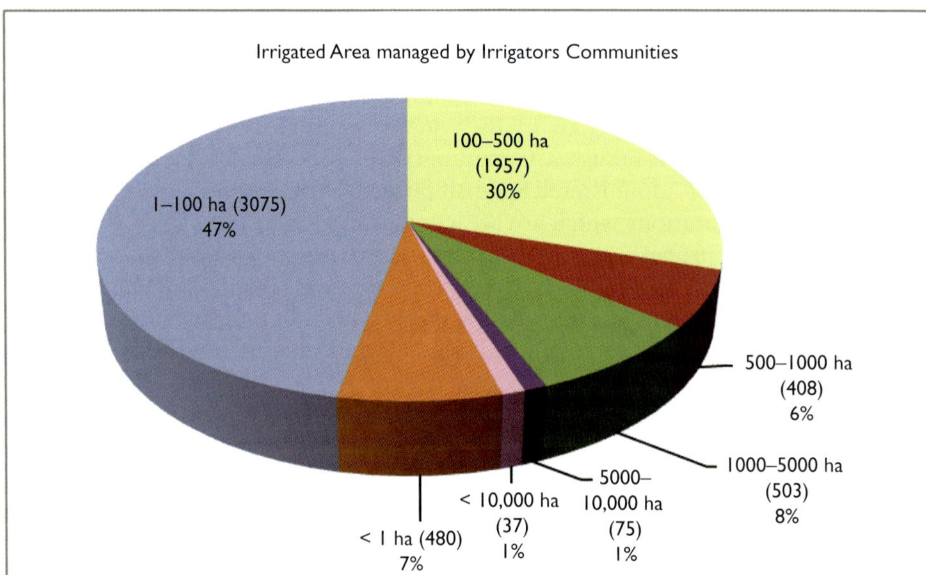

Figure 5 Areal sizes managed by Irrigators Communities (own elaboration from MAGRAMA, 2009).

Box 2: Advantages of organizing irrigation management in Users Communities

1 *Independence*: Avoiding interference between water policy and different political regimes.
2 *Versatility and flexibility*: Its functioning is able to adapt to different types of irrigation systems and other region-specific conditions.
3 *Conflict resolution*: The Irrigators Communities do great conflict resolution work on behalf of the Administration, managing and making peace in water use.
4 *Effectiveness*: The character of Public Law Corporation. It is compatible with private management in the Community which should be more efficient and effective.
5 *Data gathering and information*: Make the collection of data easier for the Government on consumption and on the operational – and exploitation costs of water.
6 *Monitoring and control*: Control and limit the abuses in the use and administration of water under the control of the Community watching over the general interest.
7 *Enforcement and sanctioning*: Allow law enforcement of national and EU laws on the daily management of water by controlling standards set by themselves and thereby also reaching a faster approval by law on partitions.
8 *Participation*: Allow users to participate in governing bodies, in management and to participate in the River Basin Authorities.
9 *Reference point*: Provide that the concession holder of public water domain is a single entity and not each individual user.
10 *Organizational flexibility*: Flexible, open and democratic model, pacifying the water use
11 *Subsidiarity*: 'small is beautiful', allowing management of resources by the person closest concerned. The Irrigators Community is a public body since it must be chosen according to the principle of subsidiarity. Only the River Basin District, the Regional Government or the State might intervene, to fund and/or subsidize the works, when the Irrigators Community fails.

In terms of typology of irrigation self-governance institutions by e.g. looking at irrigation communities, these can be divided according to three types depending on the origin of their emergence: Traditional (or historical), public initiative, and individual private irrigation in Spain.

The first type is traditional irrigation: According to the MIMAM (2000) we have to differentiate traditional or historical irrigation, for those areas that had irrigation plans that were executed prior to the year 1900. For the Segura river basin, traditional irrigation is subject to a special statutory regulation as those that existed before 1933. In a generic sense, the area in Spain irrigated according to historical methods is estimated to be around 1,075,000 ha. These tend to be located in the most fertile lowlands in rivers. Their management is done by Irrigators Communities, although important historic canals and its respective 350,000 ha under irrigation are

Box 3: Examples of traditional irrigation in different Spanish river basins

- Canal of Castilla in the Duero river basin;
- Canals of Aranjuez, the Canal of Henares in the Tagus river basin;
- Imperial Canal of Aragon in the Ebro river basin;
- Water Court of Valencia, the Royal Ditch of Moncada, Royal Ditch of Júcar, the Water Board of Plana in the Jucar river basin;
- Board of Landowners of the Huerta of Murcia, the Water Court of Orihuela, in the Segura river basin;
- Irrigators Community of Guadalmellato and the Irrigators Community Lower Valley of Guadalquivir in the Guadalquivir river basin.

still managed by the River Basin Authorities. It could be desirable to make progress in transferring the management of these canals towards the General Communities. One of the main problems this type of irrigation communities face is the small size of farmers, which tend to be smallholders (Bolea Foradada, 1998).

The second type is based on irrigation due to public initiative or public investments. This irrigation type was developed in the twentieth century thanks to the Public Administration initiative or aided by it. It covers a total of 1,518,000 ha, usually located in the most fertile valleys of the great plains and inter-fluvial areas, better suited for irrigation. The plot size is somewhat larger than in traditional irrigation, but it is still too small to be modernized according to new agricultural demands. A typical example of this type of irrigation is the area of Las Vegas of Guadiana (Badajoz Plan), the Júcar-Turia Canal, and the Cota 100 and Cota 200 Canals in the Mijares river. Between the 1960s and 1980s, the government made huge investments in large-scale public projects. Although it became apparent that the projected productivity increases were not met (Hunt, 1989), this was thought to be due to the fact that the systems were designed disregarding the problems met on a local (farmer) scale. This helped to further improve the involvement of users in the management and financial aspects of future planning (Freeman & Lowdermilk, 1991).

Finally, there is also private irrigation due to individual initiative and entrepreneurship. These irrigation plans are those that have been developed through private initiative via administrative concessions of public water or private water farms, as is for example the case particularly for groundwater. These occupy a total area of about 1,168,000 ha and irrigation water from surface and/or groundwater is usually obtained through the aid of mechanical pumping systems. Examples of these irrigation systems may be found in Castilla-La Mancha (Irrigation of the Mancha Occidental aquifer, Central Board of Irrigators of the Eastern Mancha, etc.), in Andalusia and the rest of Spain. In the table below, the character of government investments into water plans from 2001–2008 can be seen, in which over 10,000 million euros was invested for irrigation modernization and water transfer works from the Ebro basin alone (Lopez-Gunn *et al.*, 2012a).

Table 7 Budget of Hydrological Plan investments (own elaboration from MMA and EU, 2004).

Hydrological Plan investments 2001–2008	
Performance	*Investment (millions of euros)*
Irrigation methods modernization	6.150,43
Urban water supply	2.815,06
Hydraulic regulation works (reservoirs, dams...)	2.718,85
Water sewerage and treatment	2.605,46
Hydrological and forest restoration	1.859,57
River bed restoration and flood prevention	1.433,98
Water quality control programs	1.260,05
Transfer works from the Ebro river basin	4.207,08
Total	23.050,49

4 OLD AND NEW INSTITUTIONAL ENDOWMENTS: FROM LOCAL TO GLOBAL

This section will look in detail at three examples at three different levels which highlight the origins, adaptation and emergence in terms of institutional endowments in Spain. It looks at two specific cases which represent the old (and local) and the new (and regional or even inter-national) communities. It looks at the example of the Water Court of Valencia, which has been in operation for over ten centuries, a tribute to institutional resilience and social adaptation to changing conditions. Secondly, it looks at the example of a new federated institution of irrigator communities that covers the Mediterranean region, upscaling farmer demands and needs in acknowledgement of the change in the centre of gravity in terms of e.g. agricultural policy from the purely national to the European union scale and the common agricultural policy.

4.1 The case of the Water Court of Valencia's fertile lowlands

The Water Court of Valencia's fertile lowlands is without doubt one of Europe's oldest judicial institutions that still exist today. For over a thousand years, every Thursday (except from Christmas to Epiphany, January 6th), the bell rings at noon on Miguelete tower, and the eight Trustees – who form the Water Court – go to the Apostles Gate of the City Cathedral in Valencia to pass judgements.

According to historians, it was around the year 960 of the Christian era, under the reign of Abderrahman III, when it was set as we know the water court today. However, to find clear evidence of the Courts existence, we must go back to the year 1238 when King James I the Conqueror, in his jurisdiction XXXV, recognized and assigned the Valencian irrigation community the equal rights it had had in the Arab times. With this royal regulation, the irrigation in Valencia and the Water Court's permanence was ensured. This regulation involved the following features:

The Court's meeting place is established at the cathedral gate that used to be a former mosque (Figure 6). After the Reconquest, many Muslims continued cultivating our farms, but religion prevented them from entering into a Christian church, so that trials which were held inside, were held at the church's gate.

The meeting day is set on Thursdays, the eve of the feast for Muslims, and before noon, after which procedures can begin. The court is very time accurate: Trial starts at 12 o'clock If the defendant attends on time, he/she has the right to be heard but if the Court is delayed, the case is dismissed.

More recently, the existence of the Valencia Water Court has been confirmed and officially recognized by the Spanish Constitution (27/12/1978), the Statute of Autonomy of the Valencian Community (1/07/1982, as amended on 10/04/2006), the Organic Law of the Judiciary (1/07/1985) and in the Water Law (2/08/1985).

The Court's functioning is as simple as it is effective: Once a violation of the ordinances is committed, the Canal's Guardian summons the offender to appear before the Water Court next Thursday. If he does not appear, he is summoned twice and if he does not appear after a third call, he or she is judged *in absentia* and may be condemned. Once the Court is set, the Sheriff, after asking permission from the President, appoints verbally, one by one, all owners responsible for each ditch. In case there are any respondents, these enter the premises of the accused and the complainant, accompanied by the Guard's Canal they belong to. The complainant personally formulates the accusation, after which the accused defends him or herself personally and can also provide new evidence and/or witnesses. The Court's President and the other trustees can make all the necessary questions to gain information and, if necessary, the trial can even be suspended so that the members of the Court can go to see the facts on the place where they occurred in the presence of stakeholders (a visual inspection

Figure 6 Meeting of the Valencian Fertile Lowlands Water Court (Yubero, 2010).

Box 4: Water Court of Valencia's fertile lowlands and its process (Fairén-Guillen, 1988)

- *Concentration*: At the trial time, the Court has all necessary information to take legal action.
- *Oral Nature*: All trial proceedings are conducted orally; the exposure of the complaint by the complainant, the defense of the accused, the Guard's clarifications and questions of the members of the Court, and even the proclamation of the sentence (subsequently written very succinctly).
- *Speed*: Perhaps it is one of the most striking features of this Court, and perhaps the most influential in its survival. The Court meets on Thursday and if the defendant attends to the first call, which is the most typical, it is resolved that day. If the defendant does not attend, the process can be delayed by up to two weeks, but the third call, without excuse, he is being judged 'in absentia', without being present.
- *Economy*: The trial causes minimal expenses. The judges forming the Court, i.e. the Trustees receive no salary for their function, not even a *per diem honorarium*. The convicted simply pays the fine stipulated by the Ordinances and the expenditures which the Sheriff summon, which are the costs, besides payment, if any, on any financial responsibilities arising from damages that have occurred.

or 'Visura'). Without further procedures, the Court deliberates and sentences, the president saying in Valencian traditional formula: *'Este Tribunal l' absol'* ('This Court absols you') or *'Este Tribunal li condena a pena, costes, danysiperjuins en arreglo a Ordenances'* ('This Court sentences you including, costs and damages, pursuant according to Ordinance'). The rulings are final and penalties apply according to each Canal Ordinances. Besides the already mentioned judicial function, the Water Court has a governmental function focused on the administrative functioning of the complex irrigation of Valencia's fertile lowlands, ensuring proper irrigation water distribution that constitutes the historical allocation of Irrigators Communities.

4.2 Institutional organization of irrigation in the Mediterranean countries

The Euro-Mediterranean Irrigators Community (EIC) is a non-profit association which unites the different surface – and groundwater management organizations. EIC's main objective is to promote the mutual exchange of information and experience on irrigation practices among irrigators of the participating countries. This exchange of information is assumed to have a beneficial effect on the quality of the water management in the participating regions, resulting in more benefits for the irrigators. The EIC is a stakeholder for the Water Framework Directive in the Strategic Coordination Group of the European Commission. Therefore, the community can function as a spokesman, to defend the interests of irrigator communities and also individual irrigators,

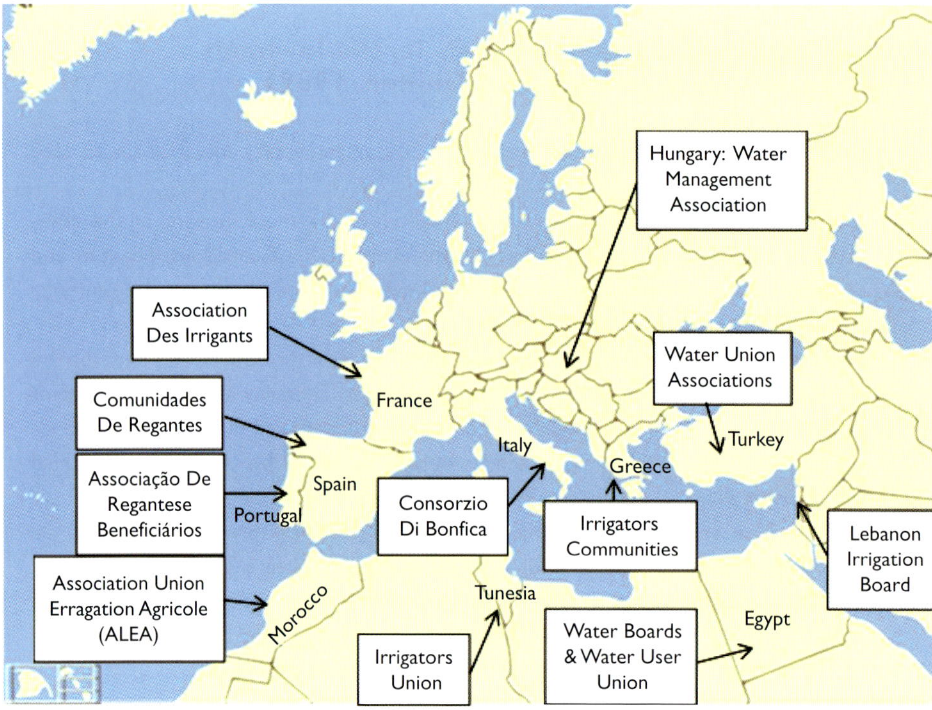

Figure 7 Irrigators communities in the other Mediterranean countries (own elaboration).

Box 5: EIC strategy and objectives

Strategy

- Exchange ideas, projects and experience to improve the institutional organization of irrigation in member countries through Irrigators Communities, Water User Associations and similar entities.
- To represent members before International Organizations and Associations related to agriculture, water and irrigation.
- To represent members and European irrigation at the European Union and its Institutions, facilitating the participation of water users in water policies of the European Union.

Objectives

- To introduce to the public the positive effects of irrigation.
- To push for R&D which strives to increase the availability of cleaner plant protection products and fertilizers.
- To promote the use of more effective and modern irrigation systems.

- To analyze and study the possible impacts of the EU Water Framework Directive on European and Mediterranean irrigation.
- To study the relations between agriculture and the environment favoring its coexistence and offer environmental training of farmers and irrigators with the goal to improve the sustainability of irrigation.

and to put forward the interests and concerns of irrigators on water policies of the European Commission and their impact on the sector (EIC, 2009).

The first basis for the EIC was laid during the celebration of the 'Corn Congress' organized by L'Association Générale des Producteurs de Maïs (AGPM) in 2000. Here, the National Federation of Irrigation Communities of Spain put forward the idea to develop an international organization to jointly represent the irrigated agricultural sector of the European and Mediterranean countries. In October 2001, at the 'International Mediterranean Irrigators Meeting' hosted by FENACORE in Murcia, concrete plans for the establishment of the EIC were made. This finally led to the inauguration of the EIC in Seville in 2002 during the celebration of the Tenth Spanish Congress of Irrigators Communities of Spain (EIC, 2009).

5 CONCLUDING REMARKS

Water resources in Spain require sophisticated institutional frameworks. This is mainly due to the great spatial and temporal variability in precipitation amounts and water availability. This leads to a growing competition amongst the different water sectors and different water users. With regard to the future, it is possible that the variability of water availability in Spain will further increase and competitiveness between the different sectors will further grow. Thus robust institutional frameworks which also include good mediation and conflict resolution are the key to accomplish this. Currently, the largest water consumer is the irrigated agricultural sector, which accounts for up to 68% of water use. About one decade ago, this figure was higher, but it has been reduced through investments in the water distribution network and increases in water efficiency through the application of modern irrigation technologies. However, these 'hard' technological innovation efforts have to be complemented with due care to existing 'soft' social institutions and social innovation (as e.g. seen in emerging user groups on desalinated water). Thus in addition to technological improvement in fields such as the growing of more drought-resistant crops, or an important shift like increasing the share of treated wastewater or desalinated water used in regions which frequently encounter water stress, it is also very important not to disregard the human infrastructure which is equally fundamental and part of the solution to cope with future water scarcity and greater uncertainty.

The long (and successful) history of irrigation associations in Spain is testimony to their ability to cope with change and uncertainty, and thus these should be at the forefront of considerations on strategies to deal with water security. Spain has a very successful history in the field of institutional organization of irrigation. Since the first Water Laws in 1866 and 1879, communities were created between water users, often farmer-irrigators, to unify and thus enable them to better represent themselves and their interests. Their success is due to the fact that they are stable communities which are financially independent from the Governments that has enabled them to function alongside different political regimes throughout history, showing a clear adaptive capacity. The community watches over the common interests according to a flexible and democratic structure, leading to relatively conflict-free water use, and therefore also ensuring law enforcement of local, national and EU laws. In addition, these communities improve the public participation of users into decision-making. Their successful history has led to the fact that similar structures have been implemented into other countries. Spain's water management is currently characterized as decentralized, since the role of the Government has seen the arrival of important new actors like the regions, which are responsible for agricultural policy and thus indirectly impact the main water use: Irrigation.

Also independent judicial organs have been formed, like the Water Court of the Valencian Fertile Lowlands (960 onwards). This court deals with irrigation-water-related-conflicts, and although the court does not make use of the national judicial instances, it resolves issues between irrigators in a very fast and economic way. Here again, the main driving force is communal pressure through social capital. Other bodies such as FENACORE are national federations which gather organizations dedicated to the administration of irrigation water use, and strive to harmonize the work of all irrigators. FENACORE has become an influential group in Spanish water policy making, putting forward the interests of irrigator communities and their individual users on a national and international scale.

An international organization of irrigators has been created – the EIC, which aims to exchange ideas of good agricultural practices, and represents members before international bodies concerned with water and agricultural issues. Its aim is to try to achieve maximal benefits for the irrigator communities of their member countries, helping to facilitate user participation in water policy making. It may be concluded that the existence of independent user communities on different levels is a key element of human capital for the success and long-livability of viable water policies. Thanks to these institutional 'soft' infrastructures, the interests of farmer-irrigators have been optimally represented resulting into increased benefits for their communities and if adequate mediation processes are in place, a key asset for water allocation and re-allocation decision-making.

REFERENCES

Barbero, A., (2008). *El programa de vigilancia ambiental del PNR: Seguimiento de la sostenibilidad del regadío.* CONAMA Congreso Nacional de Medio Ambiente, Comunicación técnica, Madrid.

Blomquist, W.A., Giansante, C., Bhat, A. & Kemper, K. (2005) *Institutional and Policy Analysis of River Basin Management: The Guadalquivir River Basin, Spain.* World Bank Policy Research Working Paper. Report number: 3526.

Bolea Foradada, J.A. (1998) *Las Comunidades de Regantes.* Canal Imperial de Aragón.

Camacho Poyato, E. (2005) Análisis de la eficiencia y el ahorro del agua en el regadío de la cuenca del Guadalquivir. *Agricultura: Revista agropecuaria*, 880, 880–887.

Del Campo García, A. (2008) Ponencia "*Las Comunidades de Regantes de España y su Federación Nacional*". Federación Nacional de Comunidades de Regantes de España (FENACORE).

EEA (European Environment Agency) (2013) *Spain freshwater assessment* [Online]. Available from: http://www.eea.europa.eu/soer/countries/es/soertopic_view?topic=freshwater [Accessed April 2013].

EIC (2009) *Our Origins* [Online]. Available from: http://www.emic.org/escaparate/portada.cgi?idempresa = 1108 [Accessed May 2013].

EU (2004) D10k: *Water Time National Context Report – Spain* [Online]. Observatorio de los Servicios Públicos. Escuela de Relaciones Laborales, Madrid. Available from: www.watertime.net/docs/WP1/NCR/D10k_Spain.doc [Accessed April 2013].

FAO, Food and Agriculture Organization of the United Nations (2002) *How design, management and policy affect the performance of irrigation.* [Online] Available from: http://www.fao.org/docrep/005/ac799e/ac799e00.htm#Contents [Accessed April 2013].

FAO (2013) *Water News: Climate Change & Water- Main findings and short – and medium-term recommendations* [Online]. Available from: http://www.fao.org/nr/water/news/climchange.html [Accessed June 2013].

Fairén-Guillén, V. (1988) *El Tribunal de las Aguas de la Vega de Valencia y su proceso.* Valencia, Ed. Caja de Ahorros de Valencia.

FENACORE (2012) *Federación Nacional de Comunidades de Regantes- Dossier De Prensa 2012.* [Online] Available from: http://www.ioncomunicacion.es/files/clientes/fenacore/dossier/DossierFenacore2012.pdf [Accessed April 2013].

Fornés, J.M. & de la Hera, A. (2007) Llamas, M.R. & Martínez-Santos P. (eds.) Legal aspects of groundwater ownership in Spain. International Water Resources Assocition. *Water International*, 32(4), 676–684.

Freeman, M. & Lowdermilk Max. K. (1991) Middle-level farmer organisations as links between farms and central irrigation systems. In: Cernea, M.M. (ed.) *Putting people First: Sociological Variables in Rural Development.* A World Bank Publication. pp. 113–144.

Giménez Casalduero, M. & Palermo, J. (2007) Organizaciones tradicionales de gestión del agua: Importancia de su reconocimiento legal como motor de su pervivencia. *Region y Sociedad*, 19(38), 3–24.

Ghazouani, W., Molle, F. & Rap, E. (2012) *Water Users Associations in the NEN Region – IFAD interventions and overall dynamics (Draft).* International Water Management Institute (IWMI).

Glick, T.F. (1970) *Irrigation and Society in Medieval Valencia.* Cambridge, Mass., Harvard University Press.

González Pérez, J., Arrieta, C. & Toledo, J. (1987) *Comentarios a la Ley de Aguas.* Ed. Civitas S.A., Madrid.

GWP (2000) *Integrated Water Resources Management.* [Online] TAC Background Paper. Report number: 4. GWP, Stockholm, Sweden. Available from: www.gwpforum.org/gwp/library/Tacno4.pdf [Accessed April 2013].

Hunt, R.C. (1989) Appropriate social organization? Water users associations in bureaucratic canal irrigation systems. *Human Organization*, 48, 79–90.

Iglesias, A. Rosenzweig, C. & Pereira, D. (2000) Agricultural impacts of climate change in Spain: Developing tools for a spatial analysis. *Global Environmental Change*, 10, 69–80.

INE, National Statistical Institute (2011) *Survey on water supply and sewerage. Environment statistics of water, year 2010.* [Online] Instituto Nacional de Estadística. Available from: www.ine.es [Accessed June 2012].

López-Gunn, E. Zorrilla, P. Prieto & Llamas, M.R. (2012a) Lost in translation? Water efficiency and productivity in Spain. *Journal of Agricultural Water management*, 108, 83–95.

López-Gunn, E., Cabrera, E., Custodio, E., Huertas, R. & Villarroya, F. (2012b) Institutional reform in Spain to address water challenges. In: De Stefano, L. & Llamas, M.R. (eds.) *Water, Agriculture and the Environment: Can we square the circle?* Taylor and Francis.

López-Gunn, E. Rica, M. & van Cauwenbergh, N. (2012c) Taming the groundwater chaos. In: De Stefano, L. & Llamas, M.R. (eds.) *Water, Agriculture and the Environment: Can we square the circle?* Taylor and Francis.

López-Gunn, E. & De Stefano, L. (in press) Between a rock and a hard place: Redefining water security under decentralization in Spain. In: Garrick, D., Anderson, G., Connell, D., Pittock, J. (eds.) (2014) *Federal Rivers: Water Management in Multi-Layered Political Systems.* Cheltenham, Edward Elgar Publishing.

López-Gunn, E., Rica, M. & Llamas, M.R. (2013) *Can cooperation lead to improved performance in groundwater management? Some reflections on the experience of Spanish groundwater user groups.* Preparing for the 2013 International Year. Water Cooperation: Making it Happen! International Annual UN-Water Zaragoza Conference 2012/2013. 8–10 January 2013, Zaragoza, Spain.

Llamas, M.R. & Garrido, A. (2007) Lessons from intensive groundwater use in Spain: Economic and social benefits and conflicts. In: Giordano, M. & Villhoth, K.G. (eds.) *The Agricultural Groundwater Revolution. Opportunities and Treath to Development.* Wallingord, CAB International Publishing.

MAGRAMA (2008) *Plan Nacional de Regadios- Horizonte.* [Online] MAGRAMA. Available from: http://www.magrama.gob.es/es/agua/temas/gestion-sostenible-de-regadios/apartado4–9_tcm7–9788.pdf [Accessed April 2013].

MAGRAMA (2009) *Subdirección General de Regadíos y Economía del Agua.* Dirección General de Desarrollo Rural. MAGRAMA.

MAGRAMA (2012) *Encuesta sobre Superficies y Rendimientos de Cultivos. Informe sobre regadíos en España 2011.* Subdirección General de Estadística. Secretaría General Técnica. MAGRAMA.

Meastu, J. (2003) *Public Participation in River Basin Management in Spain – Reflecting changes in ex-ternal and self-created context.* [Online] Workpackage 4, HarmoniCOP Project. Available from: http://www.harmonicop.uni-osnabrueck.de/_files/_down/Spain.pdf [Accessed May 2013].

Maestu, J. & Gómez, C.M. (2010) Water uses in transition. In: Garrido, A., Llamas, M. (Eds.) *Water Policy in Spain.* CRC Press.

Manero, A. (2008) *Comparative water management practices in California and Spain* (Minor Thesis)[Online].Availablefrom:http://upcommons.upc.edu/pfc/bitstream/2099.1/6053/8/07.pdf [Accessed April 2013].

MIMAM (2000) *Libro Blanco del Agua en España [White Paper on Water].* Secretaría de Estado de Aguas y Costas, Dirección General de Obras Hidráulicas y Calidad de las Aguas, Ministerio de Medio Ambiente, Spain.

MMA & EU (2004) *D10k: WaterTime National Context Report – Spain* [Online]. Observatorio de los Servicios Públicos. Escuela de Relaciones Laborales, Madrid. Available from: www.watertime.net/docs/WP1/NCR/D10k_Spain.doc [Accessed April 2013].

Moreno-Pérez M.F. & Roldán-Cañas, J. (2013) Assessment of irrigation water management in the Genil-Cabra (Córdoba, Spain) irrigation district using irrigation indicators. *Agricultural Water Management*, 120, 98–106.

Moneo, M. & Iglesias, A. (2004) *Food and Climate, Case Studies in Developed Countries: The Case Study of Spain.* [Online] Available from: www.atmosphere.mpg.de/enid/0,55a304092d09/2__Predicting_the_future/_case_study__Spain_18 s.html [Accessed April 2013].

Oñate, J.J. & Peco B. (2005) *Policy impact on desertification: Stakeholders' perceptions in southeast Spain.* Department of Ecology, Universidad Autónoma de Madrid.

Pérez Picazo, M.T. (n.d.) *Los organismos locales de gestión en los regadios mediterráneos. De los heredamientos a las comunidades de regantes.* Universidad de Murcia.

Rogers, D.H., Lamm, F.R., Alam, M., Todd, P., Trooien, G., Clark, A., Barnes, P.L. & Mankin, K. (1997) *Efficiencies and water losses of irrigation systems. Irrigation Management Series.* [Online] Kansas State University. Available from: http://cursos.uc.cl/unimit_agl_014-1/almacen/1222620929_lgurovic_sec4_pos0.pdf [Accessed April 2013].

Rojas Calderón, C. (2007) *Las Organizaciones de Usuarios de Aguas en España y Chile.* Universidad Católica del Norte (Chile) y Abogado.

Ruiz Cañete, O. & Menéndez, D. (2009) *The Water Sector in Spain.* Ciriec Working Paper.

Saleth, R.M. & Dinar, A. (2000) Institutional changes in global water sector: Trends, patterns and implications. *Water Policy*, 2(3), 175–199.

Silvestre, J. & Clar, E. (2010) The demographic impact of irrigation projects: A comparison of two case studies of the Ebro basin, Spain, 1900–2001. *J. of Historical Geography*, 36(3), 315–326.

Valero de Palma, J. (1994) *Seminario sobre Economía y Racionalización de los Usos del Agua.* Universidad Internacional Menéndez Pelayo.

Valero de Palma, J. (1996) *Régimen Jurídico de las Comunidades de Usuarios.* Colegio de Abogados de Madrid. 14 October 1996, Madrid.

Waterfootprint (2013) *National Water Footprints – Spain.* [Online] Water Footprint Network. Available from: www.waterfootprint.org [Accessed April 2013].

Yubero, F. (2010) *El Tribunal De Las Aguas De Valencia.* [Online] Available from: http://lanaveva.wordpress.com/2010/04/04/el-tribunal-de-las-aguas-de-valencia/ [Accessed March 2013].

Index